VOLUME NINETY SIX

CURRENT TOPICS IN DEVELOPMENTAL BIOLOGY

Myogenesis

VOLUME NINETY SIX

CURRENT TOPICS IN DEVELOPMENTAL BIOLOGY

Myogenesis

Edited by

GRACE K. PAVLATH
Department of Pharmacology
Emory University, Atlanta
Georgia, USA

ELSEVIER

AMSTERDAM • BOSTON • HEIDELBERG • LONDON
NEW YORK • OXFORD • PARIS • SAN DIEGO
SAN FRANCISCO • SINGAPORE • SYDNEY • TOKYO
Academic Press is an imprint of Elsevier

Academic Press is an imprint of Elsevier
525 B Street, Suite 1900, San Diego, CA 92101-4495, USA
225 Wyman Street, Waltham, MA 02451, USA
32, Jamestown Road, London NW1 7BY, UK
Linacre House, Jordan Hill, Oxford OX2 8DP, UK

First edition 2011

ISBN: 978-0-12-385940-2
ISSN: 0070-2153

For information on all Academic Press publications visit our website at elsevierdirect.com

Printed and bound by CPI Group (UK) Ltd, Croydon, CR0 4YY

Transferred to Digital Print 2011

Contents

Contributors

Esther Ardite
Department of Experimental and Health Sciences, Cell Biology Unit, CIBERNED, Pompeu Fabra University, Barcelona, Spain

Nadine Bakkar
Department of Molecular Virology, Immunology, and Medical Genetics, Human Cancer Genetics Program, Arthur G. James Comprehensive Cancer Center, The Ohio State University, Columbus, Ohio, USA

Bénédicte Chazaud
Inserm, U1016, Institut Cochin, 24 Rue du Faubourg Saint Jacques, and CNRS, UMR8104, and Univ Paris Descartes, Paris, France

Fabrice Chrétien
Institut Pasteur, Unité Histopathologie humaine et modèles animaux, 28 Rue du Docteur Roux, Paris, France

Anita H. Corbett
Department of Biochemistry, Emory University, Atlanta, Georgia, USA

Alexis Demonbreun
Department of Medicine, The University of Chicago, Chicago, Illinois, USA

Karyn A. Esser
Center for Muscle Biology, Department of Physiology, College of Medicine, University of Kentucky, Lexington, Kentucky, USA

Denis C. Guttridge
Department of Molecular Virology, Immunology, and Medical Genetics, Human Cancer Genetics Program, Arthur G. James Comprehensive Cancer Center, The Ohio State University, Columbus, Ohio, USA

Monica N. Hall
Graduate Program in Genetics and Molecular Biology, Emory University, Atlanta, Georgia, USA

Aster H. Juan
Laboratory of Muscle Stem Cell and Gene Regulation, National Institute of Arthritis and Musculoskeletal and Skin Diseases (NIAMS), National Institutes of Health, Bethesda, Maryland, USA

Gabrielle Kardon
Department of Human Genetics, University of Utah, Salt Lake City, Utah, USA

Charles Keller
Department of Pediatrics, Pape' Family Pediatric Research Institute, Oregon Health and Science University, Portland, Oregon, USA

Ken Kikuchi
Department of Pediatrics, Pape' Family Pediatric Research Institute, Oregon Health and Science University, Portland, Oregon, USA

Mellani Lefta
Center for Muscle Biology, Department of Physiology, College of Medicine, University of Kentucky, Lexington, Kentucky, USA

Christopher J. Mann
Department of Experimental and Health Sciences, Cell Biology Unit, CIBERNED, Pompeu Fabra University, Barcelona, Spain

Elizabeth M. McNally
Genomics and Systems Biology, Committee on Genetics; Department of Medicine; and Department of Human Genetics, The University of Chicago, Chicago, Illinois, USA

Rémi Mounier
Inserm, U1016, Institut Cochin, 24 Rue du Faubourg Saint Jacques, and CNRS, UMR8104, and Univ Paris Descartes, Paris, France

Pura Muñoz-Cánoves
Department of Experimental and Health Sciences, Cell Biology Unit, CIBERNED, Pompeu Fabra University, and ICREA, Barcelona, Spain

Malea Murphy
Department of Human Genetics, University of Utah, Salt Lake City, Utah, USA

Anuradha Natarajan
Biomedical Research Center, University of British Colombia, Vancouver, British Colombia, Canada

Grace K. Pavlath
Department of Pharmacology, Emory University, Atlanta, Georgia, USA

Ben Paylor
Biomedical Research Center, University of British Colombia, Vancouver, British Colombia, Canada

Eusebio Perdiguero
Department of Experimental and Health Sciences, Cell Biology Unit, CIBERNED, Pompeu Fabra University, Barcelona, Spain

Jennifer M. Peterson
Department of Molecular Virology, Immunology, and Medical Genetics, Human Cancer Genetics Program, Arthur G. James Comprehensive Cancer Center, The Ohio State University, Columbus, Ohio, USA

Avery D. Posey Jr.
Genomics and Systems Biology, Committee on Genetics, The University of Chicago, Chicago, Illinois, USA

Fabio Rossi
Biomedical Research Center, University of British Colombia, Vancouver, British Colombia, Canada

Brian P. Rubin
Pediatric Cancer Biology Program, Department of Anatomic Pathology, and Department of Molecular Genetics, Taussig Cancer Center and Lerner Research Institute, Cleveland Clinic Foundation, Cleveland, Ohio, USA

Vittorio Sartorelli
Laboratory of Muscle Stem Cell and Gene Regulation, National Institute of Arthritis and Musculoskeletal and Skin Diseases (NIAMS), National Institutes of Health, Bethesda, Maryland, USA

Antonio L. Serrano
Department of Experimental and Health Sciences, Cell Biology Unit, CIBERNED, Pompeu Fabra University, Barcelona, Spain

Berta Vidal
Department of Experimental and Health Sciences, Cell Biology Unit, CIBERNED, Pompeu Fabra University, Barcelona, Spain

Gretchen Wolff
Center for Muscle Biology, Department of Physiology, College of Medicine, University of Kentucky, Lexington, Kentucky, USA

Regan-Heng Zhang
Biomedical Research Center, University of British Colombia, Vancouver, British Colombia, Canada

Preface

Skeletal muscle is a vital tissue for movement, breathing, and metabolism. In addition, it is one of the few tissues that exhibit extensive regenerative ability in the adult due to the presence of stem cells called satellite cells. Skeletal muscle biology has engendered interest from numerous angles: sports medicine, developmental biology, gene regulation, physiology, immunology, and stem cells. In recent years, skeletal muscle research has rapidly expanded in many exciting directions. The goal of this book is to cover some key areas of muscle biology related to satellite and other progenitor cells, muscle regeneration, signal transduction, gene expression, and disease.

Key questions related to developmental origins of muscle and cancer (Chapters 1 and 2) as well as gene regulation and signal transduction (Chapters 3 and 4) are explored. Further areas that are discussed include the effects of nonmyogenic cells on satellite cells and muscle regeneration (Chapters 5 and 6) and how fibrosis develops when muscle regeneration is impaired (Chapter 7). Complementing these discussions on muscle regeneration is a consideration of how myoblast fusion is regulated by a recently described family of molecules (Chapter 8). New emerging areas of research include the effects of circadian rhythms on skeletal muscle function (Chapter 9) and the challenges of controlling nucleocytoplasmic transport in multinucleated myofibers (Chapter 10).

GRACE K. PAVLATH

CHAPTER ONE

Origin of Vertebrate Limb Muscle: The Role of Progenitor and Myoblast Populations

Malea Murphy *and* Gabrielle Kardon

Contents

Abstract

Muscle development, growth, and regeneration take place throughout vertebrate life. In amniotes, myogenesis takes place in four successive, temporally distinct, although overlapping phases. Understanding how embryonic, fetal, neonatal, and adult muscle are formed from muscle progenitors and committed myoblasts is an area of active research. In this review we examine recent expression, genetic loss-of-function, and genetic lineage studies that have been conducted in the mouse, with a particular focus on limb myogenesis. We synthesize these studies to present a current model of how embryonic, fetal, neonatal, and adult muscle are formed in the limb.

Department of Human Genetics, University of Utah, Salt Lake City, Utah, USA

Current Topics in Developmental Biology, Volume 96
ISSN 0070-2153, DOI: 10.1016/B978-0-12-385940-2.00001-2

1. Introduction

Muscle development, growth, and regeneration take place throughout vertebrate life. In amniotes, myogenesis takes place in successive, temporally distinct, although overlapping phases. Muscle produced during each of these phases is morphologically and functionally different, fulfilling different needs of the animal (reviewed in Biressi *et al.*, 2007a; Stockdale, 1992). Of intense interest is understanding how these different phases of muscle arise. Because differentiated muscle is postmitotic, muscle is generated from myogenic progenitors and committed myoblasts, which proliferate and differentiate to form muscle. Therefore, research has focused on identifying myogenic progenitors and myoblasts and their developmental origin, defining the relationship between different progenitor populations and myoblasts, and determining how these progenitors and myoblasts give rise to different phases of muscle. In this review, we will give an overview of recent expression, genetic loss-of-function, and genetic lineage studies that have been conducted in mouse, with particular focus on limb myogenesis, and synthesize these studies to present a current model of how different populations of progenitors and myoblasts give rise to muscle throughout vertebrate life.

2. Myogenesis Overview

In vertebrates, all axial and limb skeletal muscle derives from progenitors originating in the somites (Emerson and Hauschka, 2004). These progenitors arise from the dorsal portion of the somite, the dermomyotome. The limb muscle originates from limb-level somites, and cells delaminate from the ventrolateral lip of the dermomyotome and migrate into the limb, by embryonic day (E) 10.5 (in forelimb, slightly later in hindlimb). Once in the limb, these cells proliferate and give rise to two types of cells: muscle or endothelial (Hutcheson *et al.*, 2009; Kardon *et al.*, 2002). Thus, the fate of these progenitors only becomes decided once they are in the limb. Those cells destined for a muscle fate then undergo the process of myogenesis. During myogenesis, the progenitors become specified and determined as myoblasts, which in turn differentiate into postmitotic mononuclear myocytes, and these myocytes fuse to one another to form multinucleated myofibers (Emerson and Hauschka, 2004).

Myogenic progenitors, myoblasts, myocytes, and myofibers critically express either Pax or myogenic regulatory factor (MRF) transcription factors. A multitude of studies have shown that progenitors in the somites and in the limb express the paired domain transcription factors Pax3 and Pax7 (reviewed in Buckingham, 2007). Subsequently, determined myoblasts, myocytes, and myofibers in the somite and in the limb express members of the MRF family of bHLH transcription factors. The MRFs consist of four proteins: Myf5, MyoD, Mrf4 (Myf6), and Myogenin. These factors were originally identified by their *in vitro* ability to convert 10T1/2 fibroblasts to a myogenic fate (Weintraub *et al.*, 1991). Myf5, MyoD, and Mrf4 are expressed in myoblasts (Biressi *et al.*, 2007b; Kassar-Duchossoy *et al.*, 2005; Ontell *et al.*, 1993a,b; Ott *et al.*, 1991; Sassoon *et al.*, 1989), while Myogenin is expressed in myocytes (Ontell *et al.*, 1993a,b; Sassoon *et al.*, 1989). In addition, MyoD, Mrf4, and Myogenin are all expressed in the myonuclei of differentiated myofibers (Bober *et al.*, 1991; Hinterberger *et al.*, 1991; Ontell *et al.*, 1993a,b; Sassoon *et al.*, 1989; Voytik *et al.*, 1993). Identification of these molecular markers of the different stages of myogenic cells has been essential for reconstructing how myogenesis occurs.

In amniotes, there are four successive phases of myogenesis (Biressi *et al.*, 2007a; Stockdale, 1992). In the limb, embryonic myogenesis occurs between E10.5 and E12.5 in mouse and establishes the basic muscle pattern. Fetal (E14.5–P0; P, postnatal day) and neonatal (P0–P21) myogenesis are critical for muscle growth and maturation. Adult myogenesis (after P21) is necessary for postnatal growth and repair of damaged muscle. Each one of these phases involves proliferation of progenitors, determination and commitment of progenitors to myoblasts, differentiation of myocytes, and fusion of myocytes into multinucleate myofibers. The progenitors in embryonic and fetal muscle are mononuclear cells lying interstitial to the myofibers. After birth, the neonatal and adult progenitors adopt a unique anatomical position and lie in between the plasmalemma and basement membrane of the adult myofibers and thus are termed satellite cells (Mauro, 1961). During embryonic myogenesis, embryonic myoblasts differentiate into primary fibers, while during fetal myogenesis fetal myoblasts both fuse to primary fibers and fuse to one another to make secondary myofibers. During fetal and neonatal myogenesis, myofiber growth occurs by a rapid increase in myonuclear number, while in the adult myofiber hypertrophy can occur in the absence of myonuclear addition (White *et al.*, 2010).

Embryonic, fetal, and adult myoblasts and myofibers are distinctive. The different myoblast populations were initially identified based on their *in vitro* characteristics. Embryonic, fetal, and adult myoblasts differ in culture in their appearance, media requirements, response to extrinsic signaling molecules, drug sensitivity, and morphology of myofibers they generate (summarized in Table 1.1; Biressi *et al.*, 2007a; Stockdale, 1992). Recent microarray studies also demonstrate that embryonic and fetal myoblasts differ substantially in their expression of transcription factors, cell surface receptors, and extracellular matrix proteins (Biressi *et al.*, 2007b). It presently is unclear whether neonatal myoblasts differ substantially from fetal myoblasts. Differentiated primary, secondary, and adult myofibers also differ, primarily in their expression of muscle contractile proteins, including

Table 1.1 Summary of characteristics of embryonic, fetal, and adult myoblasts and myofibers

	Culture appearance and clonogenicity	Signaling molecule response	Drug sensitivity	Myofiber morphology in culture
Embryonic myoblasts	Elongated, prone to differentiate and form small colonies, do not spontaneously contract in culture	Differentiation insensitive to TGFβ-1 or BMP4	Differentiation insensitive to phorbol esters (TPA), sensitive to merocynine 540	Mononucleated myofibers or myofibers with few nuclei
Fetal myoblasts	Triangular, proliferate (to limited extent) in response to growth factors, spontaneously contract in culture	Differentiation blocked by TGFβ-1 and BMP4	Differentiation sensitive to phorbol esters (TPA)	Large, multinucleated myofibers
Satellite cells/ Adult myoblasts	Round, clonogenic, but undergo senescence after a limited number of passages, spontaneously contract in culture	Differentiation blocked by TGFβ-1 and BMP4	Differentiation sensitive to phorbol esters (TPA)	Large, multinucleated myofibers

All from Biressi *et al.* (2007b) or review of Biressi *et al.* (2007a).

	MyHCemb	MyHCperi	MyHCI	MyHCIIa	MyHCIIx	MyHCIIb
Embryonic myofibers	+	−	+	−	−	−
Fetal myofibers	+	+	+/−	+/−	+/−	+/−
Adult myofibers	−	−	−	+	+	+

Derived from Agbulut *et al.* (2003), Gunning and Hardeman (1991), Lu *et al.* (1999), Rubinstein and Kelly (2004), and Schiaffino and Reggiani (1996).

isoforms of myosin heavy chain (MyHC), myosin light chain, troponin, and tropomyosin, as well as metabolic enzymes (MyHC differences are summarized in Table 1.1; Agbulut *et al.*, 2003; Biressi *et al.*, 2007b; Gunning and Hardeman, 1991; Lu *et al.*, 1999; Rubinstein and Kelly, 2004; Schiaffino and Reggiani, 1996).

The finding that myogenesis occurs in successive phases and that embryonic, fetal, neonatal, and adult muscle are distinctive raises the question of how these different types of muscle arise. Potentially, these muscle types arise from different progenitors or alternatively from different myoblasts. Another possibility is that the differences in muscle arise during the process of differentiation of myoblasts into myocytes and myofibers. In addition, there is the overlying question of whether differences arise because of intrinsic changes in the myogenic cells or whether changes in the extrinsic environment are regulating myogenic cells.

Five theoretical, simplistic models could explain how these different types of muscle arise (Fig. 1.1). In these models, we have combined fetal and neonatal muscle into one group. (While embryonic and adult muscle are clearly distinct, the distinction between fetal and neonatal muscle is not so clear. Other than birth of the animal, fetal and neonatal muscle appear not to be discrete, but rather to be a gradually changing population of myogenic cells). In the first theoretical model, three different progenitor populations give rise to three distinct myoblast populations and these myoblasts, in turn, give rise to the different types of muscle. In this model, all differences in muscle could simply reflect initial intrinsic heterogeneities in the original progenitor populations, and it will be critical to understand the mechanisms that generate different types of progenitors. A second model is that all muscle derives from a progenitor population that changes over time to give rise to three different populations of myoblasts, and these different myoblast populations give rise to different types of muscle. In this model, the interesting question is understanding what intrinsic or extrinsic factors regulate changes in the progenitor population. In the third model there is a single invariant progenitor population which gives rise to three initially similar myoblast populations. These myoblast populations change over time such that they give rise to different muscle types. In this scenario, understanding the intrinsic or extrinsic factors that lead to differences in myoblasts will be important. In the fourth model, there is a single invariant progenitor population which gives rise to an initial myoblast population. This initial myoblast population both gives rise to embryonic muscle and gives rise to a successive series of myoblast populations. These gradually differing myoblasts then give rise to different types of muscle. Here, differences in muscle arise entirely from differences in the myoblast populations, and so it will be critical to ascertain the intrinsic and extrinsic factors altering the myoblasts. In the final model, a single invariant progenitor population gives rise to a single myoblast population. Subsequently, in the process of myoblast

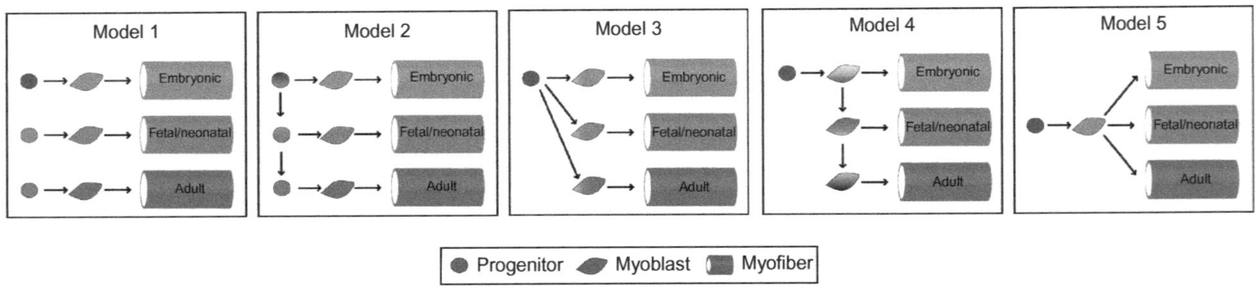

Figure 1.1 Five theoretical models describing derivation of embryonic, fetal/neonatal, and adult limb muscle in mouse. (See Color Insert.)

differentiation differences arise so that different muscle types are generated. However, this final model is unlikely to be correct because, as described above, it is well established that different myoblast populations are present and identifiable. It should be noted that a common component of all of these models is the assumption, currently made by most muscle researchers, that progenitors give rise to myoblasts and that myoblasts give rise to differentiated muscle and that this progression is irreversible. In all likelihood, myogenesis is considerably more complex than these five models. We present these models simply as a starting point to evaluate current data.

In this review, we will discuss what is known about the Pax3/7 and MRF family of transcription factors and how these data allow us to construct a model of muscle development. We focus on Pax3 and 7 and the MRFs because these both mark different myogenic populations and are functionally critical for myogenesis. We will limit our discussion to studies conducted in mouse, largely because of the availability of genetic tools available to conduct lineage, cell ablation, and conditional mutagenesis experiments (Hutcheson and Kardon, 2009). In addition, we will concentrate on myogenesis in the limb because all phase of myogenesis—embryonic, fetal/neonatal, and adult—have been studied in the limb. For discussions of myogenic progenitors in other model organisms, such as chick and zebrafish, and in the head and trunk, we refer the reader to several excellent recent reviews (Buckingham and Vincent, 2009; Kang and Krauss, 2010; Otto *et al.*, 2009; Relaix and Marcelle, 2009; Tajbakhsh, 2009)

3. Expression Analyses of Pax3/7 and MRF Transcription Factors

Multiple expression studies have established that Pax3 and Pax7 label muscle progenitors (summarized in Table 1.2). Both Pax3 and Pax7 are initially expressed in the somites. Pax3 is first expressed (beginning at E8) in the presomitic mesoderm as somites form, but is progressively restricted, first to the dermomyotome and later to dorsomedial and ventrolateral dermomyotomal lips (Bober *et al.*, 1994; Goulding *et al.*, 1994; Horst *et al.*, 2006; Schubert *et al.*, 2001; Tajbakhsh and Buckingham, 2000). Pax7 expression initiates later (beginning at E9) in the somites and is expressed in the dermomyotome, with highest levels in the central region of the dermomyotome (Horst *et al.*, 2006; Jostes *et al.*, 1990; Kassar-Duchossoy *et al.*, 2005; Relaix *et al.*, 2004). In the limb, Pax3+ progenitors are transiently present between E10.5 and E12.5 (Bober *et al.*, 1994). Although Pax3 is generally not expressed in association with muscle after E12.5, some adult satellite cells have been reported to express Pax3 (Conboy and Rando, 2002; Relaix *et al.*, 2006). Unlike Pax3 (and unlike

Table 1.2 Summary of Pax3, Pax7, Myf5, MyoD, Myogenin, and Mrf4 expression in embryonic, fetal/neonatal, and adult progenitors, myoblasts, and myofibers

	Pax3	Pax7	Myf5	MyoD	Mrf4	Myogenin
Embryonic progenitors	+ (reviewed in Buckingham, 2007; Relaix *et al.*, 2004)	+ (reviewed in Buckingham, 2007; Relaix *et al.*, 2004)	−	−	−	−
Embryonic myoblasts/ myocytes	−	−	+ (myoblasts) (Biressi *et al.*, 2007a,b; Kassar-Duchossoy *et al.*, 2005; Ott *et al.*, 1991)	+ (myoblasts) (Ontell *et al.*, 1993a; Sassoon *et al.*, 1989)	−/?	+ (myocytes) (Ontell *et al.*, 1993a; Sassoon *et al.*, 1989)
Embryonic myofibers	−	−	−	+ (Ontell *et al.*, 1993a; Sassoon *et al.*, 1989)	− (Bober *et al.*, 1991)	+ (Ontell *et al.*, 1993a; Sassoon *et al.*, 1989)
Fetal progenitors	−	+ (reviewed in Buckingham, 2007; Relaix *et al.*, 2004)	−	−	−	−
Fetal myoblasts/ myocytes	−	−	+ (myoblasts) (Biressi *et al.*, 2007a,b; Kassar-Duchossoy *et al.*, 2005)	+ (myoblasts) (Ontell *et al.*, 1993b)	−	+ (myocytes) (Ontell *et al.*, 1993b)

Fetal myofibers	–	–	–	+ (Ontell *et al.*, 1993b)	+ (Bober *et al.*, 1991)	+ (Ontell *et al.*, 1993b)
Adult progenitors	+/– (Conboy and Rando, 2002; Relaix *et al.*, 2006)	+ (Seale *et al.*, 2000)	+/– (Beauchamp *et al.*, 2000; Cornelison and Wold, 1997; Kuang *et al.*, 2007)	– (Cornelison and Wold, 1997; Kanisicak *et al.*, 2009; Yablonka-Reuveni *et al.*, 1999)	– (Cornelison and Wold, 1997; Gayraud-Morel *et al.*, 2007)	– (Cornelison and Wold, 1997)
Adult myoblasts/ myocytes	–	–	+ (myoblasts) (Cornelison and Wold, 1997; Kuang *et al.*, 2007)	+ (myoblasts) (Cornelison and Wold, 1997; Kanisicak *et al.*, 2009; Yablonka-Reuveni and Rivera, 1994)	+/– (Cornelison and Wold, 1997; Gayraud-Morel *et al.*, 2007)	+ (myocytes) (Cornelison and Wold, 1997; Yablonka-Reuveni and Rivera, 1994)
Adult myofibers	–	–	– (Hinterberger *et al.*, 1991; Voytik *et al.*, 1993)	+ (Hinterberger *et al.*, 1991; Kanisicak *et al.*, 2009; Voytik *et al.*, 1993)	++ (Gayraud-Morel *et al.*, 2007; Haldar *et al.*, 2008; Hinterberger *et al.*, 1991; Voytik *et al.*, 1993)	+ (Hinterberger *et al.*, 1991; Voytik *et al.*, 1993)

in the chick), Pax7 is not expressed in progenitors in the limb until E11.5 and then continues to be expressed in fetal and neonatal muscle (Relaix *et al.*, 2004). In the adult, Pax7 labels all satellite cells (Seale *et al.*, 2000). Much of this analysis of Pax3 and Pax7 expression has been based on RNA *in situ* hybridization and immunofluorescence. In addition, a variety of reporter alleles (both "knock-ins" and transgenes) have been developed to genetically mark Pax3+ and Pax7+ cells: $Pax3^{IRESnLacZ}$, $Pax3^{GFP}$, $Pax7^{LacZ}$, $Pax7^{nGFP}$, $Pax7^{nLacZ}$ (Mansouri *et al.*, 1996; Relaix *et al.*, 2003, 2005; Sambasivan *et al.*, 2009). These alleles have been extremely useful, as they can increase the sensitivity of detection of Pax3+ and Pax7+ cells. Nevertheless, these reporters should be used with care because, as has been often noted, the stability of the of reporter does necessarily not match the stability of the endogenous protein. For instance, the Pax3 protein is tightly regulated by ubiquitination and proteasomal degradation (Boutet *et al.*, 2007), and it has been shown that the GFP from the $Pax3^{GFP}$ allele is expressed similarly to Pax3, but perdures longer than the endogenous Pax3 protein (Relaix *et al.*, 2004).

The MRFs are expressed in myoblasts, myocytes, and myofibers in different phases oflimb myogenesis (summarized in Table 1.2). Myf5, MyoD, Mrf4, and Myogenin are all first expressed in somitic cells (Bober *et al.*, 1991; Ott *et al.*, 1991; Sassoon *et al.*, 1989; Tajbakhsh and Buckingham, 2000). However, somitic cells migrating into the limb do not initially express the MRFs (Tajbakhsh and Buckingham, 1994). Myf5 and MyoD are the earliest MRFs expressed in the developing limb. Myf5 is expressed at E10.5 in embryonic myoblasts and continues to be expressed in fetal and adult myoblasts (Biressi *et al.*, 2007b; Cornelison and Wold, 1997; Kassar-Duchossoy *et al.*, 2005; Kuang *et al.*, 2007; Ott *et al.*, 1991). Myf5 is also expressed in many, but not all adult quiescent satellite cells (Beauchamp *et al.*, 2000; Cornelison and Wold, 1997; Kuang *et al.*, 2007). Unlike the other MRFs, Myf5 expression is limited to myoblasts (or adult progenitors), as it is downregulated in differentiated myogenic cells. MyoD also begins to be expressed in the limb at E10.5 in embryonic myoblasts and myofibers (Ontell *et al.*, 1993a; Sassoon *et al.*, 1989), and subsequently is also expressed in fetal and adult myoblasts and myofibers (Cornelison and Wold, 1997; Hinterberger *et al.*, 1991; Kanisicak *et al.*, 2009; Ontell *et al.*, 1993b; Voytik *et al.*, 1993; Yablonka-Reuveni and Rivera, 1994). Unlike Myf5, MyoD rarely appears to be expressed in quiescent satellite cells (Cornelison and Wold, 1997; Yablonka-Reuveni and Rivera, 1994; Zammit *et al.*, 2002). Myogenin is expressed in the limb by E11.5 (Ontell *et al.*, 1993a; Sassoon *et al.*, 1989) and is primarily found in differentiated myocytes and myofibers of embryonic, fetal, and adult muscle (Cornelison and Wold, 1997; Hinterberger *et al.*, 1991; Ontell *et al.*, 1993a,b; Sassoon *et al.*, 1989; Voytik *et al.*, 1993; Yablonka-Reuveni and Rivera, 1994). Mrf4 is the last MRF to be expressed in the limb. It is first expressed in the limb at E13.5, with stronger expression in fetal myofibers by E16.5, and continues to be expressed as the predominant MRF in adult myofibers (Bober *et al.*, 1991; Gayraud-Morel *et al.*, 2007; Haldar *et al.*, 2008; Hinterberger *et al.*, 1991; Voytik *et al.*, 1993).

Similar to Pax3 and Pax7, expression analyses of the MRFs have been facilitated by the generation of reporter alleles $Myf5^{nLacz}$, $Myf5^{GFP\text{-}P}$, and $Mrf4^{nLacZ\text{-}P}$ (Kassar-Duchossoy *et al.*, 2004; Tajbakhsh *et al.*, 1996). These "knock-in" alleles have allowed for increased sensitivity in tracking Myf5+ and Mrf4+ cells. However, these alleles must be used with caution as Myf5 and Mrf4 are genetically linked, and the reporter constructs disrupt the expression of the linked gene to varying degrees (Kassar-Duchossoy *et al.*, 2004).

These expression studies are important both for establishing which myogenic populations are labeled by Pax3, Pax7, and MRF genes and also for describing the temporal–spatial relationship between the expression of these transcription factors and the cell populations they label. Most significantly, these studies are critical for generating testable hypotheses about gene function and cell lineage relationships. In terms of gene function, the expression of Pax3 and Pax7 in progenitors suggests that these genes are important for specification or maintenance of progenitors. The expression of MyoD and Myf5 in myoblasts suggests that these MRFs may be critical for myoblast determination. Finally, the expression of MyoD, Myogenin, and Mrf4 in myocytes or myofibers suggests that these MRFs play a role in differentiation. Thus gene expression studies strongly implicate Pax and MRF as playing roles in myogenesis and are a good starting point for designing appropriate functional experiments. However, as will be described in the following section, gene expression does not necessarily indicate critical gene function. For instance, Pax7 is strongly expressed in adult satellite cells, but is not functionally important for muscle regeneration by satellite cells (Lepper *et al.*, 2009). In terms of lineage, the finding that Pax3 is expressed before Pax7 in muscle progenitors in the limb suggests that Pax3+ cells may give rise to Pax7+ cells. In addition, the demonstration that MRFs are expressed after Pax3 also suggests that Pax3+ cells give rise to MRF+ myoblasts. However, gene expression data is not sufficient to allow us to reconstruct cell lineage. For instance, because Pax3 is only transiently expressed in progenitors, but not in myoblasts or differentiated myogenic cells, it is impossible to trace the fate of these Pax3+ progenitors. Conversely, continuity of gene expression, for example, the expression of MyoD in both myoblasts and myofibers, does not necessarily indicate continuity of cell lineage because new cells may initiate gene expression *de novo* while other cells may downregulate gene expression.

4. Functional Analysis of Pax3/7 and MRF Transcription Factors

Mouse genetic loss-of-function studies not only demonstrate that Pax3 is required for limb myogenesis, but also indicate that Pax3+ progenitors are essential to generate all the myogenic cells in the limb (Table 1.3). Pax3 function has been studied for over 50 years because of

Table 1.3 Summary of phenotypes with loss of function in mouse of Pax3, Pax7, Myf5, MyoD, Myogenin, Mrf4, and combinations of Pax3, Pax7, and MRFs

	Pax3 (Lepper *et al.*, 2009; Relaix *et al.*, 2004 and references therein)	Pax7 (Kuang *et al.*, 2006; Lepper *et al.*, 2009; Oustanina *et al.*, 2004; Relaix *et al.*, 2006; Seale *et al.*, 2000)	Pax3/Pax7 (Relaix *et al.*, 2005)	Pax3/Myf5/Mrf4 (Tajbakhsh *et al.*, 1997)	Myf5 (Gayraud-Morel *et al.*, 2007; Kassar-Duchossoy *et al.*, 2004)	MyoD (Gayraud-Morel *et al.*, 2007; Kablar *et al.*, 1997; Megeney *et al.*, 1996; Rudnicki *et al.*, 1992; White *et al.*, 2000; Yablonka-Reuveni *et al.*, 1999)	Mrf4 (Zhang *et al.*, 1995)	Myogenin or Myogenin/Myf5 or Myogenin/MyoD or Myogenin/Mrf4 (Hasty *et al.*, 1993; Nabeshima *et al.*, 1993; Rawls *et al.*, 1995, 1998; Venuti *et al.*, 1995)	Myf5/MyoD (Kassar-Duchossoy *et al.*, 2004; Kassar-Duchossoy *et al.*, 2005)	Myf5/Mrf4 (Braun and Arnold, 1995; Kassar-Duchossoy *et al.*, 2004; Tajbakhsh *et al.*, 1997)	MyoD/Mrf4 (Rawls *et al.*, 1998)	Myf5/MyoD/Mrf4 (Kassar-Duchossoy *et al.*, 2004; Rudnicki *et al.*, 1993)	MyoD/Mrf4/Myogenin (Valdez *et al.*, 2000)
Axial	Defects in somite segmentation, epaxial and hypaxial dermomyotome, trunk muscle	No phenotype observed	Only form primary myotome. No embryonic or fetal axial muscle	Defective primary myotome. No embryonic or fetal axial muscle	Delay of primary myotome formation, lack of some epaxial muscles in adult	Normal primary myotome and epaxial muscles, delay in hypaxial muscles	No phenotype observed	Embryonic axial muscle normal, no MyHCperi+ fetal axial muscle	Delay of primary myotome, embryonic axial muscle at E12.5, no fetal axial muscle	Delayed primary myotome, lack of some epaxial muscle	Embryonic axial muscle normal, no axial fetal muscle	No myotome or axial muscle	Only Myf5+ myoblasts, no myofibers
Embryonic Limb E11.5–E14.5	No limb muscle due to defects in delamination, migration, maintenance of limb progenitors	No phenotype observed	No limb muscle (see Pax3 phenotype)	No limb muscle (see Pax3 phenotype)	No phenotype observed	2.5 day delay in limb myogenesis, no limb muscle until E13.5	No phenotype observed	Normal embryonic limb myoblasts and MyHCemb+ myofibers	No limb muscle at E12.5 (MyoD phenotype), a few myofibers at E14.5	No phenotype observed	Normal embryonic limb myoblasts and myofibers	No limb muscle	Not explicitly tested

Fetal Limb E14.5–E18.5	No limb muscle (see above)	No phenotype observed	No limb muscle (see Pax3 phenotype)	No limb muscle (see Pax3 phenotype)	No phenotype observed	No phenotype observed	No phenotype observed	No MyHCperi+ fetal limb myofibers, a few residual myofibers, myoblasts present	Few myofibers at E14.5, no fetal myofibers by birth	No phenotype observed	Myoblasts present, but few residual myofibers	No limb muscle	No differentiated myofibers, no MyHCemb
Neonatal Limb P0–P21	Dead	Defects in satellite cell survival, proliferation, and differentiation (as tested by conditional deletion)	Dead	Dead	No phenotype observed	No phenotype observed	No phenotype observed	Few residual myofibers, perinatal death	No muscle at birth, perinatal death	No phenotype observed, perinatal death	Few residual myofibers, perinatal death	No limb muscle, perinatal death	No differentiated myofibers, no MyHCemb, perinatal death
Adult/ regeneration	Dead (Pax3 null mice). Not required (as tested by conditional deletion)	No effect on adult muscle regeneration (as tested by conditional deletion)	Dead	Dead	Impaired regeneration with delayed differentiation, fiber hypertrophy, increased fat and fibrosis	Delayed and impaired regeneration with increased number of satellite cells and fewer differentiated myofibers	Not explicitly tested	Dead	Dead	Dead	Dead	Dead	Dead

It should be noted that some *Myf5* null and *Mrf4* null alleles affected the expression of Mrf4 and Myf5, respectively. Thus, for instance, the $Myf5^{nlacZ/nlacZ}$ mice, originally described as *Myf5* null mice (Tajbakhsh *et al.*, 1997), are also null for *Mrf4* (Kassar-Duchossoy *et al.*, 2004). Only the $Myf5^{loxP/loxP}$ allele leaves *Mrf4* intact (Kassar-Duchossoy *et al.*, 2004). In this table, phenotypes described for both *Myf5* null and compound *Myf5* and *MyoD* null are based on $Myf5^{loxP/loxP}$ mice. Similarly, while three *Mrf4* null alleles were generated (Braun and Arnold, 1995; Patapoutian *et al.*, 1995; Zhang *et al.*, 1995), only one *Mrf4* null allele leaves *Myf5* intact (Olson *et al.*, 1996; Zhang *et al.*, 1995). The phenotype described here for the *Mrf4* null is based on this allele from the Olson lab (Zhang *et al.*, 1995).

the availability of a naturally occurring functional null allele of Pax3, the Splotch mutant (Auerbach, 1954; Epstein *et al.*, 1993). In *Pax3*^{Sp} Splotch mutants (which generally die by E14.5), as well as other splotch mutants such as *Pax3*^{SpD} (which die at E18.5), no embryonic or fetal muscle forms in the limb (Bober *et al.*, 1994; Franz *et al.*, 1993; Goulding *et al.*, 1994; Vogan *et al.*, 1993). There is a complete lack of myoblasts, myocytes, and myofibers, as indicated by the lack of expression of MRFs and muscle contractile proteins. Functional Pax3 is required for multiple aspects of somite development and limb myogenesis. In the somite, Pax3 regulates somite segmentation and formation of the dorsomedial and ventrolateral dermomyotome (Relaix *et al.*, 2004; Schubert *et al.*, 2001; Tajbakhsh and Buckingham, 2000). For limb myogenesis, Pax3 is required for maintenance of the ventrolateral somitic precursors, delamination (via activation of Met expression) from the somite of limb myogenic progenitors, migration of progenitors into the limb, and maintenance of progenitor proliferation (Relaix *et al.*, 2004). Interestingly, in the adult conditional deletion of *Pax3* in satellite cells revealed that, despite observed expression of Pax3 in satellite cells of some muscles (Relaix *et al.*, 2006), Pax3 is not required for muscle regeneration (Lepper *et al.*, 2009). Together these data show that Pax3 is required for embryonic myogenesis in the limb, but is not subsequently required in the adult. Whether Pax3 is required for fetal limb myogenesis has not been explicitly tested. These functional data also elucidate the nature of the progenitors which give rise to limb muscle. The complete absence of muscle in the limb in *Pax3* mutants, in combination with the early transient (E10.5–E12.5) expression of Pax3 in limb muscle progenitors, suggests that these early Pax3+ progenitors (present up to E12.5) give rise to all embryonic and fetal myoblasts, myocytes, and myofibers in the limb. This suggests that our theoretical Model 1, in which multiple distinct progenitors give rise to different myoblasts and myofibers, is unlikely to be correct. Instead Models 2–4 (or some variant of them), in which all muscle ultimately derives from one initial progenitor population, are more likely representations of limb myogenesis.

Functional analysis of Pax7 has established that Pax7 regulates neonatal progenitors and also reveals that there are at least two genetically distinct populations of progenitors (Table 1.3). Analysis of *Pax7* loss-of-function alleles has been complicated. Although no muscle phenotypes were initially recognized in null *Pax7*$^{LacZ/LacZ}$ (Mansouri *et al.*, 1996), subsequent analysis suggested that no satellite cells were specified in the absence of Pax7 (Seale *et al.*, 2000). Then a series of papers (Kuang *et al.*, 2006; Oustanina *et al.*, 2004; Relaix *et al.*, 2006) determined that, in fact, satellite cells were present in *Pax7* null mice. However, Pax7 was found to be critical for maintenance, proliferation, and function of satellite cells. More recently, conditional deletion of *Pax7* in satellite cells, via a tamoxifen-inducible $Pax7^{CreERT2}$ allele and a $Pax7^{fl}$ allele, has surprisingly shown that Pax7 is not required after P21 (the end of neonatal myogenesis) for effective muscle regeneration (Lepper *et al.*, 2009). However, consistent with the previous studies (Kuang *et al.*, 2006; Oustanina *et al.*, 2004; Relaix *et al.*, 2006), conditional deletion of *Pax7* between P0 and P21 did show a requirement for Pax7 in neonatal satellite cells for proper proliferation and myogenic differentiation (Lepper *et al.*, 2009). Thus, this study demonstrates that Pax7 is dispensable in the adult, but required in neonatal satellite cells for their maintenance, proliferation, and differentiation. Prior to birth, myogenesis appears not to require Pax7, as gross muscle morphology is normal (Oustanina *et al.*, 2004; Seale *et al.*, 2000). However, the reduced number of satellite cells just after birth (Oustanina *et al.*, 2004; Relaix *et al.*, 2006) suggests that proliferation and/or maintenance of fetal progenitors may be functionally dependent on Pax7. In total, these functional studies reveal that there are at least two populations of progenitors: Pax7-functionally dependent neonatal satellite cells and Pax7-functionally independent adult satellite cells. Thus, a model of myogenesis in which there is only one invariant progenitor population (as seen in Models 3, 4, and 5) is unlikely to be correct.

Compound mutants of *Myf5*, *MyoD*, and *Mrf4* demonstrate that embryonic and fetal myoblasts have different genetic requirements for their determination (Table 1.3). Over the past 20 years, multiple loss-of-function alleles of all four MRFs have been generated and allowed for detailed characterization of their function. However, analysis of Myf5 and Mrf4 function has been complicated because these two genes are genetically linked, and so many of the original *Myf5* and *Mrf4* loss-of-function alleles also affected the expression of the neighboring gene (see discussion in Kassar-Duchossoy *et al.*, 2004; Olson *et al.*, 1996). Single loss-of-function mutants of *Myf5* or *Mrf4* (in which genetically linked Mrf4 and Myf5 expression remain intact) show no defects in embryonic or fetal limb myogenesis (Kassar-Duchossoy *et al.*, 2004; Zhang *et al.*, 1995), and *MyoD* mutants have only a minor phenotype, a 2–2.5 day delay in embryonic limb myogenesis (Kablar *et al.*, 1997; Rudnicki *et al.*, 1992). Compound *Myf5* and *Mrf4* null mutants have normal embryonic and fetal limb muscle (Braun and Arnold, 1995; Kassar-Duchossoy *et al.*, 2004; Tajbakhsh *et al.*, 1997). Compound *MyoD* and *Mrf4* mutants (in which *Myf5* expression remains intact) have normal embryonic myoblasts and myofibers (but with a 2 day delay in development, reflecting the *MyoD* null phenotype) and fetal myoblasts (although fetal myofibers are absent, see below; Rawls *et al.*, 1998). Compound *Myf5* and *MyoD* loss-of-function mutants (in which *Mrf4* expression is intact) contain no fetal myoblasts or myofibers. However, a few residual embryonic myofibers are present and therefore indicate the presence of some embryonic myoblasts (Kassar-Duchossoy *et al.*, 2004). In triple *Myf5*, *MyoD*, and *Mrf4* loss-of-function mutants, no embryonic or fetal myoblasts or myofibers are present (Kassar-Duchossoy *et al.*, 2004; Rudnicki *et al.*, 1993). Together these genetic data indicate that embryonic myoblasts require Myf5, MyoD, or Mrf4 for their determination, although these MRFs differ somewhat in their function. MyoD can most efficiently determine embryonic myoblasts, as embryonic myogenesis is normal in compound *Myf5* and *Mrf4* mutants. While Myf5 can determine embryonic myoblasts, the inability of Myf5 to act as a differentiation factor leads to a delay in limb myogenesis in compound *MyoD* and *Mrf4* mutants. Mrf4 can only poorly substitute for Myf5 or MyoD as a determination factor, and so in the absence of *Myf5* and *MyoD*, limb embryonic myogenesis is only partially rescued by *Mrf4*. Unlike embryonic myoblasts, fetal myoblasts require either Myf5 or MyoD for their determination, and Mrf4 is not able to rescue this function. These data argue that embryonic and fetal myoblasts have different genetic requirements for their determination and therefore concurs with previous culture data showing that embryonic and fetal myoblasts are distinct. The presence of at least two classes of myoblasts therefore excludes Model 5, in which one myoblast population gives rise to different types of myofibers, and argues in favor of Models 1–4, in which multiple myoblast populations are important for generating different types of myofibers. It is likely that embryonic, fetal, and adult myoblasts are distinct populations. However, the genetic requirements of adult myoblasts has not been completely tested. Loss of either *Myf5* or *MyoD* leads to delayed or impaired muscle regeneration (Gayraud-Morel *et al.*, 2007; Megeney *et al.*, 1996; White *et al.*, 2000; Yablonka-Reuveni *et al.*, 1999). The role of Mrf4 in regeneration has not been explicitly tested, although the lack of Mrf4 expression in adult myoblasts suggests Mrf4 may not be required (Gayraud-Morel *et al.*, 2007). To test whether Myf5, MyoD, or Mrf4 may be acting redundantly in the adult will require conditional deletion of these MRFs in adult progenitors since compound mutants die at birth.

Compound mutants of *MyoD*, *Mrf4*, and *Myogenin* reveal that embryonic and fetal myoblasts have different genetic requirements for their differentiation (Table 1.3). Loss of *Mrf4* results in no muscle phenotype in the limbs, while loss of *MyoD* results in only a delay in embryonic limb myogenesis (Kablar *et al.*, 1997; Rudnicki *et al.*, 1992; Zhang *et al.*, 1995). Formation of embryonic myofibers (MyHCemb+) is largely unaffected with loss of *Myogenin* (although myosin levels appear lower and myofibers are less organized); however, differentiation of fetal myofibers (MyHCperi+) is completely impaired (Hasty *et al.*, 1993; Nabeshima *et al.*, 1993; Venuti *et al.*, 1995). This lack of fetal muscle is due to a defect in differentiation *in vivo*; myoblasts are still present in *Myogenin* mutant limbs and can differentiate *in vitro* (Nabeshima *et al.*, 1993). A similar phenotype is seen in compound *Myogenin/MyoD*, *Myogenin/Mrf4*, *Myogenin/Myf5*, and *Mrf4/MyoD* null mutants. In all of these mutants, embryonic muscle differentiates, but fetal muscle does not (Rawls *et al.*, 1995, 1998; Valdez *et al.*, 2000). Also, myoblasts from these compound mutants are present and *in vitro* can differentiate. In triple *Myogenin/Mrf4/MyoD* animals, no embryonic or fetal myofibers differentiate and myoblasts from these animals cannot differentiate *in vitro* (Valdez *et al.*, 2000). Together these genetic data argue that differentiation of embryonic myofibers requires Myogenin, MyoD, or Mrf4. Myf5, which is not normally expressed in differentiating myogenic cells, is not sufficient to support myofiber differentiation. The genetic requirement of fetal myofiber differentiation is more stringent and requires Myogenin *and* either Mrf4 or MyoD. Thus, the differentiation of embryonic and fetal myofibers has different genetic requirements and argues that the embryonic and fetal myoblasts (from which the myofibers derive) are genetically different. Therefore, these data support Models 1–4, in which different embryonic and fetal myoblast populations are important for the generation of embryonic and fetal myofibers.

5. Cre-Mediated Lineage and Ablation Analyses of PAX3, PAX7, and MRF+ Cells

Cre-mediated lineage analysis in mice has provided the most direct method to test the lineage relationship of progenitors and myoblasts giving rise to embryonic, fetal, neonatal, and adult muscle. These lineage studies have been enabled by the development of *Cre/loxP* technology (Branda and Dymecki, 2004; Hutcheson and Kardon, 2009). To genetically label and manipulate different populations of muscle progenitors or myoblasts, Cre lines have been created in which Cre is placed under the control of the promoter/enhancers sequences of *Pax3/7* or MRFs. Several strategies have been used to create these Cre lines. For $Pax3^{Cre}$, $Myf5^{Cre}$, and $MyoD^{Cre}$ lines, *Cre* has been placed into the ATG of the endogenous locus (Engleka *et al.*, 2005; Kanisicak *et al.*, 2009; Tallquist *et al.*, 2000). For $Pax7^{Cre}$, $Mrf4^{Cre}$, and another $Myf5^{Cre}$ line an *IRESCre* cassette was placed at the transcriptional stop (Haldar *et al.*, 2008; Keller *et al.*, 2004). $Myogenin^{Cre}$ was created as a transgene, by placing *Cre* under the control of a 1.5 kb Myogenin promoter and a 1 kb MEF2C enhancer (Li *et al.*, 2005). Recently, tamoxifen-inducible *Cre* alleles have also been created, and these *CreERT2* alleles allow for temporal control of labeling and manipulation because

Cre-mediated recombination only occurs after the delivery of tamoxifen. A tamoxifen-inducible *Pax7*^{CreERT2} allele has been created by placing a *CreERT2* cassette into the ATG of *Pax7* (Lepper and Fan, 2010; Lepper *et al.*, 2009). For each of these alleles, the ability to label and manipulate the appropriate cell requires that the Cre be faithfully expressed wherever the endogenous gene is expressed. Placing the *Cre* or *CreERT2* cassette at the endogenous ATG is the most likely strategy for ensuring that Cre expression recapitulates endogenous gene expression. However, these alleles are all "knockin/knockout" alleles in which the *Cre* disrupts expression of the targeted genes. If there is any potential issue of haplo-insufficiency, such a targeting strategy may be problematic. For the *Pax3*^{Cre}, *Myf5*^{Cre}, and *MyoD*^{Cre} lines, haplo-insufficiency has not been found to be an issue. For *Cre* alleles generated by targeted *IRESCre* to the stop or by transgenics, the fidelity of the *Cre* needs to be carefully verified. The advantage of such Cre lines, of course, is that the endogenous gene remains intact.

To follow the genetic lineage of the Pax3+, Pax7+, or MRF+ cells, these Cre lines have been crossed to various *Cre*-responsive reporter mice. In the reporter mice, reporters such as *LacZ* or *YFP* are placed under the control a ubiquitous promoter. In the absence of *Cre*, these reporters are not expressed because of the presence of a strong transcriptional stop cassette flanked by *loxP* sites, while the presence of *Cre* causes recombination of the *loxP* sites and the permanent expression of the reporter. Therefore, in mice containing both the *Cre* and the reporter, cells expressing the *Cre* and their progeny permanently express the reporter, thus allowing the fate of Pax3+, Pax7+, or MRF+ cells to be followed. The number of cells genetically labeled in response to Cre can be dramatically affected by the reporter lines used, and the utility of each reporter must be verified for each tissue and age of animal being tested. The *R26R*^{LacZ} and *R26*^{RYFP} reporters (Soriano, 1999; Srinivas *et al.*, 2001) are commonly used with good success in the embryo to label myogenic cells. In the adult, the endogenous *R26R* locus may not be sufficient to drive high levels of reporter expression, and so reporters such as *R26R*^{mTmG} (Muzumdar *et al.*, 2007) or *R26R*^{NZG} (Yamamoto *et al.*, 2009) in which a CMV β-actin promoter additionally drives reporter expression, may be necessary.

The *Cre/loxP* system can also be used to test the requirement of particular cell populations for myogenesis, by crossing *Cre* lines with *Cre*-responsive ablater lines (Hutcheson and Kardon, 2009). In these ablater lines, *Cre* activates the expression of cell-death-inducing toxins, such as *diptheria toxin* (Brockschnieder *et al.*, 2006; Wu *et al.*, 2006). The lack of receptor for diphteria toxin in mice and the expression of only the diptheria toxin fragment A (DTA, which cannot be transferred to other cells without the diptheria toxin fragment B) ensures that only cells expressing Cre, and therefore DTA, will be cell-autonomously killed. Analogous to gene loss-of-function experiments, cell ablation experiments enable the researcher to test the *necessity* of particular genetically labeled progenitors and myoblasts for myogenesis.

The expression and functional studies of Pax3 strongly suggested that Pax3+ progenitors give rise to all embryonic, fetal, neonatal, and adult muscle. Particularly because Pax3 is only transiently expressed in progenitors in the early limb bud, tracing the lineage of Pax3+ progenitors required that the cells be genetically labeled via $Pax3^{Cre}$. These Pax3 lineage studies reveal that Pax3+ cells entering the limb are initially bipotential and able to give rise to both endothelial cells and muscle (Hutcheson *et al.*, 2009; Table 1.4). Moreover, Pax3+ cells give rise to all embryonic, fetal, and adult myoblasts and myofibers (Engleka *et al.*, 2005; Hutcheson *et al.*, 2009; Schienda *et al.*, 2006). Thus, these early Pax3+ progenitors give rise to all limb muscle and exclude Model 1 of limb myogenesis, in which multiple distinct progenitors give rise to embryonic, fetal, neonatal, and adult myofibers. Of course, it is formally possible that the Pax3+ cells migrating into the limb are a heterogeneous population in which subpopulations give rise to embryonic, fetal, and adult myoblasts (and so Model 1 might be correct). However, to test this possibility, early markers of these subpopulations would be required. The necessity of Pax3+ progenitors is demonstrated by the lack of any embryonic or fetal muscle when these cells are genetically ablated (Hutcheson *et al.*, 2009). Although not formally demonstrated (because of the P0 death of $Pax3^{Cre/+};R26R^{DTA}$ mice), it is likely that the Pax3+ progenitors are also required for the formation of all adult limb muscle. In addition, these lineage studies demonstrated that all Pax7+ progenitors in the embryo and Pax7+ satellite cells in the adult are derived from the Pax3+ progenitors (Hutcheson *et al.*, 2009; Schienda *et al.*, 2006). This finding thus supports Model 2 of limb myogenesis, in which an initial progenitor population gives rise to other progenitor populations.

Genetic lineage studies of Pax7+ progenitors have established that, unlike Pax3+ progenitors, Pax7+ progenitors in the limb are restricted to a myogenic fate (Hutcheson *et al.*, 2009; Lepper and Fan, 2010). Consistent with the later expression of Pax7 (beginning at E11.5), Pax7+ progenitors do not give rise to embryonic muscle, but do give rise to all fetal and adult myoblasts and myofibers in the limb (Hutcheson *et al.*, 2009; Lepper and Fan, 2010). Pax7+ cells labeled in the early limb (via tamoxifen delivery to E11.5 $Pax7^{CreERT2/+};R26R^{LacZ/+}$ mice) also give rise to Pax7+ adult satellite cells, although it is unclear whether these labeled cells directly become satellite cells or whether their progeny give rise to satellite cells (Lepper and Fan, 2010). The loss of fetal limb muscle when Pax7+ cells are genetically ablated demonstrates that these Pax7+ progenitors are required for fetal myogenesis in the limb (Hutcheson *et al.*, 2009). The test of whether Pax7+ progenitors are *necessary* for adult myogenesis awaits the generation of $Pax7^{CreERT2/+};R26R^{DTA/+}$ mice, in which Pax7+ progenitors are genetically ablated after birth.

Recent lineage analyses of Myf5+ and MyoD+ cells have unexpectedly revealed that two populations of myoblasts may give rise to muscle (Table 1.4).

Table 1.4 Summary of genetic lineage and ablation studies in mouse

	Pax3Cre lineage	Pax3Cre-mediated ablation	Pax7Cre lineage	Pax7CreERT2 lineage	Pax7Cre-mediated ablation	Myf5Cre lineage	Myf5Cre-mediated ablation	MyoDCre lineage	MyogeninCre lineage	MyogeninCre-mediated ablation	Mrf4Cre lineage	Mrf4Cre-mediated ablation
References	Engleka *et al.* (2005), Hutcheson *et al.* (2009), Schienda *et al.* (2006)	Hutcheson *et al.* (2009)	Hutcheson *et al.* (2009)	Lepper and Fan (2010)	Hutcheson *et al.* (2009)	Gensch *et al.* (2008), Haldar *et al.* (2008), Kuang *et al.* (2007)	Gensch *et al.* (2008), Haldar *et al.* (2008)	Kanisicak *et al.* (2009)	Gensch *et al.* (2008), Li *et al.* (2005)	Gensch *et al.* (2008)	Haldar *et al.* (2008)	Haldar *et al.* (2008)
Embryonic progenitors	+	None			Present	Not analyzed	Present	Not analyzed	Not analyzed in limb	Present	Not analyzed in limb	Not analyzed in limb
Embryonic myoblasts		None			Present	+	Present	+	Not analyzed in limb	Present	Not analyzed in limb	Not analyzed in limb
Embryonic myofibers		None			Present		Present	+	+	None	Not analyzed in limb	Not analyzed in limb
Fetal/Neonatal progenitors		None	+	★	None	Not analyzed	Present	Not analyzed	Not analyzed	Not analyzed	Not analyzed	Not analyzed
Fetal/Neonatal myoblasts		None			None	Not analyzed	Present	+ (neonatal)	Not analyzed	Not analyzed	Not analyzed	Not analyzed
Fetal/Neonatal myofibers		None			None		Present	Not analyzed	+	None	Not analyzed	None
Adult progenitors	+/−	None	+		(Dead)	+/−	(dead)		Not analyzed	(Dead)	Not analyzed	(Dead)
Adult myoblasts		None			(Dead)		(Dead)	+	Not analyzed	(Dead)	Not analyzed	(Dead)
Adult myofibers		None			(Dead)		(Dead)	+	Not analyzed	(Dead)	+	(Dead)

"+" shows cells actively transcribing the gene of interest (e.g., transcribing Pax3).

Gray boxes denote progenitors, myoblasts, and myofibers entirely derived from the genetically labeled cell population (e.g., Pax3+ cells). Hatched boxes show progenitors, myoblasts, and myofibers where only some of the cells are derived from the genetically labeled cell population. Star denotes timing of tamoxifen delivery in $Pax7^{CreERT2}$ mice.

Analysis of Myf5 lineage, using two different $Myf5^{Cre}$ lines, shows that Myf5+ cells are not restricted to a muscle fate, as cells in the axial skeleton and ribs are derived from Myf5+ cells (Gensch *et al.*, 2008; Haldar *et al.*, 2008). This likely reflects early transient expression of Myf5 in the presomitic mesoderm. In contrast, MyoD+ cells appear to be restricted to a muscle fate (Kanisicak *et al.*, 2009). Interestingly, analysis of the Myf5 lineage shows that Myf5+ cells give rise to many, but not all embryonic, fetal, and adult myofibers (Gensch *et al.*, 2008; Haldar *et al.*, 2008). The distribution of Myf5-derived myofibers appears to be stochastic, as epaxial and hypaxial, slow and fast, and different anatomical muscles are randomly Myf5-derived. Unlike Myf5, analysis of MyoD lineage reveals that MyoD+ cells give rise to all embryonic and adult myofibers (fetal myofibers were not explicitly examined; Kanisicak *et al.*, 2009). Consistent with these lineage studies, ablation of Myf5+ cells did not lead to any dramatic defects in embryonic or fetal muscle (the $Myf5^{CreERT2/+};R26R^{DTA/+}$ mice die at birth from rib defects), as presumably Myf5- myoblasts compensated for the loss of Myf5+ myoblasts (Gensch *et al.*, 2008; Haldar *et al.*, 2008). Ablation of the MyoD lineage has not yet been published, but based on the lineage studies a complete loss of muscle would be expected. Together, these lineage and ablation studies argue that there are at least two populations of myoblasts, one Myf5-dependent and one Myf5-independent, thus excluding Model 5, in which only one myoblast population generates all limb muscle. It is not yet clear whether there may, in fact, be three populations of myoblasts: Myf5+MyoD−, Myf5+MyoD+, and Myf5−MyoD+. The finding that all muscle is MyoD-derived would suggest that there are no myoblasts that are Myf5+MyoD−. However, because MyoD is strongly expressed in embryonic and fetal myofibers, the finding that all muscle is YFP+ in $MyoD^{Cre/+};R26R^{YFP/+}$ mice may simply reflect MyoD expression in all myofibers, and not MyoD expression in all myoblasts. Another question yet to be resolved is whether multiple myoblast populations are present during embryonic, fetal, and neonatal myogenesis.

Analysis of the Myf5 and MyoD lineages has also revealed interesting insights about adult satellite cells. The great majority of quiescent satellite cells have been shown to be YFP labeled in $Myf5^{Cre/+};R26R^{YFP/+}$ mice (Kuang *et al.*, 2007). Given that most quiescent satellite cells express Myf5 (Beauchamp *et al.*, 2000; Cornelison and Wold, 1997), it is likely that the Myf5 lineage in satellite cells is simply reflecting active Myf5 transcription in satellite cells. However, the finding that all quiescent satellite cells are YFP labeled in $MyoD^{Cre/+};R26R^{YFP/+}$ mice was quite surprising (Kanisicak *et al.*, 2009). Multiple studies have shown that quiescent satellite cells do not express MyoD, although activated satellite cells do express MyoD (Cornelison and Wold, 1997; Yablonka-Reuveni and Rivera, 1994). Thus, the finding that quiescent satellite cells are YFP+ in $MyoD^{Cre/+};R26R^{YFP/+}$ mice suggests that all quiescent satellite cells are derived from previously activated, MyoD+ satellite cells (as suggested by Zammit *et al.*,

2004). Alternatively, all quiescent satellite cells may be derived from MyoD+ myoblasts. To definitively test whether satellite cells indeed are derived from MyoD+ myoblasts, *MyoD*$^{CreERT2/+}$;*R26R*$^{YFP/+}$ mice will need to be induced with tamoxifen in the embryo or fetus, before satellite cells are present. It will also be interesting to test using *Myf5*$^{CreERT2/+}$; *R26R*$^{YFP/+}$ mice whether Myf5+ myoblasts in the embryo or fetus give rise to satellite cells. Such a finding that MyoD+ or Myf5+ myoblasts give rise to satellite cells would profoundly change current models of myogenesis (excluding all five Models presented) because this would demonstrate that myoblasts can return to a more progenitor-like state.

Lineage analysis using *Myogenin*^{Cre} and *Mrf4*^{Cre} mice demonstrates that by birth all myofibers have expressed both Myogenin and Mrf4 (Gensch *et al.*, 2008; Haldar *et al.*, 2008; Li *et al.*, 2005; Table 1.4). A closer examination of the Myogenin lineage reveals that all embryonic and fetal muscle has derived from Myogenin+ myocyctes and/or myofibers (Gensch *et al.*, 2008; Li *et al.*, 2005). It would be worthwhile to similarly determine to what extent embryonic muscle has expressed Mrf4 since expression studies have found Mrf4 to be expressed in at least some embryonic limb muscle (Hinterberger *et al.*, 1991). Consistent with the finding that all fetal muscle has expressed Myogenin and Mrf4, ablation of Myogenin+ or Mrf4+ cells leads to a complete loss of all muscle by birth (Gensch *et al.*, 2008; Haldar *et al.*, 2008).

6. Molecular Signals Distinguishing Between Different Phases of Myogenesis

Layered on top of these expression, functional, and lineage studies concentrating on Pax3, Pax7, and MRFs are functional studies demonstrating that embryonic, fetal, and adult myogenic cells show differential sensitivity to signaling molecules. Recent microarray studies demonstrated that members of the Notch, FGF, and PDGF signaling pathways are differentially expressed in embryonic versus fetal myoblasts (Biressi *et al.*, 2007b). In addition, fetal myoblasts show upregulation of components of the TGFβ and BMP signaling pathways compared to embryonic myoblasts (Biressi *et al.*, 2007b). Such findings are consistent with *in vitro* studies demonstrating that embryonic myoblast differentiation is insensitive to treatment with TGFβ or BMP, while fetal myoblast differentiation is blocked in the presence of TGFβ or BMP (Biressi *et al.*, 2007b; Cusella-De Angelis *et al.*, 1994). Interestingly, studies examining adult myogenesis also demonstrate that BMP signaling is active in activated satellite cells and proliferating myoblasts (Ono *et al.*, 2010). Furthermore, inhibition of BMP signaling results in an increase in differentiated myocytes at the expense of proliferating myoblasts *in vitro* and smaller diameter regenerating myofibers *in vivo* (Ono *et al.*, 2010). Therefore, in mouse TGFβ and BMP signaling appear to have no effect on embryonic myoblasts, whereas they inhibit differentiation of both fetal and adult myoblasts. Thus, with respect to TGFβ and BMP signaling, fetal and adult myoblasts behave similarly. It is interesting to note that in the chick limb BMP signaling has also been shown to differentially regulate embryonic versus fetal and adult myogenesis, although BMP effects were different from those found in the mouse (Wang *et al.*, 2010).

The Wnt/β-catenin pathway also differentially regulates embryonic versus fetal and adult myogenesis. The role of β-catenin in embryonic and fetal myogenesis was tested by conditionally inactivating or activating β-catenin in embryonic muscle via $Pax3^{Cre}$ or in fetal muscle via $Pax7^{Cre}$ (Hutcheson *et al.*, 2009). After myogenic cells enter the limb, embryonic myogenic cells were found to be insensitive to perturbations in β-catenin. However, during fetal myogenesis β-catenin critically determines the number of Pax7+ progenitors and the number and fiber type of myofibers. β-catenin has also been found to positively regulate the number of Pax7+ satellite cells in the adult (Otto *et al.*, 2008; Perez-Ruiz *et al.*, 2008; but see Brack *et al.*, 2008). Thus similar to the findings for TGFβ and BMP signaling, embryonic myogenesis is insensitive to β-catenin signaling, while fetal and adult myogenesis is regulated by β-catenin.

These studies demonstrate that during embryonic myogenesis Pax3+ progenitors are insensitive to TGFβ, BMP, and Wnt/β-catenin signaling. Yet during fetal and adult myogenesis, TGFβ, BMP, and Wnt/β-catenin signaling are important for positively regulating and maintaining the population of Pax7+ progenitors. During development, postmitotic myofibers must differentiate, while proliferating progenitors must be maintained for growth. Therefore, in the same environment some progenitors must differentiate, while others must continue to proliferate. It has been hypothesized that embryonic, fetal, and adult progenitors and/or myoblasts are intrinsically different so that these cells will respond differently to similar environmental signals (Biressi *et al.*, 2007a,b). Thus, differential sensitivity to TGFβ, BMP, and Wnt/β-catenin signaling may be a molecular mechanism to allow embryonic progenitors to differentiate, but maintain a fetal and adult progenitor population.

The above examples demonstrate that embryonic, fetal, and adult myogenesis are differentially regulated by different signaling pathways. Until recently, what signals regulate the transitions from embryonic to fetal, neonatal, and adult myogenesis have been unknown. The expression of Pax7 in progenitors demarcates progenitors as being fetal/neonatal/adult progenitors, as opposed to Pax3+ embryonic progenitors. Now elegant *in vitro* and *in vivo* studies demonstrate that that the transcription Nfix is expressed in fetal and not embryonic myoblasts, and Pax7 directly binds and activates the expression Nfix (Messina *et al.*, 2010). Moreover, Nfix is critical for regulating the transition from embryonic to fetal myogenesis. Nfix both represses genes highly expressed in embryonic muscle, such as MyHCI, and activates the expression of fetal-specific genes, such as α7-integrin, β-enolase, muscle creatine kinase, and muscle sarcomeric proteins. Thus Nfix functions as an intrinsic transcriptional switch which mediates the transition from embryonic to fetal myogenesis. Recent studies have also demonstrated that extrinsic signals from the connective tissue niche, within which muscle resides, are also important for regulating muscle maturation (Mathew *et al.*, 2011). The connective tissue promotes the switch from the fetal to adult muscle by repressing developmental isoforms of myosin and promoting formation of large, multinucleate myofibers. Determining the full range of intrinsic and extrinsic factors that regulate the transitions from embryonic to fetal, neonatal, and adult myogenesis will be important areas for future research.

7. Current Model of Myogenesis

From these expression, functional, and lineage studies, a current model of myogenesis in the limb emerges that is a variant of our theoretical Models 2 and 4 (Fig. 1.2). Embryonic, fetal, neonatal, and adult muscle derive from three related, but distinct populations of progenitors. From the somite, Pax3+ progenitors migrate into the limb and are bipotential, giving rise to either endothelial cells or muscle. Myogenic Pax3+ cells require Pax3 function for their delamination from the somites, migration, and maintenance. Pax3+ cells give rise to and are required for embryonic myogenesis. In addition, Pax3+ cells give rise to Pax7+ progenitors. In turn, these Pax3-derived, Pax7+ progenitors give rise to and are required for fetal myogenesis. These Pax7+ progenitors also appear to give rise to neonatal muscle, but whether the fetal and neonatal progenitors are exactly the same population is unclear. Unlike fetal Pax7+ progenitors, neonatal Pax7+ progenitors may reside underneath the basal lamina of myofibers, similar to satellite cells. Also, while it has been shown that neonatal Pax7+ cells require Pax7 for their maintenance and proper function, it has not been explicitly tested whether fetal Pax7+ cells require Pax7. Adult muscle derives from Pax7+ progenitors, satellite cells, which reside under the myofiber basal lamina. Unlike Pax7+ neonatal progenitors, Pax7+ satellite cells do not require Pax7 for their maintenance and function. Also, the great majority of quiescent Pax7+ satellite cells express Myf5. Pax7+ satellite cells are likely to directly derive from fetal or neonatal Pax7+ progenitors. However, the finding that all quiescent Pax7+ satellite cells have expressed MyoD in their lineage suggests that satellite cells may derive indirectly from

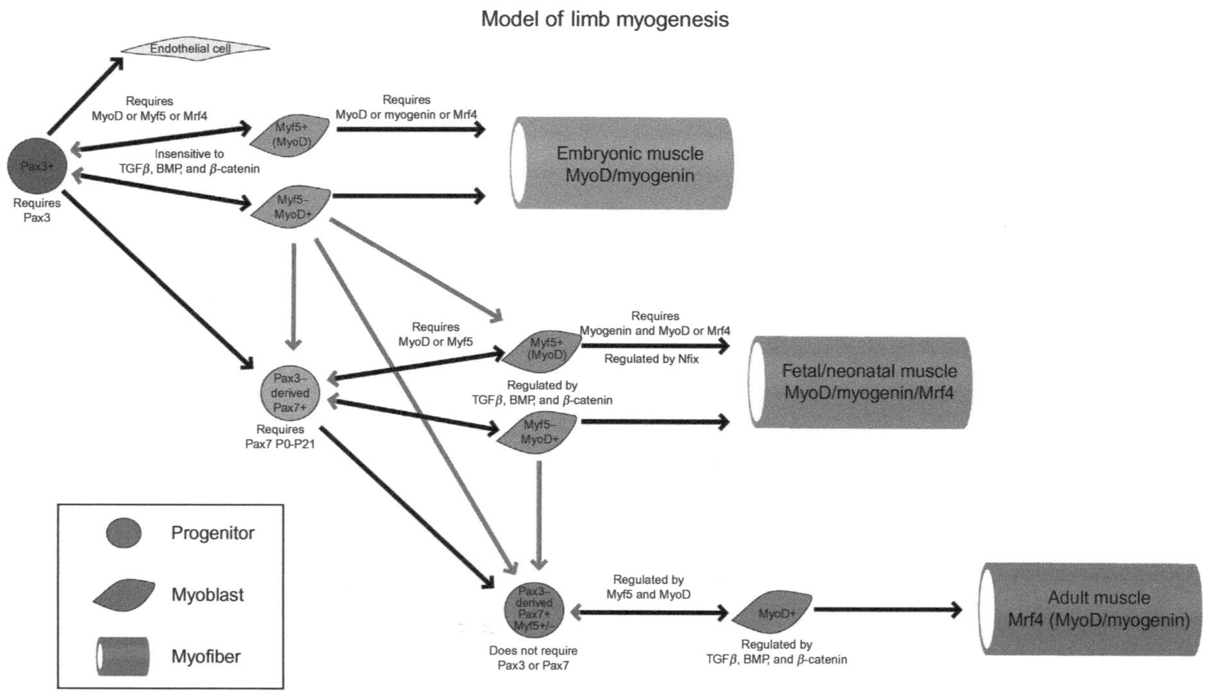

Figure 1.2 Summary of current model of embryonic, fetal/neonatal, and adult limb myogenesis in the mouse. (See Color Insert.)

Pax7+ fetal or neonatal myogenic progenitors via MyoD+ (or potentially Myf5+) myoblasts (gray arrows in Fig. 1.2). Also, some Pax7+ satellite cells may derive from adult myoblasts, generated by activated Pax7+ satellite cells. Both scenarios would suggest that the progression from progenitor to myoblasts may not be irreversible, and myoblasts may give rise to Pax7+ progenitors.

There are multiple distinct populations of myoblasts that give rise to embryonic, fetal/neonatal, and adult muscle. Embryonic myoblasts are distinct from fetal/neonatal myoblasts. Embryonic limb myoblasts require either MyoD, Myf5, or Mrf4 for their determination, while fetal myoblasts require either MyoD or Myf5 (Mrf4 cannot support fetal myoblasts). Adult myoblast function is regulated by Myf5 and MyoD, but whether Myf5 and MyoD are required has not been formally tested. Within embryonic and fetal myoblasts there appear to be at least two subpopulations, Myf5-independent and Myf5-dependent. Differentiation of embryonic and fetal myoblasts into differentiated myocytes and myofibers is differentially regulated by MRFs and signaling. Embryonic myoblasts require either MyoD, Myogenin, or Mrf4 for their differentiation, while fetal myoblasts require Myogenin and Mrf4 or MyoD. Also, while embryonic myogenesis is insensitive to TGFβ, BMP, and β-catenin signaling, fetal myogenesis is regulated by these signaling pathways. The expression of Nfix within fetal myoblasts is critical for their differentiation into fetal myofibers. Once differentiated, embryonic, fetal/neonatal, and adult myofibers express different combinations of MRFs, muscle contractile proteins (including MyHC isoforms), and metabolic enzymes.

From this model, it is clear that amniote myogenesis is complex. Multiple related, although distinct progenitor and myoblast populations give rise to embryonic, fetal, neonatal, and adult muscle. In the future, it will be important to resolve the relationships between myogenic progenitors and myoblasts and definitively answer whether myoblasts ever give rise to progenitors. Also, the extrinsic cell populations and molecular signals differentially regulating the different phases of myogenesis are largely unknown. Finally, a critical question is the identification of the intrinsic and extrinsic factors that maintain the populations of myogenic progenitors, particularly in the embryo and fetus where progenitors reside alongside actively differentiating myogenic cells.

ACKNOWLEDGMENTS

We thank D.D. Cornelison, D. Goldhamer, F. Relaix, and S. Tajbakhsh for discussion and review of the manuscript. We also thank members of the Kardon lab (particularly D.A. Hutcheson), S. Biressi, and G. Messina for many helpful discussions. Research in the Kardon lab is supported by the Pew Foundation, MDA, and NIH R01 HD053728.

REFERENCES

Agbulut, O., Noirez, P., Beaumont, F., and Butler-Browne, G. (2003). Myosin heavy chain isoforms in postnatal muscle development of mice. *Biol. Cell* **95,** 399–406.

Auerbach, R. (1954). Analysis of the developmental effects of a lethal mutation in the house mouse. *J. Exp. Zool.* **127,** 305–329.

Beauchamp, J. R., Heslop, L., Yu, D. S., Tajbakhsh, S., Kelly, R. G., Wernig, A., Buckingham, M. E., Partridge, T. A., and Zammit, P. S. (2000). Expression of CD34 and Myf5 defines the majority of quiescent adult skeletal muscle satellite cells. *J. Cell Biol.* **151,** 1221–1234.

Biressi, S., Molinaro, M., and Cossu, G. (2007a). Cellular heterogeneity during vertebrate skeletal muscle development. *Dev. Biol.* **308,** 281–293.

Biressi, S., Tagliafico, E., Lamorte, G., Monteverde, S., Tenedini, E., Roncaglia, E., Ferrari, S., Cusella-De Angelis, M. G., Tajbakhsh, S., and Cossu, G. (2007b). Intrinsic phenotypic diversity of embryonic and fetal myoblasts is revealed by genome-wide gene expression analysis on purified cells. *Dev. Biol.* **304,** 633–651.

Bober, E., Lyons, G. E., Braun, T., Cossu, G., Buckingham, M., and Arnold, H. H. (1991). The muscle regulatory gene, Myf-6, has a biphasic pattern of expression during early mouse development. *J. Cell Biol.* **113,** 1255–1265.

Bober, E., Franz, T., Arnold, H. H., Gruss, P., and Tremblay, P. (1994). Pax-3 is required for the development of limb muscles: a possible role for the migration of dermomyotomal muscle progenitor cells. *Development* **120,** 603–612.

Boutet, S. C., Disatnik, M. H., Chan, L. S., Iori, K., and Rando, T. A. (2007). Regulation of Pax3 by proteasomal degradation of monoubiquitinated protein in skeletal muscle progenitors. *Cell* **130,** 349–362.

Brack, A. S., Conboy, I. M., Conboy, M. J., Shen, J., and Rando, T. A. (2008). A temporal switch from notch to Wnt signaling in muscle stem cells is necessary for normal adult myogenesis. *Cell Stem Cell* **2,** 50–59.

Branda, C. S., and Dymecki, S. M. (2004). Talking about a revolution: The impact of site-specific recombinases on genetic analyses in mice. *Dev. Cell* **6,** 7–28.

Braun, T., and Arnold, H. H. (1995). Inactivation of Myf-6 and Myf-5 genes in mice leads to alterations in skeletal muscle development. *EMBO J.* **14,** 1176–1186.

Brockschnieder, D., Pechmann, Y., Sonnenberg-Riethmacher, E., and Riethmacher, D. (2006). An improved mouse line for Cre-induced cell ablation due to diphtheria toxin A, expressed from the Rosa26 locus. *Genesis* **44,** 322–327.

Buckingham, M. (2007). Skeletal muscle progenitor cells and the role of Pax genes. *C. R. Biol.* **330,** 530–533.

Buckingham, M., and Vincent, S. D. (2009). Distinct and dynamic myogenic populations in the vertebrate embryo. *Curr. Opin. Genet. Dev.* **19,** 444–453.

Conboy, I. M., and Rando, T. A. (2002). The regulation of Notch signaling controls satellite cell activation and cell fate determination in postnatal myogenesis. *Dev. Cell* **3,** 397–409.

Cornelison, D. D., and Wold, B. J. (1997). Single-cell analysis of regulatory gene expression in quiescent and activated mouse skeletal muscle satellite cells. *Dev. Biol.* **191,** 270–283.

Cusella-De Angelis, M. G., Molinari, S., Le Donne, A., Coletta, M., Vivarelli, E., Bouche, M., Molinaro, M., Ferrari, S., and Cossu, G. (1994). Differential response of embryonic and fetal myoblasts to TGF beta: A possible regulatory mechanism of skeletal muscle histogenesis. *Development* **120,** 925–933.

Emerson, C. P., Jr., and Hauschka, S. D. (2004). Embryonic origins of skeletal muscles. *In* "Myology," (A. G. Engel and C. Franzini-Armstrong, eds.), pp. 3–44. McGraw-Hill, New York.

Engleka, K. A., Gitler, A. D., Zhang, M., Zhou, D. D., High, F. A., and Epstein, J. A. (2005). Insertion of Cre into the Pax3 locus creates a new allele of Splotch and identifies unexpected Pax3 derivatives. *Dev. Biol.* **280,** 396–406.

Epstein, D. J., Vogan, K. J., Trasler, D. G., and Gros, P. (1993). A mutation within intron 3 of the Pax-3 gene produces aberrantly spliced mRNA transcripts in the splotch (Sp) mouse mutant. *Proc. Natl Acad. Sci. USA* **90,** 532–536.

Franz, T., Kothary, R., Surani, M. A., Halata, Z., and Grim, M. (1993). *Anat. Embryol. (Berl.).* The Splotch mutation interferes with muscle development in the limbs. **187,** 153–160.

Gayraud-Morel, B., ChrÈtien, F., Flamant, P., GomËs, D., Zammit, P. S., and Tajbakhsh, S. (2007). A role for the myogenic determination gene Myf5 in adult regenerative myogenesis. *Dev. Biol.* **312,** 13–28.

Gensch, N., Borchardt, T., Schneider, A., Riethmacher, D., and Braun, T. (2008). Different autonomous myogenic cell populations revealed by ablation of Myf5-expressing cells during mouse embryogenesis. *Development* **135,** 1597–1604.

Goulding, M., Lumsden, A., and Paquette, A. J. (1994). Regulation of Pax-3 expression in the dermomyotome and its role in muscle development. *Development* **120,** 957–971.

Gunning, P., and Hardeman, E. (1991). Multiple mechanisms regulate muscle fiber diversity. *FASEB J.* **5,** 3064–3070.

Haldar, M., Karan, G., Tvrdik, P., and Capecchi, M. R. (2008). Two cell lineages, myf5 and myf5-independent, participate in mouse skeletal myogenesis. *Dev. Cell* **14,** 437–445.

Hasty, P., Bradley, A., Morris, J. H., Edmondson, D. G., Venuti, J. M., Olson, E. N., and Klein, W. H. (1993). Muscle deficiency and neonatal death in mice with a targeted mutation in the myogenin gene. *Nature* **364,** 501–506.

Hinterberger, T. J., Sassoon, D. A., Rhodes, S. J., and Konieczny, S. F. (1991). Expression of the muscle regulatory factor MRF4 during somite and skeletal myofiber development. *Dev. Biol.* **147,** 144–156.

Horst, D., Ustanina, S., Sergi, C., Mikuz, G., Juergens, H., Braun, T., and Vorobyov, E. (2006). Comparative expression analysis of Pax3 and Pax7 during mouse myogenesis. *Int. J. Dev. Biol.* **50,** 47–54.

Hutcheson, D. A., and Kardon, G. (2009). Genetic manipulations reveal dynamic cell and gene functions: Cre-ating a new view of myogenesis. *Cell Cycle* **8,** 3675–3678.

Hutcheson, D. A., Zhao, J., Merrell, A., Haldar, M., and Kardon, G. (2009). Embryonic and fetal limb myogenic cells are derived from developmentally distinct progenitors and have different requirements for beta-catenin. *Genes Dev.* **23,** 997–1013.

Jostes, B., Walther, C., and Gruss, P. (1990). The murine paired box gene, Pax7, is expressed specifically during the development of the nervous and muscular system. *Mech. Dev.* **33,** 27–37.

Kablar, B., Krastel, K., Ying, C., Asakura, A., Tapscott, S. J., and Rudnicki, M. A. (1997). MyoD and Myf-5 differentially regulate the development of limb versus trunk skeletal muscle. *Development* **124,** 4729–4738.

Kang, J. S., and Krauss, R. S. (2010). Muscle stem cells in developmental and regenerative myogenesis. *Curr. Opin. Clin. Nutr. Metab. Care* **13,** 243–248.

Kanisicak, O., Mendez, J. J., Yamamoto, S., Yamamoto, M., and Goldhamer, D. J. (2009). Progenitors of skeletal muscle satellite cells express the muscle determination gene, MyoD. *Dev. Biol.* **332,** 131–141.

Kardon, G., Campbell, J. K., and Tabin, C. J. (2002). Local extrinsic signals determine muscle and endothelial cell fate and patterning in the vertebrate limb. *Dev. Cell* **3,** 533–545.

Kassar-Duchossoy, L., Gayraud-Morel, B., Gomes, D., Rocancourt, D., Buckingham, M., Shinin, V., and Tajbakhsh, S. (2004). Mrf4 determines skeletal muscle identity in Myf5: Myod double-mutant mice. *Nature* **431,** 466–471.

Kassar-Duchossoy, L., Giacone, E., Gayraud-Morel, B., Jory, A., Gomes, D., and Tajbakhsh, S. (2005). Pax3/Pax7 mark a novel population of primitive myogenic cells during development. *Genes Dev.* **19,** 1426–1431.
Keller, C., Hansen, M. S., Coffin, C. M., and Capecchi, M. R. (2004). Pax3:Fkhr interferes with embryonic Pax3 and Pax7 function: implications for alveolar rhabdomyosarcoma cell of origin. *Genes Dev.* **18,** 2608–2613.
Kuang, S., Charge, S. B., Seale, P., Huh, M., and Rudnicki, M. A. (2006). Distinct roles for Pax7 and Pax3 in adult regenerative myogenesis. *J. Cell Biol.* **172,** 103–113.
Kuang, S., Kuroda, K., Le Grand, F., and Rudnicki, M. A. (2007). Asymmetric self-renewal and commitment of satellite stem cells in muscle. *Cell* **129,** 999–1010.
Lepper, C., and Fan, C. M. (2010). Inducible lineage tracing of Pax7-descendant cells reveals embryonic origin of adult satellite cells. *Genesis* **48,** 424–436.
Lepper, C., Conway, S. J., and Fan, C. M. (2009). Adult satellite cells and embryonic muscle progenitors have distinct genetic requirements. *Nature* **460,** 627–631.
Li, S., Czubryt, M. P., McAnally, J., Bassel-Duby, R., Richardson, J. A., Wiebel, F. F., Nordheim, A., and Olson, E. N. (2005). Requirement for serum response factor for skeletal muscle growth and maturation revealed by tissue-specific gene deletion in mice. *Proc. Natl. Acad. Sci. USA* **102,** 1082–1087.
Lu, B. D., Allen, D. L., Leinwand, L. A., and Lyons, G. E. (1999). Spatial and temporal changes in myosin heavy chain gene expression in skeletal muscle development. *Dev. Biol.* **216,** 312–326.
Mansouri, A., Stoykova, A., Torres, M., and Gruss, P. (1996). Dysgenesis of cephalic neural crest derivatives in Pax7-/- mutant mice. *Development* **122,** 831–838.
Mathew, S. J., Hansen, J. M., Merrell, A. J., Murphy, M. M., Lawson, J. A., Hutcheson, D. A., Hansen, M. S., Angus-Hill, M., and Kardon, G. (2011). Connective tissue fibroblasts and Tcf4 regulate myogenesis. *Development* **138**(2), 371–384.
Mauro, A. (1961). Satellite cell of skeletal muscle fibers. *J. Biophys. Biochem. Cytol.* **9,** 493–495.
Megeney, L. A., Kablar, B., Garrett, K., Anderson, J. E., and Rudnicki, M. A. (1996). MyoD is required for myogenic stem cell function in adult skeletal muscle. *Genes Dev.* **10,** 1173–1183.
Messina, G., Biressi, S., Monteverde, S., Magli, A., Cassano, M., Perani, L., Roncaglia, E., Tagliafico, E., Starnes, L., Campbell, C. E., Grossi, M., Goldhamer, D. J., *et al.* (2010). Nfix regulates fetal-specific transcription in developing skeletal muscle. *Cell* **140,** 554–566.
Muzumdar, M. D., Tasic, B., Miyamichi, K., Li, L., and Luo, L. (2007). A global double-fluorescent Cre reporter mouse. *Genesis* **45,** 593–605.
Nabeshima, Y., Hanaoka, K., Hayasaka, M., Esumi, E., Li, S., and Nonaka, I. (1993). Myogenin gene disruption results in perinatal lethality because of severe muscle defect. *Nature* **364,** 532–535.
Olson, E. N., Arnold, H. H., Rigby, P. W. J., and Wold, B. J. (1996). Three phenotypes in null mutants of the myogenic bHLH gene MRF4. *Cell* **85,** 1–4.
Ono, Y., Calbaheu, F., Morgan, J. E., Katagiri, T., Amthor, H., and Zammit, P. S. (2010). BMP signalling permits population expansion by preventing premature myogenic differentiation in muscle satellite cells. *Cell Death Differ.* **18**(2), 222–234.
Ontell, M., Ontell, M. P., Sopper, M. M., Mallonga, R., Lyons, G., and Buckingham, M. (1993a). Contractile protein gene expression in primary myotubes of embryonic mouse hindlimb muscles. *Development* **117,** 1435–1444.
Ontell, M. P., Sopper, M. M., Lyons, G., Buckingham, M., and Ontell, M. (1993b). Modulation of contractile protein gene expression in fetal murine crural muscles: Emergence of muscle diversity. *Dev. Dyn.* **198,** 203–213.

Ott, M. O., Bober, E., Lyons, G., Arnold, H., and Buckingham, M. (1991). Early expression of the myogenic regulatory gene, myf-5, in precursor cells of skeletal muscle in the mouse embryo. *Development* **111,** 1097–1107.

Otto, A., Schmidt, C., Luke, G., Allen, S., Valasek, P., Muntoni, F., Lawrence-Watt, D., and Patel, K. (2008). Canonical Wnt signalling induces satellite-cell proliferation during adult skeletal muscle regeneration. *J. Cell Sci.* **121,** 2939–2950.

Otto, A., Collins-Hooper, H., and Patel, K. (2009). The origin, molecular regulation and therapeutic potential of myogenic stem cell populations. *J. Anat.* **215,** 477–497.

Oustanina, S., Hause, G., and Braun, T. (2004). Pax7 directs postnatal renewal and propagation of myogenic satellite cells but not their specification. *EMBO J.* **23,** 3430–3439.

Patapoutian, A., Yoon, J. K., Miner, J. H., Wang, S., Stark, K., and Wold, B. (1995). Disruption of the mouse MRF4 gene identifies multiple waves of myogenesis in the myotome. *Development* **121,** 3347–3358.

Perez-Ruiz, A., Ono, Y., Gnocchi, V. F., and Zammit, P. S. (2008). beta-Catenin promotes self-renewal of skeletal-muscle satellite cells. *J. Cell Sci.* **121,** 1373–1382.

Rawls, A., Morris, J. H., Rudnicki, M., Braun, T., Arnold, H. H., Klein, W. H., and Olson, E. N. (1995). Myogenin's functions do not overlap with those of MyoD or Myf-5 during mouse embryogenesis. *Dev. Biol.* **172,** 37–50.

Rawls, A., Valdez, M. R., Zhang, W., Richardson, J., Klein, W. H., and Olson, E. N. (1998). Overlapping functions of the myogenic bHLH genes MRF4 and MyoD revealed in double mutant mice. *Development* **125,** 2349–2358.

Relaix, F., and Marcelle, C. (2009). Muscle stem cells. *Curr. Opin. Cell Biol.* **21,** 748–753.

Relaix, F., Polimeni, M., Rocancourt, D., Ponzetto, C., Schafer, B. W., and Buckingham, M. (2003). The transcriptional activator PAX3-FKHR rescues the defects of Pax3 mutant mice but induces a myogenic gain-of-function phenotype with ligand-independent activation of Met signaling in vivo. *Genes Dev.* **17,** 2950–2965.

Relaix, F., Rocancourt, D., Mansouri, A., and Buckingham, M. (2004). Divergent functions of murine Pax3 and Pax7 in limb muscle development. *Genes Dev.* **18,** 1088–1105.

Relaix, F., Rocancourt, D., Mansouri, A., and Buckingham, M. (2005). A Pax3/Pax7-dependent population of skeletal muscle progenitor cells. *Nature* **435**(7044), 948–953.

Relaix, F., Montarras, D., Zaffran, S., Gayraud-Morel, B., Rocancourt, D., Tajbakhsh, S., Mansouri, A., Cumano, A., and Buckingham, M. (2006). Pax3 and Pax7 have distinct and overlapping functions in adult muscle progenitor cells. *J. Cell Biol.* **172,** 91–102.

Rubinstein, N. A., and Kelly, A. M. (2004). The diversity of muscle fiber types and its origin during development. *In* "Myology," (A. G. Engel and C. Franzini-Armstrong, eds.), pp. 87–103. McGraw-Hill, New York.

Rudnicki, M. A., Braun, T., Hinuma, S., and Jaenisch, R. (1992). Inactivation of MyoD in mice leads to up-regulation of the myogenic HLH gene Myf-5 and results in apparently normal muscle development. *Cell* **71,** 383–390.

Rudnicki, M. A., Schnegelsberg, P. N., Stead, R. H., Braun, T., Arnold, H. H., and Jaenisch, R. (1993). MyoD or Myf-5 is required for the formation of skeletal muscle. *Cell* **75,** 1351–1359.

Sambasivan, R., Gayraud-Morel, B., Dumas, G., Cimper, C., Paisant, S., Kelly, R. G., and Tajbakhsh, S. (2009). Distinct regulatory cascades govern extraocular and pharyngeal arch muscle progenitor cell fates. *Dev. Cell* **16,** 810–821.

Sassoon, D., Lyons, G., Wright, W. E., Lin, V., Lassar, A., Weintraub, H., and Buckingham, M. (1989). Expression of two myogenic regulatory factors myogenin and MyoD1 during mouse embryogenesis. *Nature* **341,** 303–307.

Schiaffino, S., and Reggiani, C. (1996). Molecular diversity of myofibrillar proteins: Gene regulation and functional significance. *Physiol. Rev.* **76,** 371–423.

Schienda, J., Engleka, K. A., Jun, S., Hansen, M. S., Epstein, J. A., Tabin, C. J., Kunkel, L. M., and Kardon, G. (2006). Somitic origin of limb muscle satellite and side population cells. *Proc. Natl. Acad. Sci. USA* **103,** 945–950.
Schubert, F. R., Tremblay, P., Mansouri, A., Faisst, A. M., Kammandel, B., Lumsden, A., Gruss, P., and Dietrich, S. (2001). Early mesodermal phenotypes in splotch suggest a role for Pax3 in the formation of epithelial somites. *Dev. Dyn.* **222,** 506–521.
Seale, P., Sabourin, L. A., Girgis-Gabardo, A., Mansouri, A., Gruss, P., and Rudnicki, M. A. (2000). Pax7 is required for the specification of myogenic satellite cells. *Cell* **102,** 777–786.
Soriano, P. (1999). Generalized lacZ expression with the ROSA26 Cre reporter strain. *Nat. Genet.* **21,** 70–71.
Srinivas, S., Watanabe, T., Lin, C. S., William, C. M., Tanabe, Y., Jessell, T. M., and Costantini, F. (2001). Cre reporter strains produced by targeted insertion of EYFP and ECFP into the ROSA26 locus. *BMC Dev. Biol.* **1,** 4.
Stockdale, F. E. (1992). Myogenic cell lineages. *Dev. Biol.* **154,** 284–298.
Tajbakhsh, S. (2009). Skeletal muscle stem cells in developmental versus regenerative myogenesis. *J. Intern. Med.* **266,** 372–389.
Tajbakhsh, S., and Buckingham, M. E. (1994). Mouse limb muscle is determined in the absence of the earliest myogenic factor myf-5. *Proc. Natl. Acad. Sci. USA* **91,** 747–751.
Tajbakhsh, S., and Buckingham, M. (2000). The birth of muscle progenitor cells in the mouse: spatiotemporal considerations. *Curr. Top. Dev. Biol.* **48,** 225–268.
Tajbakhsh, S., Bober, E., Babinet, C., Pournin, S., Arnold, H., and Buckingham, M. (1996). Gene targeting the myf-5 locus with nlacZ reveals expression of this myogenic factor in mature skeletal muscle fibres as well as early embryonic muscle. *Dev. Dyn.* **206,** 291–300.
Tajbakhsh, S., Rocancourt, D., Cossu, G., and Buckingham, M. (1997). Redefining the genetic hierarchies controlling skeletal myogenesis: Pax-3 and Myf-5 act upstream of MyoD. *Cell* **89,** 127–138.
Tallquist, M. D., Weismann, K. E., Hellstrom, M., and Soriano, P. (2000). Early myotome specification regulates PDGFA expression and axial skeleton development. *Development* **127,** 5059–5070.
Valdez, M. R., Richardson, J. A., Klein, W. H., and Olson, E. N. (2000). Failure of Myf5 to support myogenic differentiation without myogenin, MyoD, and MRF4. *Dev. Biol.* **219,** 287–298.
Venuti, J. M., Morris, J. H., Vivian, J. L., Olson, E. N., and Klein, W. H. (1995). Myogenin is required for late but not early aspects of myogenesis during mouse development. *J. Cell Biol.* **128,** 563–576.
Vogan, K. J., Epstein, D. J., Trasler, D. G., and Gros, P. (1993). The splotch-delayed (Spd) mouse mutant carries a point mutation within the paired box of the Pax-3 gene. *Genomics* **17,** 364–369.
Voytik, S. L., Przyborski, M., Badylak, S. F., and Konieczny, S. F. (1993). Differential expression of muscle regulatory factor genes in normal and denervated adult rat hindlimb muscles. *Dev. Dyn.* **198,** 214–224.
Wang, H., Noulet, F., Edom-Vovard, F., Le Grand, F., and Duprez, D. (2010). Bmp signaling at the tips of skeletal muscles regulates the number of fetal muscle progenitors and satellite cells during development. *Dev. Cell* **18,** 643–654.
Weintraub, H., Davis, R., Tapscott, S., Thayer, M., Krause, M., Benezra, R., Blackwell, T. K., Turner, D., Rupp, R., Hollenberg, S., *et al.* (1991). The myoD gene family: Nodal point during specification of the muscle cell lineage. *Science* **251,** 761–766.
White, J. D., Scaffidi, A., Davies, M., McGeachie, J., Rudnicki, M. A., and Grounds, M. D. (2000). Myotube formation is delayed but not prevented in MyoD-deficient skeletal muscle: Studies in regenerating whole muscle grafts of adult mice. *J. Histochem. Cytochem.* **48,** 1531–1544.

White, R. B., Bierinx, A. S., Gnocchi, V. F., and Zammit, P. S. (2010). Dynamics of muscle fibre growth during postnatal mouse development. *BMC Dev. Biol.* **10,** 21.

Wu, S., Wu, Y., and Capecchi, M. R. (2006). Motoneurons and oligodendrocytes are sequentially generated from neural stem cells but do not appear to share common lineage-restricted progenitors in vivo. *Development* **133,** 581–590.

Yablonka-Reuveni, Z., and Rivera, A. J. (1994). Temporal expression of regulatory and structural muscle proteins during myogenesis of satellite cells on isolated adult rat fibers. *Dev. Biol.* **164,** 588–603.

Yablonka-Reuveni, Z., Rudnicki, M. A., Rivera, A. J., Primig, M., Anderson, J. E., and Natanson, P. (1999). The transition from proliferation to differentiation is delayed in satellite cells from mice lacking MyoD. *Dev. Biol.* **210,** 440–455.

Yamamoto, M., Shook, N. A., Kanisicak, O., Yamamoto, S., Wosczyna, M. N., Camp, J. R., and Goldhamer, D. J. (2009). A multifunctional reporter mouse line for Cre- and FLP-dependent lineage analysis. *Genesis* **47,** 107–114.

Zammit, P. S., Heslop, L., Hudon, V., Rosenblatt, J. D., Tajbakhsh, S., Buckingham, M. E., Beauchamp, J. R., and Partridge, T. A. (2002). Kinetics of myoblast proliferation show that resident satellite cells are competent to fully regenerate skeletal muscle fibers. *Exp. Cell Res.* **281,** 39–49.

Zammit, P. S., Golding, J. P., Nagata, Y., Hudon, V., Partridge, T. A., and Beauchamp, J. R. (2004). Muscle satellite cells adopt divergent fates: A mechanism for self-renewal? *J. Cell Biol.* **166,** 347–357.

Zhang, W., Behringer, R. R., and Olson, E. N. (1995). Inactivation of the myogenic bHLH gene MRF4 results in up-regulation of myogenin and rib anomalies. *Genes Dev.* **9,** 1388–1399.

CHAPTER TWO

Developmental Origins of Fusion-Negative Rhabdomyosarcomas

Ken Kikuchi,* Brian P. Rubin,†,‡ *and* Charles Keller*

Contents

Abstract

Rhabdomyosarcomas (RMS) are very heterogeneous tumors that can be divided into three major groups: alveolar rhabdomyosarcoma, embryonal rhabdomyosarcoma, and pleomorphic rhabdomyosarcoma. Concerted efforts over the past a decade have led to an understanding of the genetic underpinnings of many human tumors through genetically engineered models; however, left largely behind in this effort have been rare tumors with poorly understood chromosomal abnormalities including the vast majority of RMS lacking a pathognomonic translocation, i.e. fusion-negative RMS. In this chapter, we review the characteristic genetic abnormalities associated with human RMS and the genetically engineered animal models for these fusion-negative RMS. We explore not only how specific combinations of mutations and cell of origin give rise to different histologically and biologically distinguishable pediatric and adult

* Department of Pediatrics, Pape' Family Pediatric Research Institute, Oregon Health and Science University, Portland, Oregon, USA
† Pediatric Cancer Biology Program, Department of Anatomic Pathology, Taussig Cancer Center and Lerner Research Institute, Cleveland Clinic Foundation, Cleveland, Ohio, USA
‡ Department of Molecular Genetics, Taussig Cancer Center and Lerner Research Institute, Cleveland Clinic Foundation, Cleveland, Ohio, USA

Current Topics in Developmental Biology, Volume 96
ISSN 0070-2153, DOI: 10.1016/B978-0-12-385940-2.00002-4

RMS subtypes, but we also examine how tumor cell phenotype (and tumor "stem" cell phenotype) can vary markedly from the cell of origin.

1. Introduction

Soft-tissue sarcomas are malignant tumors believed to be of the mesodermal lineage, and therefore derived from nonepithelial, nonhematopoietic tissues. While extremely rare in adults, rhabdomyosarcomas (RMS) are one of the more common neoplasms in children and adolescents (Parham and Ellison, 2006). RMS, as the name suggests, are presumed to be associated with the skeletal muscle lineage because of myogenic marker expression. The names of many soft-tissue sarcomas imply a particular line of differentiation (i.e., the tissue that the tumor most closely resembles), without any information about the cell of origin.

Histologically, RMS are very heterogeneous tumors that can be divided into three major groups: alveolar RMS (ARMS), embryonal RMS (ERMS), and pleomorphic RMS (Parham and Ellison, 2006). Pleomorphic RMS affect mainly adults, while ARMS and ERMS affect children and adolescents. Clinically, ARMS is more common in older children, involves the trunk and extremities, and has a worse prognosis, while ERMS presents at an earlier age, mainly in the orbit, head, and neck and retroperitoneum, and is associated with a better prognosis (Parham and Ellison, 2006); nevertheless, metastatic ERMS portends only 40% overall survival (Breneman *et al.*, 2003).

Beside clinicopathological differences, these tumors also differ at the molecular level. In approximately one-third of soft-tissue sarcomas, a specific translocation drives sarcomagenesis (Borden *et al.*, 2003). This subset of RMS, which accounts for 80–90% of ARMS, is characterized by specific chromosomal translocations having either a *t*(2; 13) or a *t*(1; 13) translocation, involving *PAX3:FKHR* or *PAX7:FKHR* fusion genes, respectively. Further, the fusion type correlates strongly with outcome, since *PAX3: FKHR* is associated clinically with more aggressive tumors than *PAX7: FKHR* (Sorensen *et al.*, 2002). The remaining two-thirds of RMS (including ERMS and pleomorphic RMS) do not have a pathogenetic translocation. Instead, these tumors are often characterized by complex karyotypes with inactivation of the p53 tumor suppressor pathway (Borden *et al.*, 2003).

A hypothesis that is currently in vogue in the field of carcinogenesis is that a small, but identifiable population of cells within a tumor provides the "self-renewal" phenotype of cancer. Many names have been used to identify this population but the term cancer stem cell has received broad acceptance. Cancer stem cells have been defined as "a cell within a tumor that possesses the capacity to self-renew and to cause the heterogeneous lineages of tumor cells that comprise the tumor" (Clark *et al.*, 2005). These two definitive biological properties are what make the cancer stem cell the prime candidate for initiation of relapse, thereby becoming a crucial target for the development of novel therapies. An extension of this hypothesis is that these cancer stem cells most closely resemble the cell of origin for a given tumor (Reya *et al.*, 2001). Based on this hypothesis, the originating cell of a tumor is increasingly believed to be a primitive cell with the ability to undergo division with high-phenocopy fidelity and retention of multipotency. Meanwhile, there is an ongoing debate over whether cancer stem cells represent a mature tissue stem cell which has undergone malignant change or whether more differentiated cells reinitiate a "stemness" programme as part of, or following, malignant transformation. Until we have this information, it is important to consider independently the concepts of cell of origin and cancer stem cell, as defined purely by self-renewal and capacity to differentiate.

We previously reported a conditional mouse model of ARMS in which *Pax3:Fkhr* was activated and *Trp53* was inactivated in maturing myoblasts (Keller *et al.*, 2004). In these mice, cells expressing high levels of *Pax3:Fkhr* were most capable of repopulating tumors at metastatic sites (Nishijo *et al.*, 2009). These results suggested that *Pax3:Fkhr* fusion-positive ARMS tumors display heterogeneity in *Pax3:Fkhr* expression, and that *Pax3:Fkhr* overexpressing cells might be tumor repopulating cells, that is, cancer stem cells. Meanwhile, cancer stem cells of fusion-negative RMS, including ERMS and pleomorphic RMS, have not yet either described in great detail either. In this review, we will consider the genetically engineered tumor models of fusion-negative RMS using transgenic and germline mice with oncogenic mutations, and conditional mice that express oncogenes. Based on these models and corresponding human RMS biology, we will make an effort to understand the cell of origin, tumor phenotype, and cancer stem cell of fusion-negative RMS.

2. Mutations Seen in Fusion-Negative RMS

2.1. Congenital syndromes associated with fusion-negative RMS and mutations detected in sporadic fusion-negative RMS

Although no consistent chromosomal rearrangement have been identified in ERMS and the majority of ERMS occur as sporadic cases, these tumors have been reported in hereditable conditions with tumor predisposition such as Li-Fraumeni syndrome (LFS), Beckwith-Wiedemann syndrome (BWS), Costello syndrome, Neurofibromatosis type-1 (NF1), Noonan Syndrome, and Gorlin syndrome (Tables 2.1 and 2.2).

Table 2.1 Congenital syndromes associated with fusion-negative RMS

Cancersyndrome	Locus	Gene	Characteristic malignancies	Frequency	Reference
Li-Fraumeni syndrome	17p13.1	*TP53*	Sarcomas, breast cancer, brain tumors, adrenocortical carcinoma	1–10% of RMS patients	Le Bihan *et al.* (1995), Schneider and Garber (2010), Xia *et al.* (2002)
Beckwith-Wiedemann syndrome	11p15.5	*KCNQ1OT1, H19/IGF2, CDKN1C*	Wilms tumor, hepatoblastoma, neuroblastoma, adrenocortical carcinoma, rhabdomyosarcoma	7.5% childhood risk for tumor development	Shuman *et al.* (2010)
Costello syndrome	11p15.5	*HRAS*	Rhabdomyosarcoma, neuroblastoma, transitional cell carcinoma of the bladder	15% lifetime risk for tumor development	Gripp (2005), Gripp and Lin (2009)
Neurofibromatosis type 1	17q1.2	*NF1*	Glioma, malignant peripheral nerve sheath tumor, rhabdomyosarcoma	1–6% of RMS patients	Ferrari *et al.* (2007), Xia *et al.* (2002)
Noonan Syndrome	12q24, 12p12.1, 3p25, 2p21	*PTPN11, SOS1, RAF1, KRAS*	JMML, AML, ALL	RMS: three cases	Moschovi *et al.* (2007)
Gorlin syndrome	9q22.3	*PTCH1*	Basal cell carcinoma, medulloblastoma, fibroma	RMS: three cases	Gorlin (2004), Xia *et al.* (2002)
Hereditary retinoblastoma	13q14.2	*RB1*	Osteosarcoma, soft-tissue sarcomas, melanoma	Standardized incidence ratio of RMS = 279	Kleinerman *et al.* (2007)

Table 2.2 Mutations detected in sporadic fusion-negative RMS

Locus	Gene	Mutation	Frequency	Percentage (%)	Reference
17p13.1	*TP53*	Point mutation	5/36	8–50	Diller *et al.* (1995), Felix *et al.* (1992), Taylor *et al.* (2000)
12q14.3-q15	*MDM2*	Amplification	1/12	8	Taylor *et al.* (2000)
9p21	*CDKN2A/ARF*	Deletion	3/10	25–33	Gil-Benso *et al.* (2003), Iolascon *et al.* (1996)
9q22.3	*PTCH*	Deletion	7/20	33–38	Bridge *et al.* (2000), Tostar *et al.* (2006)
12p12.1	*KRAS*	Point mutation	2/65	0–14	Chen *et al.* (2006), Martinelli *et al.* (2009), Stratton *et al.* (1989)
1p13.2	*NRAS*	Point mutation	10/65	5–21	Chen *et al.* (2006), Martinelli *et al.* (2009), Stratton *et al.* (1989)
11p15.5	*HRAS*	Point mutation	2/54	0–33	Chen *et al.* (2006), Martinelli *et al.* (2009), Wilke *et al.* (1993)
12q24	*PTPN11*	Point mutation	2/51	3–5	Chen *et al.* (2006), Martinelli *et al.* (2009)
5q35.1-qter	*FGFR4*	Point mutation	5/58	9	Taylor *et al.* (2009)
11p15.5	*H19/IGF2, CDKN1C*	LOH	30/39	77	Davicioni *et al.* (2009)
13q14.2	*RB1*	Allelic imbalance	13/27	48	Kohashi *et al.* (2008)

TP53, a transcription factor, is a tumor suppressor gene product involved in apoptosis, cell cycle withdrawal, differentiation, and cellular senescence. p53 is activated by a number of cellular stresses such as DNA damage. Germline mutations of *TP53* are the underlying etiology of LFS, which is associated with high risk of a diverse spectrum of childhood and adult-onset malignancies. One study, based on five families with LFS, estimated age-specific cancer risks as 42% at ages 0–16 years, 38% at ages 17–45 years, and 63% after age 45 years; overall lifetime cancer risk was calculated at 85% (Le Bihan *et al.*, 1995; Schneider and Garber, 2010). Individuals with LFS are at increased risk of developing soft-tissue (i.e., muscle and connective tissue) sarcomas (e.g., RMS, liposarcoma) and sarcomas of bone (e.g., osteosarcoma, chondrosarcoma). Almost all types of sarcomas have been noted in families with *TP53* mutations with the exception of Ewing sarcoma, which is not associated with LFS. The median age of soft-tissue sarcomas in individuals with LFS is 14 years by comparison with sporadic sarcomas, which is 61.3 years (Olivier *et al.*, 2003). Somatic mutations of *TP53* (Diller *et al.*, 1995; Felix *et al.*, 1992; Taylor *et al.*, 2000) and dysregulation of its associated regulatory proteins have also been implicated in the development of ERMS. Loss of p53 tumor suppressor function in sporadic RMS may result from overexpression of *MDM2*, since *MDM2* encodes a protein capable of binding and inactivating p53. *MDM2*, along with other genes at the 12q13–15 locus, have been found to be amplified in ERMS (Taylor *et al.*, 2000). Several studies have also found that MDM2 is overexpressed at both mRNA and protein levels (Keleti *et al.*, 1996; Miyachi *et al.*, 2009), again suggesting a role for MDM2 in ERMS. Another member of the p53 pathway is ARF, which blocks MDM2 function and stabilizes the p53 protein. Although mutations of *ARF* have not been reported in RMS, homozygous deletions of the 9p21 region, which contains the *CDKN2A* common locus for both $P16^{\mathrm{INK4A}}$ and *ARF*, were found in 25–33% of ERMS cases (Gil-Benso *et al.*, 2003; Iolascon *et al.*, 1996).

BWS is a disorder of growth characterized by macrosomia, macroglossia, visceromegaly, omphalocele, neonatal hypoglycemia, ear creases/pits, adrenocortical cytomegaly, renal abnormalities, and embryonal tumors. Children with BWS have an increased risk of mortality associated with neoplasia, particularly Wilms' tumor and hepatoblastoma, but also neuroblastoma, adrenocortical carcinoma, and RMS (ERMS and fusion-negative ARMS; Smith *et al.*, 2001). Also seen are a wide variety of other tumors. The estimated risk for tumor development in children with BWS is 7.5%. The increased risk for neoplasia seems to be concentrated in the first 8 years of life. Tumor development is uncommon in affected individuals older than 8 years of age (Shuman *et al.*, 2010). The diagnosis of BWS relies primarily on clinical findings. Clinically available molecular genetic testing can identify several different types of 11p15 abnormalities in individuals with BWS: (1) loss of methylation at *KCNQ1OT1* (*DMR2*) is observed in 50% of individuals; (2) gain of methylation at *H19/IGF2* (*DMR1*) is observed in 2–7%; (3) paternal uniparental disomy for chromosome 11p15 is observed in 10–20%. Testing reveals mutations in the *CDKN1C* gene in 40% of familial cases and 5–10% of sporadic cases. Molecular analyses of polymorphic loci in sporadic cases of ERMS revealed frequent allelic loss of 11p15 (Davicioni *et al.*, 2009; Koufos *et al.*, 1985). Comparison of the allelic loss pattern in ERMS tumors to the allelic status of the patients' parents revealed that ERMS tumors preferentially maintain the paternally inherited allele and lose the maternal allele (Scrable *et al.*, 1989). *H19* and *CDKN1C* are preferentially expressed from the maternally inherited alleles and *IGF2* is imprinted in the opposite direction so that the paternally inherited alleles are preferentially expressed. Some alterations lead to overexpression of the *IGF2* fetal growth factor, while others serve to mutate or inactivate expression of growth suppressive genes such as *H19* and *CDKN1C*. However, definitive genetic alteration in this region has not been identified in ERMS, and whether BWS and ERMS share common genetic mutations is yet to be proven.

The RAS-mitogen-activated protein kinase (MAPK) pathway is a signal transduction cascade that has been studied extensively during the last decades due to its role in human oncogenesis. Mutations in *RAS* genes have been found in numerous malignancies, including sporadic ERMS (Table 2.2; Chen *et al.*, 2006; Martinelli *et al.*, 2009; Stratton *et al.*, 1989; Wilke *et al.*, 1993). Recently, germline mutations in genes coding for different components of the RAS signaling cascade have been recognized as the cause of several phenotypically overlapping disorders, referred to as the neuro-cardio-facial-cutaneous syndromes, Costello (*HRAS* 80–90%), NF1, Noonan (*PTPN11* 50%, *RAF1* 3–17%, *SOS1* 10–13%, and *KRAS* $<5\%$), LEOPARD (*PTPN11* 90% and *RAF1* 3%), and cardiofaciocutaneous (*BRAF* 75–80%, *MAP2K1, 2* 10–15%, and *KRAS* $<5\%$) syndromes all present with variable degrees of psychomotor delay, congenital heart defects, facial dysmorphism, short stature, skin abnormalities, and a predisposition for malignancy including ERMS (Ferrari *et al.*, 2007; Gripp, 2005; Gripp and Lin, 2009; Moschovi *et al.*, 2007; Xia *et al.*, 2002). Costello syndrome is a rare congenital anomaly syndrome and characterized by severe postnatal failure to thrive and short stature. Individuals with Costello syndrome have an approximately 15% lifetime risk for developing a malignant tumor. The most common tumor in Costello syndrome is ERMS, followed by neuroblastoma and bladder carcinoma (Gripp, 2005; Gripp and Lin, 2009).

Gorlin syndrome, an autosomal dominant disorder associated with a predisposition for multiple basal cell carcinomas and other neoplasms such as medulloblastoma and ovarian fibrosarcoma results from a germline mutation in *PTCH1*, a tumor suppressor gene and a negative regulator of the sonic hedgehog signaling (Shh) pathway, which plays an integral role in nervous system development and anterior–posterior limb patterning (Gorlin, 2004). Although rare cases of ERMS have been found in association with Gorlin syndrome, a disease-associated locus has been mapped to 9q22, a region of relatively frequent genomic loss in some series (Bridge *et al.*, 2000; Tostar *et al.*, 2006), and additional supporting evidence for Shh playing a role in a subset of ERMS is provided by observational studies using human tumors wherein expression of the Shh signaling cascade are altered (Oue *et al.*, 2010; Tostar *et al.*, 2006).

Survivors of hereditary retinoblastoma have substantially increased risks of developing subsequent primary malignancies including osteosarcoma, melanoma, and soft-tissue sarcomas. The predisposition to malignancies in retinoblastoma survivors has been attributed to a germline mutation in the *RB1* gene, which encodes the cell cycle regulatory retinoblastoma protein (pRb; Lohmann and Gallie, 2010). Early studies suggested that pRb abnormalities rarely occur in ERMS or ARMS (De Chiara *et al.*, 1993), but in a contemporary series, 6 of 36 ERMS lacked pRb staining on an immunohistochemical study (Takahashi *et al.*, 2004). A recent study reported that *Rb1* allelic imbalance occurred in 13 of 27 ERMS tested (but much less frequently than in ARMS; Kohashi *et al.*, 2008). However, among patients with hereditary *RB1* mutations, RMS reported to have occurred were more often categorized as RMS "Not Otherwise Specified (NOS)" than ERMS (note that NOS may imply inadequate tissue rather than diagnostic uncertainty), and associated with in the field of radiation (Kleinerman *et al.*, 2007). Indeed, Rubin *et al.* (2011) indicate that while *Rb1* loss of function alone does not lead to tumor initiation, *Rb1* loss of function in combination with other oncogenic factors such as *Trp53* nullizygous with/without *PTCH1* haploinsufficiency is strongly associated with an undifferentiated phenotype, wherein myogenic marker expression is reduced or absent. Tumor cell proliferation (Ki67 positivity) and myodifferentiation capacity under low serum conditions were also severely altered. Therefore, it might be suggested that *Rb1* is best characterized as a "modifier" of phenotype in ERMS.

2.2. Results of RMS subtype by mutation type using transgenic mice

Genetic analyses of sporadically occurring RMS have pinpointed several common alterations (Table 2.3). The role of *TP53* in heritable and sporadic RMS has been confirmed by results of transgenic mouse experiments. *Trp53* nullizygous mice have increased tumor susceptibility in adulthood, and the majority of tumors are lymphomas, whereas RMS is very rare in these mice (Harvey *et al.*, 1993a,b). The low incidence of RMS in individuals carrying *Trp53* alterations indicates that additional genetic lesions and/ or appropriate temporal and tissue specific alterations are required to cause

Table 2.3 Results of rhabdomyosarcoma type by mutation type using transgenic mice

Mutation type	Tumor type	Mouse strain	Onset	Frequency (%)	Reference
Trp53+/−	RMS	129Sv X C57Bl6	21–78 weeks	3	Harvey *et al.* (1993b)
Trp53−/−	RMS	129Sv	17 weeks	6	Harvey *et al.* (1993a)
Trp53+/− HER-2 tg	Embryonal RMS	BALB/c	11 weeks	100	Nanni *et al.* (2003)
Trp53−/− Fos−/−	Embryonal RMS	C57Bl6 X 129Sv	10 weeks	93	Fleischmann *et al.* (2003)
Ink4a/Arf −/− HGF/SF tg	Embryonal RMS	C57Bl6 X FVB	3 months	100	Sharp *et al.* (2002)
Ink4a/Arf −/− Lig4+/−	Sarcomas (including RMS)	129SvJ X C57Bl6	22.9 weeks (average)	47 (sarcomas)	Sharpless *et al.* (2001)
Ptch1+/−	Embryonal RMS	129Sv X CD-1	6–13 weeks	9	Hahn *et al.* (1998)
Ptch1+/− Ptch2−/−	Embryonal RMS	C57Bl6 X 129Sv	2–14 months	40	Lee *et al.* (2006)

this tumor. Interestingly, however, the frequencies of RMS decrease in mice with dominant negative (R175H) *Trp53* mutation (Lang *et al.*, 2004).

The HER family of genes also appear to play a role in RMS tumor initiation and progression. *HER-1/EGF-R* sustains RMS cell growth whereas *HER-3* induces myogenic differentiation *in vitro*, and both *HER-1* and *HER-3* heterodimerize with *HER-2*. Activation of *HER-2* can lead to transformation *in vitro* and *in vivo* in many cell types, is required for myoblast cell survival and is expressed in approximately one-half of human RMS (Andrechek *et al.*, 2002; Ricci *et al.*, 2000). Nanni *et al.* (2003) described remarkable results in which the combination of *Trp53* inactivation and *HER-2* activation resulted in the induction of ERMS in all male mice. These tumors which arose between 11 and 21 weeks exclusively in the genitourinary tract expressed not only desmin and myosin but also insulin-like growth factor-II.

Fos is a zinc finger transcription factor consisting of AP-1 and regulates various biological processes by transcriptional activation of a number of target genes downstream of signaling pathways such as protein kinase C. *Fos−/−* mice display osteopetrosis due to lack of osteoclasts. Overexpression of *Fos* in transgenic mice causes them to develop chondrosarcoma and osteosarcoma (Grigoriadis *et al.*, 1993; Wang *et al.*, 1991). Fos is also known to regulate apoptotic-signaling pathways. Fleischmann *et al.* (2003) generated *Trp53/Fos* compound mutant mice (*Trp53−/− Fos−/−*) and these mice initially developed osteopetrosis similar to the phenotype seen in *Fos−/−* mice (Grigoriadis *et al.*, 1994). However, *Trp53−/− Fos−/−* mice started to develop tumors of the facial and orbital regions at 10 weeks of age and the tumor penetrance was 93% at 25 weeks. These tumors contained cells with polygonal or elongated shapes and expressed cell cycle-associated proteins and a number of skeletal muscle markers such as desmin and MyoD. These characteristics are very similar to human embryonal RMS. Cell lines isolated from these tumors expressed *Pax7*, which was significantly reduced by *Fos* reexpression. In addition, overexpression of *Fos* in the primary myoblasts also downregulated muscle specific gene expression, and *Pax7* gene expression was decreased to below the detection limit. These results suggest an interesting molecular mechanism by which *Fos* is the repressor for *Pax7* gene transcription and upregulation of *Pax7* gene expression may result in increased myoblast proliferation and prevent apoptosis in this experimental context. Recently, Singh *et al.* investigated Wnt signaling for these *Trp53−/− Fos−/−* mouse tumors, and demonstrated a critical role for suppression of the canonical Wnt pathway in human ERMS cell lines that was mediated by AP-1 (Singh *et al.*, 2010).

Another RMS-associated mutant mouse strain carries a germline knockout of the *Cdkn2a* locus (Sharp *et al.*, 2002; Sharpless *et al.*, 2001). This locus encodes two unrelated proteins with tumor suppressor function, $p16^{Ink4a}$ and $p19^{Arf}$. p16 regulates the cell cycle pathway through modulation of

CDK4 and CDK6 and p19 regulates p53 checkpoint function. It is not surprising that *Ink4a/Arf*-knockout mice are prone to cancer, but they develop mainly hematopoietic tumors, some melanomas and fibrosarcomas and, rarely, RMS (Serrano *et al.*, 1996).

Activation of the *c-MET* signaling pathway occurs in several types of human cancers, and affects cell motility, proliferation and survival in experimental systems. Activation of *c-MET* can occur through overexpression or activating mutations of the *c-MET* receptor itself or by autocrine expression of *HGF/SF*. Transgenic *HGF/SF* mice develop tumors rarely, and these are mostly melanomas, although a few RMS also develop (Takayama *et al.*, 1997). The surprise came from mice that expressed the *HGF/SF* transgene and, in addition, were deficient for *Ink4a/Arf*. The doubly mutant mice developed ERMS with nearly a 100% frequency with a mean age of onset of 3.3 months. Intriguingly, this study demonstrated the presence of preneoplastic hyperplastic satellite cells in the skeletal muscle of *HGF/SF* Ink4a/Arf−/− mice aged 6–10 weeks, which suggested that these myogenic precursors were the source of RMS.

LIG4 syndrome is a rare autosomal recessive disorder arising from mutations in the *LIG4* gene, which plays a critical role in the repair of DNA double-strand breaks by nonhomologous end-joining. It is characterized by chromosomal instability, immunodeficiency, developmental delay, and an increased risk for lymphoid malignancies. Most commonly, *Lig4/Trp53* null mice die from pre-B cell lymphomas (Frank *et al.*, 2000). Sharpless *et al.* described results in which *Ink4a/Arf−/− Lig4* haploinsufficient mice developed soft-tissue sarcomas including RMS, which possess clonal amplifications, deletions, and translocations. Because reduced *Lig4* gene dosage is sufficient to facilitate transformation, this data suggests that tumors with *Lig4* mutations would have increased genomic instability even in the setting of a retained and expressed wild-type allele. *LIG4* appears to constitute a new class of tumor suppressor protein that does not conform to the Knudson paradigm. The authors concluded that either germline or somatic loss of NHEJ proteins might underlie chromosomal aberrations known to be responsible for human malignancies.

PTCH1, a member of the patched gene family, is the receptor for SHH, a secreted molecule implicated in the formation of embryonic structures and in tumorigenesis, as well as the desert hedgehog and Indian hedgehog proteins. This gene is thought to function as a tumor suppressor. The experimental demonstration that *Ptch1* haploinsufficiency is linked to ERMS tumor initiation has been reported. Second allele inactivation is thought to occur by methylation (Calzada-Wack *et al.*, 2002; Hahn *et al.*, 1998). Lee *et al.* investigated the role of *Ptch2*, which is highly similar to *Ptch1* in tumor suppression by generating *Ptch2*-deficient mice. In striking contrast to *Ptch1−/−* mice, *Ptch2−/−* mice are born alive and show no obvious defects and are not cancer prone. However, loss of *Ptch2* markedly affects tumor formation in combination with *Ptch1* haploinsufficiency. These data suggest that *Ptch2* mutations enhance tumorigenesis, resulting from defects in Shh signaling due to *Ptch1* heterozygosity, suggesting a cooperation of *Ptch2* loss with *Ptch1* mutation (Lee *et al.*, 2006).

In our own studies, we have examined the influence of *Ptch1* haploinsufficiency on *Trp53* nullizygous muscle and found *Ptch1* loss to be both important as a cooperative initiating mutation for soft-tissue sarcomas, but also as a modifier of myogenic differentiation that depends upon the cellular context. *Rb1* loss in the context of *Trp53* nullizygous (with or without *Ptch1* loss) also appears to be a modifier, resulting uniformly in loss of differentiation. The work of others affirms the assertion that *Trp53* phenotype can be modified by cooperating mutations: *Trp53−/− Fos−/−*mice develop ERMS in facial and orbital regions only. In contrast, *Trp53+/− HER-2* transgenic mice develop ERMS in the male genitourinary tract only.

3. Cell of Origin of RMS

3.1. Pleomorphic RMS

Pleomorphic RMS typically arises in the skeletal musculature of adults and is distinctly uncommon in children. As the name implies, pleomorphic RMS is a high-grade pleomorphic sarcoma composed of pleomorphic spindle cells containing irregular, hyperchromatic nuclei and numerous mitoses, admixed with varying numbers of rhabdomyoblasts, which demonstrate muscle filaments by electron microscopy. Immunohistochemistry confirms the myogenic nature of these tumors, which demonstrate positivity for desmin, MyoD1, and myogenin. Cytogenetic studies reveal complex karyotypes, with rearrangements and evidence of gene amplification, that offer no distinctions from other adult pleomorphic sarcomas (Parham and Ellison, 2006).

The cell of origin for pleomorphic RMS remains to be defined. Doyle *et al.* developed a mouse model of pleomorphic RMS in which *Cytochrome P450* promoter-driven Cre (AhCre)-activated oncogenic *KRasG12V* can cooperate with *Trp53* mutation and/or loss in pleomorphic RMS development. Frequencies of pleomorphic RMS seen in AhCre *KRasG12V Trp53+/−* mice, AhCre *KRasG12V Trp53R172H/+* mice, and AhCre *KRasG12V Trp53−/−* were 6%, 88%, and 94%, respectively. AhCre, under the control of the *Cyp1A1* promoter, was originally described as showing expression in the small intestine, liver, and colon following induction with β-naphthoflavone (Ireland *et al.*, 2004). However, spontaneous Cre expression has been seen in other organs (Sansom *et al.*, 2005). Mice expressing AhCre were crossed to mice expressing conditional *GFP*, and

GFP expression was seen throughout skeletal muscle. This result indicated that cell of origin for pleomorphic RMS might be a muscle resident cell early in the myogenic lineage (Doyle *et al.*, 2010). Tsumura *et al.* (2006) utilized genetically engineered mice with Cre-inducible oncogenic *Kras* and loss of the *Trp53* allele to generate a spatially and temporally restricted mouse model. In this model, a Cre expression vector is electroporated into the muscle and 10 weeks later pleomorphic RMS develops at the site of electroporation. The electroporation clearly inflicts tissue damage, followed by mononuclear skeletal muscle cell accumulation, and this subsequently disappears in control mice, but numerous mononuclear skeletal muscle cells continue to proliferate in intramuscular tissue of mice harboring conditional *Kras* and *Trp53* genes. It is presumed that activated *KrasG12V* and loss of *Trp53* creates a premalignant population of myogenic precursors incapable of withdrawing from the cell cycle. As Cre recombinase was locally introduced into muscular tissue by electroporation, cells might have been derived from residential stem cells in the muscle, including satellite cells. Loosely interpreted, these results suggest that the cell of origin for pleomorphic RMS might be an adult satellite cell, and that *Trp53* function impairment and *Kras* activation are crucial for tumor initiation (Fig. 2.1).

3.2. ERMS

ERMS are so-named because of their remarkable resemblance to developing embryonic and fetal skeletal muscle. As such, these tumors are characterized by variable zones of condensation that produce alternating foci of hypocellularity and hypercellularity. Like embryonic muscle, the dense zones typically contain areas of more overt myogenesis, whereas the loose areas more closely resemble primitive mesenchyme and lie in a loose gelatinous matrix (Hall and Miyake, 2000).

Using Zebrafish, Langenau *et al.* (2007) discovered that a transgene which expressed human *KrasG12D* by a *rag2* promoter rapidly induced tumors that appeared to be skeletal muscle in origin, based on the presence of multinucleated striated muscle fibers and a battery of diagnostic markers (i.e., desmin, myod, met, myf5, mcadherin; Merlino and Khanna, 2007). Gene set enrichment analysis (GSEA), successful in classifying human, mouse, and even zebrafish cancer (Lam *et al.*, 2006), was used to determine whether a gene set derived from their zebrafish model (tumor vs. normal muscle) was enriched in human data sets of a variety of tumors versus their corresponding normal tissues. Notably, gene sets upregulated in the zebrafish tumors were significantly associated with human ERMS. The *rag2* promoter is expressed in immature T and B cell lineages, olfactory rosettes, and sperm (Jessen *et al.*, 2001). In this chapter, using *rag2-EGFP-bcl2* and *rag2-dsRED2* transgenic animals revealed that transgene-expressing cells

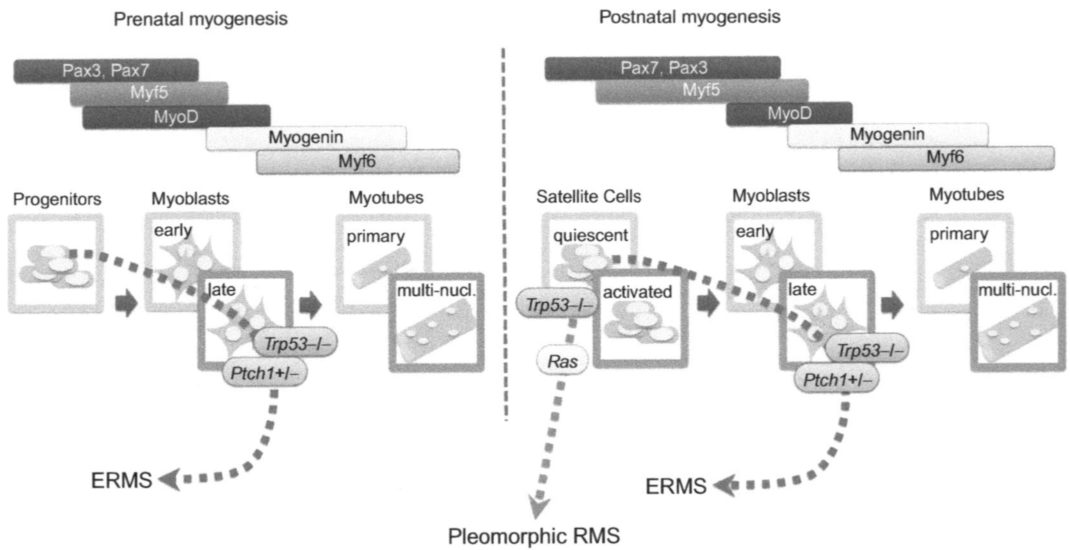

Figure 2.1 Model of skeletal myogenesis and possible cellular origins of fusion-negative RMS. Prenatal muscle development and postnatal muscle maintenance and regeneration are regulated by Pax3, Pax7, and muscle regulatory factors (MyoD, Myf5, Myogenin, and Myf6). Cell of origin for pleomorphic RMS might be an adult satellite cell. Meanwhile, cell of origins for ERMS might be a differentiating late myoblast. (See Color Insert.)

were also detected in the mononuclear component of the skeletal musculature, comprising mononuclear satellite cells, differentiating myoblasts, and rare fusing myoblasts, but not multinucleated terminally differentiated muscle fibers.

Linardic *et al.* (2005) established ERMS model mice with xenografts of human skeletal muscle cell precursors and muscle myoblasts infected with retroviruses encoding SV40 large-T and small-t oncoproteins, the hTERT catalytic subunit of telomerase and HRasG12V. Interestingly, introducing genetic changes characteristic of RMS in cultures of human fetal skeletal muscle cell precursors led to a broad spectrum of sarcomas, ranging from undifferentiated small round blue cell tumors (sarcomas, NOS) to tumors exhibiting differentiation markers characteristic of rhabdomyoblasts, but lacking frank histopathologic features of either ERMS or ARMS. On the other hand, transformation of human adolescent skeletal muscle myoblasts generated an ERMS phenotype.

Recently, Rubin *et al.* generate mouse models of ERMS, through lineage-specific homozygous deletion of *Trp53* with or without heterozygous *Ptch1* deletion by interbreeding *Ptch1* or *Trp53* conditional lines with the myogenic Cre mouse line MCre (Brown *et al.*, 2005), Myf5Cre (Haldar *et al.*, 2008), Pax7CreER (Nishijo *et al.*, 2009), or Myf6Cre (Keller *et al.*, 2004). MCre is specific for the prenatal and postnatal hypaxial lineage of Pax3 that includes postnatal satellite cells (Brown *et al.*, 2005; Relaix *et al.*, 2006). Myf5Cre is specific for the prenatal and postnatal lineage of Myf5 that includes quiescent and activated satellite cells and early myoblasts (Beauchamp *et al.*, 2000; Cornelison and Wold, 1997; Kuang *et al.*, 2007). Pax7CreER is specific for the postnatal lineage of Pax7 that includes quiescent and activated satellite cells (Nishijo *et al.*, 2009), and Myf6Cre is specific for the prenatal and postnatal lineage of Myf6 that includes maturing myoblasts (Keller *et al.*, 2004). When *Trp53* was homozygously inactivated with or without heterozygous *Ptch1* deletion, tumors in all mouse lines developed at a penetrance rate of 13–56% within a 600-day follow-up period. They developed several types of tumors including RMS (embryonal, alveolar, and pleomorphic) and non-RMS soft-tissue sarcoma including undifferentiated pleomorphic sarcoma and osteosarcoma. Interestingly, Myf6Cre *Trp53−/−* lineage gave rise to the highest percentage of ERMS (100%; Rubin *et al.*, 2011).

These results suggest that cell of origin for ERMS is most commonly the differentiating late myoblast (Fig. 2.1). The final stage of transformation might be at this differentiating stage. Indeed, the development of a viable germline *Trp53−/−* mouse model demonstrates that the *Trp53* molecule is not critical during development. Furthermore, White *et al.* (2002) indicated that p53 is not required for the regulation of myoblast proliferation, differentiation, or myotube formation during myogenesis of adult skeletal muscle *in vivo*. Thus, in the context of *p53* mutations, complementary mutations are required to generate RMS and Rubin *et al.* indicate that *PTCH1* and *RB1* mutations are not uncommon in human RMS.

A common impression is that RMS arise in skeletal muscle, but in fact many pediatric examples arise in viscera such as prostate, urinary bladder, and gallbladder, which are devoid of striated muscle fibers. However, revealing new reports of the presence of satellite cells in the urethral rhabdosphincter are emerging (Sumino *et al.*, 2007), and the presence of similar cells in other regions of the bladder, the prostate, and biliary tract are intriguing. RMS occasionally present with diffuse bone marrow involvement and no clear primary tumor suggests that RMS can arise from non-muscle cells, such as a mesenchymal stem cell with the capacity to be pushed down the skeletal muscle lineage; however, almost all of the reported cases associated with marrow involvement, including those presenting as possible acute leukemia, have been of the alveolar, not embryonal subtype (Chen *et al.*, 2004; Lisboa *et al.*, 2008; Sandberg *et al.*, 2001).

Are cell of origin and cancer stem cell the same? Certainly the tumor initiating cell may be able to become a cancer stem cell, but the biology of the "tumor when conceived" may be distinct from the final tumor product. Several different cell types may be capable of becoming RMS by processes of differentiation, dedifferentiation (Odelberg *et al.*, 2000) or trans-differentiation (Lagha *et al.*, 2009; Messina *et al.*, 2009; Wiggan *et al.*, 2002). Will tumors retain characteristics of their originating linage? It may be dangerous to generalize, but at least in the context of certain mutations, Rubin *et al.* (2011) suggest this to be the case. We speculate that cell of origin (tumor initiation cell) may be able to become a cancer stem cell, but by way of a process that may significantly change its character.

4. Tumor Phenotype and Cancer Stem Cells

The hypothesis that malignant tumors grow and progress as a result of rare subsets of tumor repopulating cells that are more tumorigenic than other cancer cells has increasingly gained credence. Presuming that a single cell of origin gives rise to any individual tumor, the basis for this functional heterogeneity has been explained by one of two models, the hierarchy model and the stochastic model (Bomken *et al.*, 2010). The hierarchy model predicts that a malignancy is organized in a manner analogous to the normal tissue hierarchy with tumor/tissue stem cells able to produce identical daughter stem cells with self-renewal capacity, and committed progenitor daughter cells with limited, although potentially still significant, potential to divide. This model, with a rare cancer stem cell at the apex, is essentially synonymous with the cancer stem cell model (Quintana *et al.*, 2008). The stochastic model predicts that a malignancy is composed of a homogeneous population of cells, which generate their heterogeneity in response to particular combinations of endogenous and exogenous factors. Endogenous effects include gene dosage effects and transcriptional and translational control mechanisms, whereas exogenous effects include cytokine concentrations, cell–cell interactions and niche environment. This model has stemness as a functional phenotype, and does not yet predict whether stemness is found truly within each population, or whether cells first undergo a process of dedifferentiation to a more tissue specific stem cell-like phenotype, reacquiring stemness in the process (Gupta *et al.*, 2009). However, the current definitive test for a cancer stem cell is the capacity to propagate tumors as xenografts in immunocompromised mice (Clarke *et al.*, 2006). What factors are responsible for ERMS tumor repopulating cells? Hirotsu *et al.* (2009) reported xenoengraftment of single FGFR3-positive ERMS cells yielded tumor formation. In this study, cancer stem cells were enriched in RMS subpopulations defined not by side population characteristics (Komuro *et al.*, 2007) or CD133 positive-cells, but FGFR3 alone. It has been reported that bFGF promotes proliferation and inhibits differentiation of muscle satellite cells (Guthridge *et al.*, 1992; Lefaucheur and Sebille, 1995), and while FGF ligand and receptor interactions are still being defined, the binding of bFGF to FGFR3 is known to activate FGF signaling (Maric *et al.*, 2007). The mRNA expression of FGFR3 and FGFR4 are higher in quiescent than activated satellite cells (Fukada *et al.*, 2007) suggesting a commonality between tumor repopulating cell and normal muscle stem cell maintenance mechanisms. Further, identification of new FGFR4-activating mutations in 9% human ERMS were identified, and this mutation promoted metastasis and "remote" tumor repopulation in xenotransplanted models (Taylor *et al.*, 2009).

As mentioned previously, Langenau *et al.* (2007) discovered that expression of human *KrasG12D* driven by a *rag2* promoter rapidly induced ERMS in Zebrafish (Merlino and Khanna, 2007). By coinjecting both *rag2-dsRED2* and *rag2-KrasG12D* constructs into α-actin-GFP transgenic zebrafish embryos, they were able to differentially label RMS cells based on their differentiation status; *rag2* promoter-directed red fluorescence marked mononuclear myogenic precursors (R+), while α-actin promoter-driven green fluorescence labeled more mature muscle cells (G+). R+G+ cells represented cells of intermediate muscle differentiation, while double-negative cells were positive for blood cell markers. Using serial transplantation and limiting-dilution transplantation, functional hallmarks for cancer stem cell behavior, Langenau *et al.* found greater stem-like potential among the less-well-differentiated myogenic population (R+), likely representing the target cells for *Kras* transformation. Based on microarray analysis, the R+ population found in zebrafish ERMS was similar to activated satellite cells and shared pathways involved in normal satellite cell self-renewal. Recently, Rubin *et al.* also found that murine ERMS cells share features of satellite cells. Perhaps the most fascinating aspect of this study is the high concordance of gene expression between murine ERMS and activated rather than quiescent satellite cells (Fukada *et al.*, 2007; Rubin *et al.*, 2011).

Taken together, these data suggest that ERMS cancer stem cells may represent a myogenic precursor state, akin to a satellite cell, and express FGFRs. The notion that the tumor is driven by cancer stem cells has obvious therapeutic implications. The efficacy of tumor response to systemic therapy has traditionally been assessed based on the bulk of tumor cells by monitoring of changes in tumor size. However, if only a small fraction of cancer stem cells are capable of initiating tumor formation, then curative therapy should be designed to target these rare cancer stem cells rather than the bulk of nontumorigenic cells. Analysis of RMS cancer stem cells might thus yield novel and important therapeutic targets.

REFERENCES

Andrechek, E. R., Hardy, W. R., Girgis-Gabardo, A. A., Perry, R. L., Butler, R., Graham, F. L., Kahn, R. C., Rudnicki, M. A., and Muller, W. J. (2002). ErbB2 is required for muscle spindle and myoblast cell survival. *Mol. Cell. Biol.* **22,** 4714–4722.

Beauchamp, J. R., Heslop, L., Yu, D. S., Tajbakhsh, S., Kelly, R. G., Wernig, A., Buckingham, M. E., Partridge, T. A., and Zammit, P. S. (2000). Expression of CD34 and Myf5 defines the majority of quiescent adult skeletal muscle satellite cells. *J. Cell Biol.* **151,** 1221–1234.

Bomken, S., Fiser, K., Heidenreich, O., and Vormoor, J. (2010). Understanding the cancer stem cell. *Br. J. Cancer* **103,** 439–445.

Borden, E. C., Baker, L. H., Bell, R. S., Bramwell, V., Demetri, G. D., Eisenberg, B. L., Fletcher, C. D., Fletcher, J. A., Ladanyi, M., Meltzer, P., O'Sullivan, B., Parkinson, D. R., *et al.* (2003). Soft tissue sarcomas of adults: State of the translational science. *Clin. Cancer Res.* **9,** 1941–1956.

Breneman, J. C., Lyden, E., Pappo, A. S., Link, M. P., Anderson, J. R., Parham, D. M., Qualman, S. J., Wharam, M. D., Donaldson, S. S., Maurer, H. M., Meyer, W. H., Baker, K. S., *et al.* (2003). Prognostic factors and clinical outcomes in children and adolescents with metastatic rhabdomyosarcoma–A report from the Intergroup Rhabdomyosarcoma Study IV. *J. Clin. Oncol.* **21,** 78–84.

Bridge, J. A., Liu, J., Weibolt, V., Baker, K. S., Perry, D., Kruger, R., Qualman, S., Barr, F., Sorensen, P., Triche, T., and Suijkerbuijk, R. (2000). Novel genomic imbalances in embryonal rhabdomyosarcoma revealed by comparative genomic hybridization and fluorescence in situ hybridization: An intergroup rhabdomyosarcoma study. *Genes Chromosomes Cancer* **27,** 337–344.

Brown, C. B., Engleka, K. A., Wenning, J., Min Lu, M., and Epstein, J. A. (2005). Identification of a hypaxial somite enhancer element regulating Pax3 expression in migrating myoblasts and characterization of hypaxial muscle Cre transgenic mice. *Genesis* **41,** 202–209.

Calzada-Wack, J., Kappler, R., Schnitzbauer, U., Richter, T., Nathrath, M., Rosemann, M., Wagner, S. N., Hein, R., and Hahn, H. (2002). Unbalanced overexpression of the mutant allele in murine patched mutants. *Carcinogenesis* **23,** 727–733.

Chen, L., Shah, H. O., and Lin, J. H. (2004). Alveolar rhabdomyosarcoma with concurrent metastases to bone marrow and lymph nodes simulating acute hematologic malignancy. *J. Pediatr. Hematol. Oncol.* **26,** 696–697.
Chen, Y., Takita, J., Hiwatari, M., Igarashi, T., Hanada, R., Kikuchi, A., Hongo, T., Taki, T., Ogasawara, M., Shimada, A., and Hayashi, Y. (2006). Mutations of the PTPN11 and RAS genes in rhabdomyosarcoma and pediatric hematological malignancies. *Genes Chromosomes Cancer* **45,** 583–591.
Clark, M. A., Fisher, C., Judson, I., and Thomas, J. M. (2005). Soft-tissue sarcomas in adults. *N. Engl. J. Med.* **353,** 701–711.
Clarke, M. F., Dick, J. E., Dirks, P. B., Eaves, C. J., Jamieson, C. H., Jones, D. L., Visvader, J., Weissman, I. L., and Wahl, G. M. (2006). Cancer stem cells—Perspectives on current status and future directions: AACR workshop on cancer stem cells. *Cancer Res.* **66,** 9339–9344.
Cornelison, D. D., and Wold, B. J. (1997). Single-cell analysis of regulatory gene expression in quiescent and activated mouse skeletal muscle satellite cells. *Dev. Biol.* **191,** 270–283.
Davicioni, E., Anderson, M. J., Finckenstein, F. G., Lynch, J. C., Qualman, S. J., Shimada, H., Schofield, D. E., Buckley, J. D., Meyer, W. H., Sorensen, P. H., and Triche, T. J. (2009). Molecular classification of rhabdomyosarcoma-genotypic and phenotypic determinants of diagnosis: A report from the Children's Oncology Group. *Am. J. Pathol.* **174,** 550–564.
De Chiara, A., T'Ang, A., and Triche, T. J. (1993). Expression of the retinoblastoma susceptibility gene in childhood rhabdomyosarcomas. *J. Natl. Cancer Inst.* **85,** 152–157.
Diller, L., Sexsmith, E., Gottlieb, A., Li, F. P., and Malkin, D. (1995). Germline p53 mutations are frequently detected in young children with rhabdomyosarcoma. *J. Clin. Invest.* **95,** 1606–1611.
Doyle, B., Morton, J. P., Delaney, D. W., Ridgway, R. A., Wilkins, J. A., and Sansom, O. J. (2010). p53 mutation and loss have different effects on tumourigenesis in a novel mouse model of pleomorphic rhabdomyosarcoma. *J. Pathol.* **222,** 129–137.
Felix, C. A., Kappel, C. C., Mitsudomi, T., Nau, M. M., Tsokos, M., Crouch, G. D., Nisen, P. D., Winick, N. J., and Helman, L. J. (1992). Frequency and diversity of p53 mutations in childhood rhabdomyosarcoma. *Cancer Res.* **52,** 2243–2247.
Ferrari, A., Bisogno, G., Macaluso, A., Casanova, M., D'Angelo, P., Pierani, P., Zanetti, I., Alaggio, R., Cecchetto, G., and Carli, M. (2007). Soft-tissue sarcomas in children and adolescents with neurofibromatosis type 1. *Cancer* **109,** 1406–1412.
Fleischmann, A., Jochum, W., Eferl, R., Witowsky, J., and Wagner, E. F. (2003). Rhabdomyosarcoma development in mice lacking Trp53 and Fos: Tumor suppression by the Fos protooncogene. *Cancer Cell* **4,** 477–482.
Frank, K. M., Sharpless, N. E., Gao, Y., Sekiguchi, J. M., Ferguson, D. O., Zhu, C., Manis, J. P., Horner, J., DePinho, R. A., and Alt, F. W. (2000). DNA ligase IV deficiency in mice leads to defective neurogenesis and embryonic lethality via the p53 pathway. *Mol. Cell* **5,** 993–1002.
Fukada, S., Uezumi, A., Ikemoto, M., Masuda, S., Segawa, M., Tanimura, N., Yamamoto, H., Miyagoe-Suzuki, Y., and Takeda, S. (2007). Molecular signature of quiescent satellite cells in adult skeletal muscle. *Stem Cells* **25,** 2448–2459.
Gil-Benso, R., Lopez-Gines, C., Carda, C., Lopez-Guerrero, J. A., Ferrer, J., Pellin-Perez, A., and Llombart-Bosch, A. (2003). Cytogenetic and molecular findings related to rhabdomyosarcoma. An analysis of seven cases. *Cancer Genet. Cytogenet.* **144,** 125–133.
Gorlin, R. J. (2004). Nevoid basal cell carcinoma (Gorlin) syndrome. *Genet. Med.* **6,** 530–539.
Grigoriadis, A. E., Schellander, K., Wang, Z. Q., and Wagner, E. F. (1993). Osteoblasts are target cells for transformation in c-fos transgenic mice. *J. Cell Biol.* **122,** 685–701.

Grigoriadis, A. E., Wang, Z. Q., Cecchini, M. G., Hofstetter, W., Felix, R., Fleisch, H. A., and Wagner, E. F. (1994). c-Fos: A key regulator of osteoclast-macrophage lineage determination and bone remodeling. *Science* **266,** 443–448.

Gripp, K. W. (2005). Tumor predisposition in Costello syndrome. *Am. J. Med. Genet. C Semin. Med. Genet.* **137C,** 72–77.

Gripp, K. W., and Lin, A. E. (2009). Costello Syndrome. *In* "*GeneReviews* at GeneTests: Medical Genetics Information. Resource (database online)," (R. A. Pagon, T. D. Bird, C. R. Dolan, and K. Stephens, eds.), Copyright, University of Washington, Seattle. 1993-2011. Available at http://www.genetests.org. Accessed March 21, 2011.

Gupta, P. B., Chaffer, C. L., and Weinberg, R. A. (2009). Cancer stem cells: Mirage or reality? *Nat. Med.* **15,** 1010–1012.

Guthridge, M., Wilson, M., Cowling, J., Bertolini, J., and Hearn, M. T. (1992). The role of basic fibroblast growth factor in skeletal muscle regeneration. *Growth Factors* **6,** 53–63.

Hahn, H., Wojnowski, L., Zimmer, A. M., Hall, J., Miller, G., and Zimmer, A. (1998). Rhabdomyosarcomas and radiation hypersensitivity in a mouse model of Gorlin syndrome. *Nat. Med.* **4,** 619–622.

Haldar, M., Karan, G., Tvrdik, P., and Capecchi, M. R. (2008). Two cell lineages, myf5 and myf5-independent, participate in mouse skeletal myogenesis. *Dev. Cell* **14,** 437–445.

Hall, B. K., and Miyake, T. (2000). All for one and one for all: Condensations and the initiation of skeletal development. *Bioessays* **22,** 138–147.

Harvey, M., McArthur, M. J., Montgomery, C. A., Jr., Bradley, A., and Donehower, L. A. (1993a). Genetic background alters the spectrum of tumors that develop in p53-deficient mice. *FASEB J.* **7,** 938–943.

Harvey, M., McArthur, M. J., Montgomery, C. A., Jr., Butel, J. S., Bradley, A., and Donehower, L. A. (1993b). Spontaneous and carcinogen-induced tumorigenesis in p53-deficient mice. *Nat. Genet.* **5,** 225–229.

Hirotsu, M., Setoguchi, T., Matsunoshita, Y., Sasaki, H., Nagao, H., Gao, H., Sugimura, K., and Komiya, S. (2009). Tumour formation by single fibroblast growth factor receptor 3-positive rhabdomyosarcoma-initiating cells. *Br. J. Cancer* **101,** 2030–2037.

Iolascon, A., Faienza, M. F., Coppola, B., Rosolen, A., Basso, G., Della Ragione, F., and Schettini, F. (1996). Analysis of cyclin-dependent kinase inhibitor genes (CDKN2A, CDKN2B, and CDKN2C) in childhood rhabdomyosarcoma. *Genes Chromosomes Cancer* **15,** 217–222.

Ireland, H., Kemp, R., Houghton, C., Howard, L., Clarke, A. R., Sansom, O. J., and Winton, D. J. (2004). Inducible Cre-mediated control of gene expression in the murine gastrointestinal tract: Effect of loss of beta-catenin. *Gastroenterology* **126,** 1236–1246.

Jessen, J. R., Jessen, T. N., Vogel, S. S., and Lin, S. (2001). Concurrent expression of recombination activating genes 1 and 2 in zebrafish olfactory sensory neurons. *Genesis* **29,** 156–162.

Keleti, J., Quezado, M. M., Abaza, M. M., Raffeld, M., and Tsokos, M. (1996). The MDM2 oncoprotein is overexpressed in rhabdomyosarcoma cell lines and stabilizes wild-type p53 protein. *Am. J. Pathol.* **149,** 143–151.

Keller, C., Arenkiel, B. R., Coffin, C. M., El-Bardeesy, N., DePinho, R. A., and Capecchi, M. R. (2004). Alveolar rhabdomyosarcomas in conditional Pax3:Fkhr mice: Cooperativity of Ink4a/ARF and Trp53 loss of function. *Genes Dev.* **18,** 2614–2626.

Kleinerman, R. A., Tucker, M. A., Abramson, D. H., Seddon, J. M., Tarone, R. E., and Fraumeni, J. F., Jr. (2007). Risk of soft tissue sarcomas by individual subtype in survivors of hereditary retinoblastoma. *J. Natl. Cancer Inst.* **99,** 24–31.

Kohashi, K., Oda, Y., Yamamoto, H., Tamiya, S., Takahira, T., Takahashi, Y., Tajiri, T., Taguchi, T., Suita, S., and Tsuneyoshi, M. (2008). Alterations of RB1 gene in embryonal and alveolar rhabdomyosarcoma: Special reference to utility of pRB immunoreactivity in differential diagnosis of rhabdomyosarcoma subtype. *J. Cancer Res. Clin. Oncol.* **134,** 1097–1103.

Komuro, H., Saihara, R., Shinya, M., Takita, J., Kaneko, S., Kaneko, M., and Hayashi, Y. (2007). Identification of side population cells (stem-like cell population) in pediatric solid tumor cell lines. *J. Pediatr. Surg.* **42,** 2040–2045.

Koufos, A., Hansen, M. F., Copeland, N. G., Jenkins, N. A., Lampkin, B. C., and Cavenee, W. K. (1985). Loss of heterozygosity in three embryonal tumours suggests a common pathogenetic mechanism. *Nature* **316,** 330–334.

Kuang, S., Kuroda, K., Le Grand, F., and Rudnicki, M. A. (2007). Asymmetric self-renewal and commitment of satellite stem cells in muscle. *Cell* **129,** 999–1010.

Lagha, M., Brunelli, S., Messina, G., Cumano, A., Kume, T., Relaix, F., and Buckingham, M. E. (2009). Pax3:Foxc2 reciprocal repression in the somite modulates muscular versus vascular cell fate choice in multipotent progenitors. *Dev. Cell* **17,** 892–899.

Lam, S. H., Wu, Y. L., Vega, V. B., Miller, L. D., Spitsbergen, J., Tong, Y., Zhan, H., Govindarajan, K. R., Lee, S., Mathavan, S., Murthy, K. R., Buhler, D. R., *et al.* (2006). Conservation of gene expression signatures between zebrafish and human liver tumors and tumor progression. *Nat. Biotechnol.* **24,** 73–75.

Lang, G. A., Iwakuma, T., Suh, Y. A., Liu, G., Rao, V. A., Parant, J. M., Valentin-Vega, Y. A., Terzian, T., Caldwell, L. C., Strong, L. C., El-Naggar, A. K., and Lozano, G. (2004). Gain of function of a p53 hot spot mutation in a mouse model of Li-Fraumeni syndrome. *Cell* **119,** 861–872.

Langenau, D. M., Keefe, M. D., Storer, N. Y., Guyon, J. R., Kutok, J. L., Le, X., Goessling, W., Neuberg, D. S., Kunkel, L. M., and Zon, L. I. (2007). Effects of RAS on the genesis of embryonal rhabdomyosarcoma. *Genes Dev.* **21,** 1382–1395.

Le Bihan, C., Moutou, C., Brugieres, L., Feunteun, J., and Bonaiti-Pellie, C. (1995). ARCAD: A method for estimating age-dependent disease risk associated with mutation carrier status from family data. *Genet. Epidemiol.* **12,** 13–25.

Lee, Y., Miller, H. L., Russell, H. R., Boyd, K., Curran, T., and McKinnon, P. J. (2006). Patched2 modulates tumorigenesis in patched1 heterozygous mice. *Cancer Res.* **66,** 6964–6971.

Lefaucheur, J. P., and Sebille, A. (1995). Basic fibroblast growth factor promotes in vivo muscle regeneration in murine muscular dystrophy. *Neurosci. Lett.* **202,** 121–124.

Linardic, C. M., Downie, D. L., Qualman, S., Bentley, R. C., and Counter, C. M. (2005). Genetic modeling of human rhabdomyosarcoma. *Cancer Res.* **65,** 4490–4495.

Lisboa, S., Cerveira, N., Vieira, J., Torres, L., Ferreira, A. M., Afonso, M., Norton, L., Henrique, R., and Teixeira, M. R. (2008). Genetic diagnosis of alveolar rhabdomyosarcoma in the bone marrow of a patient without evidence of primary tumor. *Pediatr. Blood Cancer* **51,** 554–557.

Lohmann, D. R., and Gallie, B. L.(2010). Retinoblastoma. *In* "*GeneReviews* at GeneTests: Medical Genetics Information Resource (database online)," (R. A. Pagon,T.D. Bird, C. R. Dolan, and K. Stephens, eds.), Copyright, University of Washington, Seattle. 1993–2011. Available at http://www.genetests.org. Accessed March 21, 2011.

Maric, D., Fiorio Pla, A., Chang, Y. H., and Barker, J. L. (2007). Self-renewing and differentiating properties of cortical neural stem cells are selectively regulated by basic fibroblast growth factor (FGF) signaling via specific FGF receptors. *J. Neurosci.* **27,** 1836–1852.

Martinelli, S., McDowell, H. P., Vigne, S. D., Kokai, G., Uccini, S., Tartaglia, M., and Dominici, C. (2009). RAS signaling dysregulation in human embryonal rhabdomyosarcoma. *Genes Chromosomes Cancer* **48,** 975–982.

Merlino, G., and Khanna, C. (2007). Fishing for the origins of cancer. *Genes Dev.* **21,** 1275–1279.

Messina, G., Sirabella, D., Monteverde, S., Galvez, B. G., Tonlorenzi, R., Schnapp, E., De Angelis, L., Brunelli, S., Relaix, F., Buckingham, M., and Cossu, G. (2009). Skeletal muscle differentiation of embryonic mesoangioblasts requires pax3 activity. *Stem Cells* **27,** 157–164.

Miyachi, M., Kakazu, N., Yagyu, S., Katsumi, Y., Tsubai-Shimizu, S., Kikuchi, K., Tsuchiya, K., Iehara, T., and Hosoi, H. (2009). Restoration of p53 pathway by nutlin-3 induces cell cycle arrest and apoptosis in human rhabdomyosarcoma cells. *Clin. Cancer Res.* **15,** 4077–4084.

Moschovi, M., Touliatou, V., Papadopoulou, A., Mayakou, M. A., Nikolaidou-Karpathiou, P., and Kitsiou-Tzeli, S. (2007). Rhabdomyosarcoma in a patient with Noonan syndrome phenotype and review of the literature. *J. Pediatr. Hematol. Oncol.* **29,** 341–344.

Nanni, P., Nicoletti, G., De Giovanni, C., Croci, S., Astolfi, A., Landuzzi, L., Di Carlo, E., Iezzi, M., Musiani, P., and Lollini, P. L. (2003). Development of rhabdomyosarcoma in HER-2/neu transgenic p53 mutant mice. *Cancer Res.* **63,** 2728–2732.

Nishijo, K., Hosoyama, T., Bjornson, C. R., Schaffer, B. S., Prajapati, S. I., Bahadur, A. N., Hansen, M. S., Blandford, M. C., McCleish, A. T., Rubin, B. P., Epstein, J. A., Rando, T. A., *et al.* (2009). Biomarker system for studying muscle, stem cells, and cancer in vivo. *FASEB J.* **23,** 2681–2690.

Odelberg, S. J., Kollhoff, A., and Keating, M. T. (2000). Dedifferentiation of mammalian myotubes induced by msx1. *Cell* **103,** 1099–1109.

Olivier, M., Goldgar, D. E., Sodha, N., Ohgaki, H., Kleihues, P., Hainaut, P., and Eeles, R. A. (2003). Li-Fraumeni and related syndromes: Correlation between tumor type, family structure, and TP53 genotype. *Cancer Res.* **63,** 6643–6650.

Oue, T., Yoneda, A., Uehara, S., Yamanaka, H., and Fukuzawa, M. (2010). Increased expression of the hedgehog signaling pathway in pediatric solid malignancies. *J. Pediatr. Surg.* **45,** 387–392.

Parham, D. M., and Ellison, D. A. (2006). Rhabdomyosarcomas in adults and children: An update. *Arch. Pathol. Lab. Med.* **130,** 1454–1465.

Quintana, E., Shackleton, M., Sabel, M. S., Fullen, D. R., Johnson, T. M., and Morrison, S. J. (2008). Efficient tumour formation by single human melanoma cells. *Nature* **456,** 593–598.

Relaix, F., Montarras, D., Zaffran, S., Gayraud-Morel, B., Rocancourt, D., Tajbakhsh, S., Mansouri, A., Cumano, A., and Buckingham, M. (2006). Pax3 and Pax7 have distinct and overlapping functions in adult muscle progenitor cells. *J. Cell Biol.* **172,** 91–102.

Reya, T., Morrison, S. J., Clarke, M. F., and Weissman, I. L. (2001). Stem cells, cancer, and cancer stem cells. *Nature* **414,** 105–111.

Ricci, C., Landuzzi, L., Rossi, I., De Giovanni, C., Nicoletti, G., Astolfi, A., Pupa, S., Menard, S., Scotlandi, K., Nanni, P., and Lollini, P. L. (2000). Expression of HER/erbB family of receptor tyrosine kinases and induction of differentiation by glial growth factor 2 in human rhabdomyosarcoma cells. *Int. J. Cancer* **87,** 29–36.

Rubin, B. P., Nishijo, K., Chen, H.-I. H., Yi, X., Schuetze, D. P., Pal, R., Prajapati, S. I., Abraham, J., Arenkiel, B. R., Chen, Q.-R., Davis, S., McCleish, A. T., *et al.* (2011). Evidence for an unanticipated relationship between undifferentiated pleomorphic sarcoma and embryonal rhabdomyosarcoma. *Cancer Cell* **19,** 177–791.

Sandberg, A. A., Stone, J. F., Czarnecki, L., and Cohen, J. D. (2001). Hematologic masquerade of rhabdomyosarcoma. *Am. J. Hematol.* **68,** 51–57.

Sansom, O. J., Griffiths, D. F., Reed, K. R., Winton, D. J., and Clarke, A. R. (2005). Apc deficiency predisposes to renal carcinoma in the mouse. *Oncogene* **24,** 8205–8210.

Schneider, K., and Garber, J. (2010). Li-Fraumeni Syndrome. *In* "*GeneReviews* at GeneTests: Medical Genetics Information Resource (database online)," (R. A. Pagon, T. D. Bird, C. R. Dolan, K. Stephens, eds.), Copyright, University of Washington, Seattle. 1993-2011. Available at http://www.genetests.org. Accessed March 21, 2011.

Scrable, H., Cavenee, W., Ghavimi, F., Lovell, M., Morgan, K., and Sapienza, C. (1989). A model for embryonal rhabdomyosarcoma tumorigenesis that involves genome imprinting. *Proc. Natl. Acad. Sci. USA* **86,** 7480–7484.

Serrano, M., Lee, H., Chin, L., Cordon-Cardo, C., Beach, D., and DePinho, R. A. (1996). Role of the INK4a locus in tumor suppression and cell mortality. *Cell* **85,** 27–37.

Sharp, R., Recio, J. A., Jhappan, C., Otsuka, T., Liu, S., Yu, Y., Liu, W., Anver, M., Navid, F., Helman, L. J., DePinho, R. A., and Merlino, G. (2002). Synergism between INK4a/ARF inactivation and aberrant HGF/SF signaling in rhabdomyosarcomagenesis. *Nat. Med.* **8,** 1276–1280.

Sharpless, N. E., Ferguson, D. O., O'Hagan, R. C., Castrillon, D. H., Lee, C., Farazi, P. A., Alson, S., Fleming, J., Morton, C. C., Frank, K., Chin, L., Alt, F. W., *et al.* (2001). Impaired nonhomologous end-joining provokes soft tissue sarcomas harboring chromosomal translocations, amplifications, and deletions. *Mol. Cell* **8,** 1187–1196.

Shuman, C., Smith, A. C., and Weksberg, R. (2010). Beckwith-Wiedemann Syndrome. *In* "*GeneReviews* at GeneTests: Medical Genetics Information Resource (database online)," (R. A. Pagon, T. D. Bird, C. R. Dolan, and K. Stephens, eds.), Copyright, University of Washington, Seattle. 1993–2011. Available at http://www.genetests.org. Accessed March 21, 2011.

Singh, S., Vinson, C., Gurley, C. M., Nolen, G. T., Beggs, M. L., Nagarajan, R., Wagner, E. F., Parham, D. M., and Peterson, C. A. (2010). Impaired Wnt signaling in embryonal rhabdomyosarcoma cells from p53/c-fos double mutant mice. *Am. J. Pathol.* **177,** 2055–2066.

Smith, A. C., Squire, J. A., Thorner, P., Zielenska, M., Shuman, C., Grant, R., Chitayat, D., Nishikawa, J. L., and Weksberg, R. (2001). Association of alveolar rhabdomyosarcoma with the Beckwith-Wiedemann syndrome. *Pediatr. Dev. Pathol.* **4,** 550–558.

Sorensen, P. H., Lynch, J. C., Qualman, S. J., Tirabosco, R., Lim, J. F., Maurer, H. M., Bridge, J. A., Crist, W. M., Triche, T. J., and Barr, F. G. (2002). PAX3-FKHR and PAX7-FKHR gene fusions are prognostic indicators in alveolar rhabdomyosarcoma: A report from the children's oncology group. *J. Clin. Oncol.* **20,** 2672–2679.

Stratton, M. R., Fisher, C., Gusterson, B. A., and Cooper, C. S. (1989). Detection of point mutations in N-ras and K-ras genes of human embryonal rhabdomyosarcomas using oligonucleotide probes and the polymerase chain reaction. *Cancer Res.* **49,** 6324–6327.

Sumino, Y., Hirata, Y., Sato, F., and Mimata, H. (2007). Growth mechanism of satellite cells in human urethral rhabdosphincter. *Neurourol. Urodyn.* **26,** 552–561.

Takahashi, Y., Oda, Y., Kawaguchi, K., Tamiya, S., Yamamoto, H., Suita, S., and Tsuneyoshi, M. (2004). Altered expression and molecular abnormalities of cell-cycle-regulatory proteins in rhabdomyosarcoma. *Mod. Pathol.* **17,** 660–669.

Takayama, H., LaRochelle, W. J., Sharp, R., Otsuka, T., Kriebel, P., Anver, M., Aaronson, S. A., and Merlino, G. (1997). Diverse tumorigenesis associated with aberrant development in mice overexpressing hepatocyte growth factor/scatter factor. *Proc. Natl. Acad. Sci. USA* **94,** 701–706.

Taylor, A. C., Shu, L., Danks, M. K., Poquette, C. A., Shetty, S., Thayer, M. J., Houghton, P. J., and Harris, L. C. (2000). P53 mutation and MDM2 amplification frequency in pediatric rhabdomyosarcoma tumors and cell lines. *Med. Pediatr. Oncol.* **35,** 96–103.

Taylor, J. G. t., Cheuk, A. T., Tsang, P. S., Chung, J. Y., Song, Y. K., Desai, K., Yu, Y., Chen, Q. R., Shah, K., Youngblood, V., Fang, J., Kim, S. Y., *et al.* (2009). Identification of FGFR4-activating mutations in human rhabdomyosarcomas that promote metastasis in xenotransplanted models. *J. Clin. Invest.* **119,** 3395–3407.

Tostar, U., Malm, C. J., Meis-Kindblom, J. M., Kindblom, L. G., Toftgard, R., and Unden, A. B. (2006). Deregulation of the hedgehog signalling pathway: A possible role for the PTCH and SUFU genes in human rhabdomyoma and rhabdomyosarcoma development. *J. Pathol.* **208,** 17–25.

Tsumura, H., Yoshida, T., Saito, H., Imanaka-Yoshida, K., and Suzuki, N. (2006). Cooperation of oncogenic K-ras and p53 deficiency in pleomorphic rhabdomyosarcoma development in adult mice. *Oncogene* **25,** 7673–7679.

Wang, Z. Q., Grigoriadis, A. E., Mohle-Steinlein, U., and Wagner, E. F. (1991). A novel target cell for c-fos-induced oncogenesis: Development of chondrogenic tumours in embryonic stem cell chimeras. *EMBO J.* **10,** 2437–2450.

White, J. D., Rachel, C., Vermeulen, R., Davies, M., and Grounds, M. D. (2002). The role of p53 in vivo during skeletal muscle post-natal development and regeneration: Studies in p53 knockout mice. *Int. J. Dev. Biol.* **46,** 577–582.

Wiggan, O., Fadel, M. P., and Hamel, P. A. (2002). Pax3 induces cell aggregation and regulates phenotypic mesenchymal–epithelial interconversion. *J. Cell Sci.* **115,** 517–529.

Wilke, W., Maillet, M., and Robinson, R. (1993). H-ras-1 point mutations in soft tissue sarcomas. *Mod. Pathol.* **6,** 129–132.

Xia, S. J., Pressey, J. G., and Barr, F. G. (2002). Molecular pathogenesis of rhabdomyosarcoma. *Cancer Biol. Ther.* **1,** 97–104.

CHAPTER THREE

Sculpting Chromatin Beyond the Double Helix: Epigenetic Control of Skeletal Myogenesis

Vittorio Sartorelli *and* Aster H. Juan

Contents

Abstract

Satellite cells (SCs) are the main source of adult skeletal muscle stem cells responsible for muscle growth and regeneration. By interpreting extracellular cues, developmental regulators control quiescence, proliferation, and differentiation of SCs by influencing coordinate gene expression. The scope of this review is limited to the description and discussion of protein complexes that

Laboratory of Muscle Stem Cell and Gene Regulation, National Institute of Arthritis and Musculoskeletal and Skin Diseases (NIAMS), National Institutes of Health, Bethesda, Maryland, USA

Current Topics in Developmental Biology, Volume 96
ISSN 0070-2153, DOI: 10.1016/B978-0-12-385940-2.00003-6

introduce and decode heritable histone and chromatin modifications and how these modifications are relevant for SC biology.

1. Introduction

Epigenetics can be defined as the ensemble of heritable changes in gene function that occur without modifications of primary DNA sequence (Bird, 2007; Russo *et al.*, 1996) to include DNA methylation and chromatin structural changes introduced by histone modifications and nucleosome positioning. During specification, proliferation, and differentiation of skeletal muscle cells (myogenesis), the epigenetic marks deposited by chromatin-modifying enzymes at specific loci determine whether different subsets of genes will be repressed or activated, effectively controlling the fate of muscle progenitors and the transition between each developmental phase. Therefore, the balance between cell self-renewal and differentiation is ultimately regulated by coordination of stage-specific transcriptional machineries through epigenetic mechanisms. Many excellent and comprehensive reviews on satellite cells (SCs) have been published (Buckingham, 2007; Charge and Rudnicki, 2004; Dhawan and Rando, 2005; Kang and Krauss, 2010) and the reader is referred to them.

2. Satellite Cells

SCs represent the primary source of multipotent stem cells responsible for postnatal skeletal muscle growth and regeneration (Charge and Rudnicki, 2004; Partridge, 2004). They originate from Pax3/7-expressing muscle progenitors in the embryonic dermomyotome (Gros *et al.*, 2005; Relaix *et al.*, 2005). In the mouse, SCs expressing Pax7 appear at approximately embryonic day 16.5 (E16.5) in the developing limb muscle (Kassar-Duchossoy *et al.*, 2005) and acquire their characteristic position under the basal lamina of mature muscle fibers (Mauro, 1961). SCs comprise about 30% of total muscle nuclei in neonatal mice and 2–7% of muscle nuclei in the adult mice. They express a distinct profile of surface markers and transcription factors (TFs; Charge and Rudnicki, 2004). Recent studies have revealed that SCs are a heterogeneous population (Beauchamp *et al.*, 2000; Collins *et al.*, 2005; Kuang *et al.*, 2007; Sherwood *et al.*, 2004) composed of two groups of cells: noncommitted stem cells which maintain self-renewal ability and committed myogenic progenitors which undergo lineage-specific differentiation (Kuang *et al.*, 2007). Because of these two distinguished characteristics, SCs qualify as a *bona fide* tissue-specific adult stem cell. Under normal circumstances, adult SCs are in a quiescent state.

In response to environmental cues, such as mechanical stress, muscle injury, or degenerative muscle diseases, SCs exit quiescence, enter cell cycle, and actively proliferate (activated state). Following activation, a subset of SC returns to its niche under the basal lamina to replenish the stem cell reservoir. Majority of SCs activate expression of the myogenic factor MyoD and give rise to committed myoblasts, which undergo rapid proliferation and subsequently exit from the cell cycle to terminally differentiate and fuse to form new muscle fibers or repair damaged muscle (Dhawan and Rando, 2005). The identification of mesangioblasts and other non-SC muscle resident and circulating progenitors expand the repertoire of cells potentially participating to muscle growth and regeneration (Benchaouir *et al.*, 2007; Corbel *et al.*, 2003; Cossu and Bianco, 2003; Dellavalle *et al.*, 2007; Joe *et al.*, 2010; Mitchell *et al.*, 2010; Sampaolesi *et al.*, 2003, 2006; reviewed in Tedesco *et al.*, 2010).

3. Repressing Muscle Gene Expression: Skeletal Muscle Developmental Regulators and Polycomb Proteins in Embryonic Stem (ES) and Nonmuscle Cells

Totipotent ES cells retain the ability of self-renewing and generating all adult cell types. These two properties are mutually exclusive: an individual ES cell either renews or differentiates into a given cell lineage. Understanding the molecular mechanisms that regulate this dichotomic decision is not only of scientific interest to developmental biologists but also holds great promise for regenerative medicine. Under defined culture conditions, ES cells can be indefinitely propagated in an undifferentiated state. ES cell differentiation into specialized cell lineages occurs in culture or during development when appropriate signals are provided and correctly interpreted. For these processes to be properly executed, transcription and translation of developmental regulators (DRs) need to be tightly controlled. DRs are TFs or related molecules that coordinate temporal and spatial expression of battery of genes involved in the specification and maintenance of a given cell state. DRs favoring ES self-renewal (such as Oct4, Nanog, and Sox2) are expressed in undifferentiated ES cells but extinguished upon lineage commitment and differentiation (Chambers *et al.*, 2003, 2007; Masui *et al.*, 2007; Nichols *et al.*, 1998). Conversely, cell type-specific DRs, such as MyoD (Davis *et al.*, 1987), are repressed in undifferentiated ES cells and their expression is activated once a given cell fate is specified. Polycomb group (PcG) proteins are pivotal regulators of DRs expression in ES cells (Bernstein *et al.*, 2006; Boyer *et al.*, 2006; Lee *et al.*, 2006). PcG proteins are assembled in multiple Polycomb Repressive Complexes

(PRCs) that exert their regulatory functions by introducing epigenetic modifications to chromatin structure conducive to transcriptional repression (Simon and Kingston, 2009). PRC2—and related PRC3 and PRC4—contains the core components Ezh2, Suz12, and Eed. Ezh2 is the PRC2 catalytic component with methyltransferase activity directed at lysine 27 of histone H3 (H3K27; Cao *et al.*, 2002; Czermin *et al.*, 2002; Kirmizis *et al.*, 2004; Kuzmichev *et al.*, 2002; Muller *et al.*, 2002), with Suz12 being required for the enzymatic activity (Cao and Zhang, 2004; Pasini *et al.*, 2004). Eed specifically recognizes and binds to trimethylated repressive marks, including H3K27me3. This event leads to the allosteric activation of the methyltransferase activity of PRC2, resulting in further propagation of the H3K27me3 mark to extended chromatin blocks and possibly transmission and retention of this histone modification through DNA replication (Margueron *et al.*, 2009). H3K27me3 serves, in most cases, as a docking site for PRC1 recruitment, which promotes further chromatin condensation to ensure gene silencing (Francis *et al.*, 2004).

The paired domain- and homeobox-containing TFs Pax3 and Pax7 and the myogenic determinant factor MyoD are nodal DRs expressed in quiescent and activated SCs (Kassar-Duchossoy *et al.*, 2005; Megeney *et al.*, 1996; Relaix *et al.*, 2005; Rudnicki *et al.*, 1993; Seale *et al.*, 2000). Myf6 (also referred as to MRF4) is transcribed during SC differentiation. In ES cells, Pax3, Pax7, MyoD, and MRF4 are not expressed and their respective loci are occupied by Suz12 and nucleosomes with H3K27me3 repressive mark (Lee *et al.*, 2006). However, in addition to the repressive H3K27me3 mark, the regulatory regions of Pax3, Pax7, MyoD, and MRF4 also contain the H3K4me3 mark (Zhao *et al.*, 2007), a histone modification associated with transcriptional activation (Barski *et al.*, 2007). The coexistence of negative and positive histone marks (bivalent domains; Bernstein *et al.*, 2006) would serve the purpose of silencing DRs in ES cells while keeping them poised for immediate activation once signals are received and decoded to initiate cell lineage specific gene expression. Consistent with this proposed function, bivalent domains are resolved in monovalent regions retaining only H3K4me3 when they experience gene activation or H3K27me3 when remaining repressed. The Pax3, Pax7, MyoD, and myogenin—another pivotal regulator of SC differentiation—loci engage Suz12, are H3K27me3 marked, and repressed also in human embryonic fibroblasts (Bracken *et al.*, 2006) and numerous nonepidermal "master" DRs , including MyoD, are marked by H3K27me3 and transcriptionally repressed in committed embryonic basal epidermal progenitors (Ezhkova *et al.*, 2009). Pax3 and Pax7 are also occupied by Suz12 and marked with H3K27me3 in F9 embryonic carcinoma cells (Squazzo *et al.*, 2006). Thus, DRs may need to be continuously repressed not only in ES cells but also in cells that have entered a cell lineage-specific fate, which is incompatible with expression of the repressed DRs. Alternatively, H3K27me3 may be a remnant mark introduced by PcG in ES cells and self-propagated (Hansen *et al.*, 2008) in specified cells. The presence of PRC2 at some DRs is indicative of an active and continuously repressive role exerted by PcG in specified cell lineages. ES cells in which either Suz12 or Eed has been genetically ablated (Boyer *et al.*, 2006; Lee *et al.*, 2006) or Ezh2 has been reduced by miR-214 overexpression (Juan *et al.*, 2009) display an aberrant expression of certain cell lineage-specific DRs, including Pax7, Gata1-6, and Sox17. Neither MyoD nor MRF4 appear to be expressed in Suz12- or Eed-null ES cells nor in ES cells with miR-214-mediated reduced Ezh2, indicating that derepression does not necessarily coincides with transcriptional activation, at least for certain loci. Conditional Ezh2 inactivation in committed embryonic basal epidermal progenitors is not sufficient to activate expression of nonepidermal PcG targets (Ezhkova *et al.*, 2009). Interestingly, mouse embryonic fibroblasts (MEFs) derived from mice carrying an hypomorphic allele and expressing ~25% of the normal complement for the TF YY1—a protein known to participate in PcG-mediated repression (Atchison *et al.*, 2003; Satijn *et al.*, 2001; Srinivasan and Atchison, 2004; Woo *et al.*, 2010; Yue *et al.*, 2009)—display increased expression of several muscle-specific transcripts (actins, myosins, troponins), which are physiologically absent in MEFs (Affar el *et al.*, 2006).

4. Sculpting Chromatin for Transcription in Skeletal Muscle Cells

Chromatin accessibility to enzymes and TFs characterizes different cell types (Weintraub and Groudine, 1976) and different developmental stages within the same cell lineage (Groudine and Weintraub, 1981), including skeletal muscle cells (Carmon *et al.*, 1982). This is a reflection of different conformations that nucleosomes adopt as a consequence of histone and DNA modifications and positioning introduced by specific enzymatic complexes. These enzymatic protein complexes are recruited at discrete DNA regions via interaction with proteins (TFs) that recognize sequence-specific DNA modules. Chromatin-modifying complexes are transferred to stalled RNA polymerases or travel in association with elongating RNA polymerases. Noncoding RNAs can also recruit enzymatic complexes at defined chromatin regions (Mohammad *et al.*, 2008; Rinn *et al.*, 2007; Zhao *et al.*, 2008).

4.1. Pax3/Pax7 and alternative SC fates

As already mentioned, Pax3 and Pax7 are important regulators of SC biology (Oustanina *et al.*, 2004; Relaix *et al.*, 2005; Seale *et al.*, 2000). They both regulate expression of numerous genes including the myogenic determinant factor Myf5 (Maroto *et al.*, 1997; Tajbakhsh *et al.*, 1997). In C2C12 cell and satellite-derived primary myoblasts, Pax7 interacts with and engages the Wdr5-Ash2-MLL2 histone methyltransferase at the Myf5 locus, promoting H3K4me3 and gene activation (McKinnell *et al.*, 2008). Pax3 activates Myf5 through the intermediate induction of Dmrt2, a TF that interacts with one of the Myf5 enhancers (Sato *et al.*, 2010). Differences were noted in transcriptional potency, with Pax3 being a poor activator compared to Pax7 (McKinnell *et al.*, 2008; Relaix *et al.*, 2003), perhaps reflective of recruitment of distinct regulatory protein complexes. An important question relates to the inability of Pax7 to activate Myf5 in SCs returning to their niche. In these cells ($Pax7^+/Myf5^-$; Kuang *et al.*, 2007), the Myf5 locus may retain H3K27me3 introduced by PcG in ES cells (epigenetic memory; Lee *et al.*, 2006) or be occupied by histones, such as histone linker H1, or histone variants that may preclude Pax7 chromatin access. Alternatively, the Pax7 protein may be modified to avoid association with regulatory protein complexes required to activate Myf5 or such regulatory complexes may be absent in the self-renewing daughter cell. The hypothesized H3K27me3-marked Myf5 alleles may segregate with $Pax7^+/Myf5^-$ cells and bear methylated DNA. Indeed, Ezh2 directly controls DNA methylation by recruiting DNA methyltransferase proteins (Vire *et al.*, 2006). Preferential asymmetric segregation of histone- and/or DNA-methylated Myf5 alleles in the self-renewing cell maybe achieved by DNA cosegregation with the immortal DNA strand (Conboy *et al.*, 2007; Shinin *et al.*, 2006).

4.2. Myogenic bHLH protein binding modalities

MyoD, Myf5, myogenin, and MRF4 heterodimerize with ubiquitously expressed bHLH E-proteins and recognize the DNA core consensus sequence CANNTG (the E-box; reviewed in Tapscott, 2005). Chromatin immunoprecipitation (ChIP) combined with either microarray hybridization or high-throughput sequencing and gene expression profiling (Bergstrom *et al.*, 2002; Blais *et al.*, 2005; Cao *et al.*, 2006, 2010) has allowed identification of myogenic bHLH targets on a genome-wide scale. As anticipated, MyoD and myogenin bind to regulatory regions of genes whose transcription is regulated during skeletal muscle differentiation (Blais *et al.*, 2005; Cao *et al.*, 2006). Unexpectedly, MyoD binding is evident at several sites before differentiation (Blais *et al.*, 2005) and majority of MyoD binding occurs at either introns or intergenic regions, coinciding with local histone acetylation (Cao *et al.*, 2010). One interpretation of these findings is that, in addition to regulating gene expression, MyoD may have a role in the epigenetic reprogramming of skeletal muscle cells (Cao *et al.*, 2010). Importantly, genome-wide MyoD binding distribution in the C2C12 cell line, primary muscle cells, and fibroblasts expressing MyoD are extremely similar (Cao *et al.*, 2010), a reassuring finding further confirming coincidence of regulatory transcriptional modalities in different cellular model of skeletal muscle differentiation.

4.3. Marking chromatin for repression

The composition of the transcripitonal machinery (protein complexes) and the chromatin asset (histone and DNA) at muscle-specific genes reflects the different states of the cell. In the vicinity of transcriptional start sites (TSS) of genes not expressed in undifferentiated myoblasts, core histones H3 and H4 are mainly deacetylated and methylated at specific lysines (H3K9,14,18 deAc, H4K5,8,16 deAc, H3K9me2, and H3K27me3) by histone deaceylases (HDACs; Fulco *et al.*, 2003; Lu *et al.*, 2000b; Mal *et al.*, 2001; Ohkawa *et al.*, 2006; Puri *et al.*, 2001) and methyltransferases (HMTs; Suv39h1/KMT1A and PcG; Caretti *et al.*, 2004; Mal, 2006). These enzymes have developed the ability to function as cooperative modules (Margueron and Reinberg, 2010). This is a property deriving by mutually exclusive histone modifications. Acetylation and methylation cannot occur on the same lysine residue and deacetylation is often a prerequisite for subsequent methylation. Thus, HDACs are often found in protein complexes containing HMTs. Class I, II, and III HDACs, SUv39h1/KMT1A, the chromodomain-containing methyl-binding protein HP1, and PcG Ezh2 associate in different protein complexes (Caretti *et al.*, 2004; Kuzmichev *et al.*, 2005; van der Vlag and Otte, 1999; Vaquero *et al.*, 2007; Vaute *et al.*, 2002) and are recruited, along with the PcG-related TF YY1, at chromatin regulatory regions of inactive muscle genes (Fig. 3.1A; Caretti *et al.*, 2004; Fulco *et al.*, 2003; Mal, 2006; Ohkawa *et al.*, 2006; Wang *et al.*, 2007; Zhang *et al.*, 2002). Initially identified as histone deacetylases and methyltransferases, HDACs, and HMTs were subsequently found to act also on nonhistone proteins, including TFs such as MyoD, MEF2, and p53 (Chuikov *et al.*, 2004; Fulco *et al.*, 2003; Huang *et al.*, 2006, 2010; Mal *et al.*, 2001; Puri *et al.*, 2001; Zhao *et al.*, 2005).

Specific histone isoforms participate in shaping repressive chromatin structure. Through specific interaction with the homeoprotein Msx1, histone H1b is deposited at the core enhancer region of the MyoD locus within a repressive chromatin region characterized by H3K9 methylation, decreased H3K9 and 14 acetylation, and reduced H3Ser10 phosphorylation (Lee *et al.*, 2004). H1b depletion abrogated the ability of Msx1 to inhibit muscle cell differentiation, directly linking the repressive function of Msx1 to histone H1b.

4.4. SIRT1: A multifaceted regulator

SIRT1 is an NAD$^+$-dependent protein deacetylase (Imai *et al.*, 2000) expressed in quiescent and activated SCs (J. G. Ryall, A. Pasut, M. A. Rudnicki, and V. S., unpublished results). Reducing SIRT1 in the C2C12

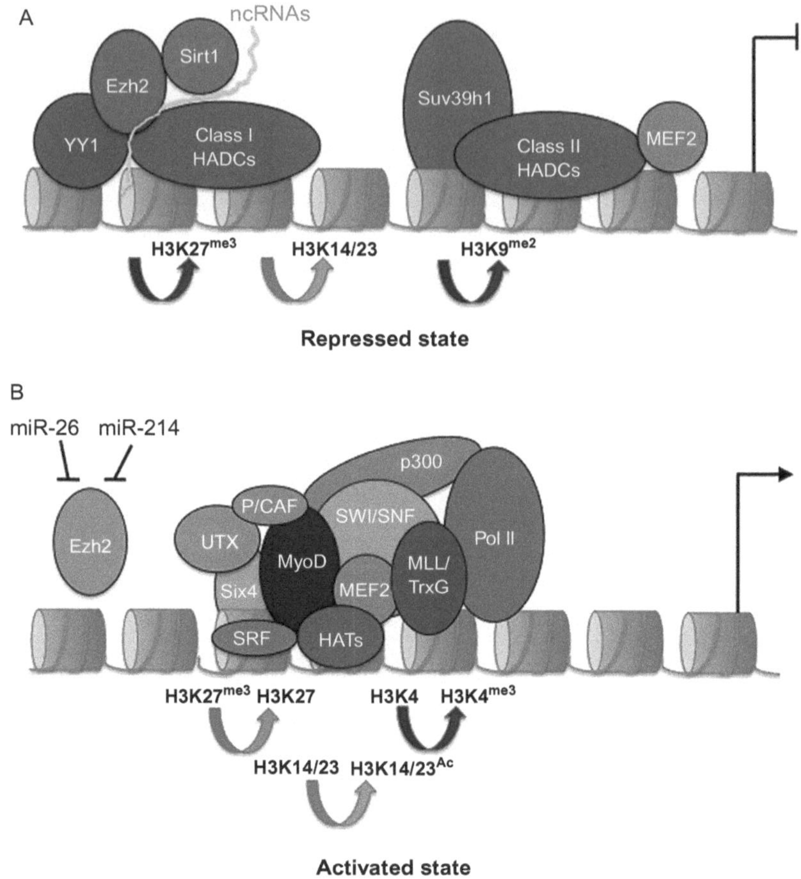

Figure 3.1 (A) Repressive histone marks are introduced by protein complexes composed by class I, II, and III (SIRT1) deacetylases and histone methyltransferases (Suv39h1 and Polycomb Ezh2). Ezh2 may be recruited by the Pleiohomeotic (PHO)-related YY1 protein or noncoding RNAs (ncRNAs). Lysines 14–23 of histone H3 are deacetylated and lysine 27 (Ezh2) and lysine 9 (Suv39h1) are methylated. (B) Repressors are replaced by a new set of protein complexes at transcribed genes. Via interaction with the transcriptional activators MyoD, MEF2, SRF, and Six4, acetyltransferases (HATs, p300, P/CAF) and nucleosome remodeling machines (SWI/SNF) promote lysine acetylation (H3K14-23^{AC}) and nucleosome remodeling, respectively. Demethylation of lysine 27 is achieved by transcriptional downregulation and microRNA-mediated (miR-26 and miR-214) repression of Ezh2. In addition, the lysine 27-specific UTX actively removes methyl groups from lysine 27. The Trithorax-related mixed lineage leukemia (MLL) proteins promote methylation of histone H3 lysine 4, a mark associated with polymerase II (PolII) recruitment and gene activation. (See Color Insert.)

cell line or in primary myoblasts results in their premature differentiation (Fulco *et al.*, 2003) and, analogously, primary myoblasts derived from heterozygous SIRT1$^{+/-}$ exhibit early differentiation and are resistant to antidifferentiative cues, such as glucose restriction (Fulco *et al.*, 2008). Overexpression experiments have revealed that SIRT1 increases SC proliferation (Rathbone *et al.*, 2008). It is likely that the effects of SIRT1 on muscle cell differentiation results from its deacetylase activity on different substrates, including histones, acetyltransferases, and transcriptional regulators (Brunet *et al.*, 2004; Fulco *et al.*, 2003; Motta *et al.*, 2004). Indeed, SIRT1 deacetylates MyoD and P/CAF (Fulco *et al.*, 2003), an acetyltransferase that activates MyoD (Dilworth *et al.*, 2004; Kuninger *et al.*, 2006; Sartorelli *et al.*, 1999). MEF2D, an important TF that collaborates with MyoD to activate genes critical for muscle differentiation, is also deacetylated by SIRT1 (Zhao *et al.*, 2005). In addition to controlling cell differentiation, SIRT1, and related sirtuins, may have other functions in SCs that remain to be explored. For instance, SIRT1 interacts with and mediates transcriptional repression induced by hairy/Hes1 and HEY2 (Bianchi-Frias *et al.*, 2004; Rosenberg and Parkhurst, 2002; Takata and Ishikawa, 2003), two transcriptional repressors whose activation is promoted by the Notch pathway. Given the central role played by the Notch pathway in regulating SC activation and cell fate determination in postnatal myogenesis (Brack *et al.*, 2008; Conboy and Rando, 2002), asymmetric cell division (Conboy *et al.*, 2007; Kuang *et al.*, 2007; Shinin *et al.*, 2006), restoring the regenerative potential of aged muscle (Conboy *et al.*, 2003; Luo *et al.*, 2005), and the ability of SIRT1 to control SC proliferation and differentiation (Fulco *et al.*, 2003; Rathbone *et al.*, 2008), to retard degenerative processes and increase lifespan across species (reviewed in Imai and Guarente, 2010), it will be of great interest to determine whether the Notch and SIRT1 pathways cross-talk. Another process that may be potentially influenced by SIRT1 or other sirtuins is the cell response to oxidative stress. It is intriguing that quiescent SCs have developed strategies that protect them from xenobiotics, genotoxics, and oxidative stress (Pallafacchina *et al.*, 2010), likely to maintain genome integrity. Since SIRT1 is activated by and protects from the deleterious effects induced by oxidative stress , DNA damage, and hypoxia (Brunet *et al.*, 2004; Dioum *et al.*, 2009; Motta *et al.*, 2004; Oberdoerffer *et al.*, 2008), it may contribute to SC resistance to environmetal insults. In principle, SIRT1 transcription needs not to be regulated in quiescent versus activated SCs, as its enzymatic activity can be modulated by the redox state (Fulco *et al.*, 2003).

4.5. Turning on transcription: Erasing and writing

Reception and decodification of prodifferentiative signals induce modification of the chromatin structure conducive to gene activation. Repressive marks need to be erased and substituted with histone modifications and nucleosome restructuring compatible with RNA polymerase II (PolII) engagement and elongation. A pivotal step in this process is the recruitment of histone acetyltransferases (HATs), ATP-dependent chromatin remodeling complexes, methyltransferases, demethylases, and specific subunits of the basal transcriptional machinery.

H3K27me3 is reduced at specific muscle regulatory regions upon gene activation (Caretti *et al.*, 2004). Several independent, but functionally coherent, mechanisms contribute to this phenomenon (Fig. 3.1B). H3K27me3 deposition is curtailed as Ezh2 transcription is reduced (Caretti *et al.*, 2004) and the residual Ezh2 transcripts are targeted for translational inhibition by the microRNA miR-26 and miR-214 (Juan *et al.*, 2009; Wong and Tellam, 2008). Active H3K27 demethylation is concomitantly brought about by Six4-mediated recruitment of the UTX demethylase (Seenundun *et al.*, 2010). The presence of the H3K4 methyltransferases mixed lineage leukemia (MLL) proteins within the UTX complex provides a parsimonious and elegant mechanism whereby a repressive mark is erased and one with activation property is introduced by the same protein complex (Fig. 3.1B; Rampalli *et al.*, 2007). Jmjd1, a specific H3K9me2 demethylase (Yamane *et al.*, 2006) may be involved in erasing H3K9me2 from muscle regulatory regions. During the differentiation process, class I HDACs are disengaged from MyoD and possibly other transcriptional regulators and redistributed to alternative protein complexes via cell cycle-regulated events involving pRb hypophosphorylation (Puri *et al.*, 2001) and calcium-mediated activation of the calmodulin-dependent protein kinase (CaMK) stimulates MEF2 activity by dissociating it from class II HDACs (Lu *et al.*, 2000a). Pharmacological HDACs inhibition promotes, in a temporal-specific manner, muscle gene expression (Iezzi *et al.*, 2002, 2004), favoring a functional and morphological amelioration in two different animal models of muscle dystrophies (Minetti *et al.*, 2006; reviewed in (Mozzetta *et al.*, 2009). Removal of HDACs from transcribed regions has been classically seen as a prerequisite for gene activation and HATs and HDACs cooccupancy mutually exclusive. However, in genome-wide mapping studies, HDACs have been detected with HATs at active genes with acetylated histones and their recruitment imputed to phosphorylated PolII, providing the basis for dynamic cycles of acetylation/deacetylation required to reset chromatin after PolII elongation (Wang *et al.*, 2009). These findings are consistent with the observation that SIRT1 and PCAF cooccupy transcribed chromatin regions (Fulco *et al.*, 2003). However, it cannot be excluded that, at activated genes, SIRT1 deacetylase activity may be tempered by unfavorable [NAD^+]/[NADH] ratio (Sartorelli and Caretti, 2005). The HATs p300, PCAF, and related GNC5 are engaged at muscle-activated genes by different TFs—including myogenic bHLH, MEF2 factors, SRF, six proteins, and ubiquitous transcriptional regulators (reviewed in Guasconi and Puri, 2009; Perdiguero *et al.*, 2009). Association of MyoD with the HATs p300 and PCAF (Puri *et al.*, 1997b) is promoted by AKT1 and 2 kinases via direct phosphorylation of p300 (Serra *et al.*, 2007). Several

TFs regulating muscle gene expression are acetylated by the same HATs they recruit on the chromatin (Avantaggiati *et al.*, 1997; Duquet *et al.*, 2006; Gu *et al.*, 1997; Lill *et al.*, 1997; Polesskaya *et al.*, 2000; Sartorelli *et al.*, 1999; Simone *et al.*, 2004). In some instances, that is, p53 and MyoD, acetylation increases DNA binding affinity (Gu *et al.*, 1997; Sartorelli *et al.*, 1999), providing a positive feedback mechanism to increase HATs recruitment and consequent histone acetylation.

4.6. Moving obstacles

Nucleosomes containing epigenetic marks compatible with transcriptional activation need to be accordingly redistributed as to avoid occluding access of regulatory DNA regions to TFs. Nucleosome positioning can be achieved by ATP-dependent chromatin remodeling complexes or by TFs in the absence of additional factors (reviewed in Radman-Livaja and Rando, 2010). Binding of a TF near the entry/exit point of a nucleosome modifies the chromatin structure making accessible other sites located toward the center of the nucleosome (Polach and Widom, 1996). With a related mechanism, an NFkB-p65 subunit mutant without activation domains continues to regulate transcription of a subset of target genes by permitting access to secondary TFs (van Essen *et al.*, 2009). Pbx1/Meis1 is a "pioneer" TF capable of penetrating repressive chromatin. While the detailed mechanisms endowing Pbx1/Meis with such ability are not known, it may be that its binding sites are preferentially situated in nucleosome free-regions, as in the case of the IFN-b promoter, or that Pbx1/Meis1 may directly interact with and displace histones, as it occurs for HNF3 (Sagerstrom, 2004). Pbx1/Meis1 is constitutively bound to muscle regulatory regions before transcriptional activation and in the absence of MyoD. Pbx1/Meis interacts with MyoD (Knoepfler *et al.*, 1999) and can position it and tether it even at imperfect MyoD binding sites (Berkes *et al.*, 2004). Interaction of Pbx1/Meis1 with MyoD occurs via two domains required for initiation of chromatin remodeling (Bergstrom and Tapscott, 2001; Gerber *et al.*, 1997). These studies provide a mechanistic explanation of how MyoD can penetrate repressive chromatin to initiate remodeling. Pbx1/Meis1 may need to be somehow modified or interact with transcriptional repressors to repel MyoD and avoid premature gene activation. Alternatively, regulated interaction of Pbx1/Meis1 with SWI/SNF may permit productive MyoD recruitment (de la Serna *et al.*, 2005). Recruitment of the ATP-dependent SWI/SNF BRM and brahma-like 1 (BRG-1) chromatin remodeling complex through interaction with MyoD (de la Serna *et al.*, 2001; reviewed in de la Serna *et al.*, 2006) is regulated via direct phosphorylation of the SWI/SNF BAF60c subunit mediated by the MAPK p38 (mitogen-activated kinases; Simone *et al.*, 2004; reviewed in Albini and Puri, 2010) and required for binding of MyoD, myogenin, and MEF2 (de la

Serna *et al.*, 2005) and sustained gene expression (Ohkawa *et al.*, 2007). At the myogenin promoter, MyoD induces histone hyperacetylation before and independently of BRG-1 activity (de la Serna *et al.*, 2005), suggesting the formation of a temporally regulated transcriptional protein complex where MyoD-mediated HATs recruitment (Eckner *et al.*, 1996; Puri *et al.*, 1997a,b; Sartorelli *et al.*, 1997; Yuan *et al.*, 1996) precedes BRG-1-dependent chromatin remodeling. BRG-1-meditaed remodeling at different muscle loci is temporally regulated by sequential activities of the arginine methyltransferases PRMT5 and CARM1/PRMT4 (Dacwag *et al.*, 2009; Mallappa *et al.*, 2010). An interesting relation between SWI/SNF and PcG proteins was uncovered when brahma was identified as a suppressor of PcG mutants in homeotic gene expression, suggesting that SWI/SNF may regulate gene expression by antagonizing repressive chromatin structure organized by PcG proteins (Tamkun *et al.*, 1992). A dedicated role of BAF60c to myogenesis is suggested by its preferential expression in developing heart and somites and has been confirmed by its genetic ablation, which results in cardiac and skeletal muscle defects (Lickert *et al.*, 2004). Intriguingly, and consistent with its chromatin remodeling ability, BRG-1 has been found to facilitate iPS formation (Singhal *et al.*, 2010).

4.7. RNA helicases: Connecting histone acetylation, chromatin remodeling, and microRNA maturation

The highly related RNA helicases p68 and p72 (p68/p72) and the long noncoding RNA (lncRNA) steroid receptor activator (SRA) associate with MyoD and coregulate its transcriptional activity (Caretti *et al.*, 2006). They also serve as coregulators of other TFs, including p53, estrogen, and androgen receptors. Their mechanistic role is not fully understood as the RNA helicase activity seems dispensable to regulate transcription (Bates *et al.*, 2005; Caretti *et al.*, 2006). However, several other features endow p68/p72 with regulatory properties. P68 interacts with HDAC1, repressing transcription in a promoter-specific manner (Wilson *et al.*, 2004). In other biological circumstances, p68 can interact with and is acetylated by p300. Acetylation increases p68/p72 stability, thus stimulating its ability to coactivate estrogen receptor-mediated transcription (Mooney *et al.*, 2010). Sumoylation may constitute a regulatory switch as it enhances p68 transcriptional repression by favoring its association with HDAC1 (Fig. 3.2A; Jacobs *et al.*, 2007). In differentiated skeletal muscle cells, p68/p72 were detected on chromatin regulatory regions in association with the lncRNA SRA and MyoD (Fig. 3.2B). Reducing p68 level by RNAi in either C2C12 cell line or satellite-cell-derived primary muscle cells hampered differentiation, transcription of selected muscle genes, and chromatin engagement of BRG-1, TATA-binding protein (TBP), and PolII without affecting either MyoD or p300 recruitment (Caretti *et al.*, 2006). Genes whose expression

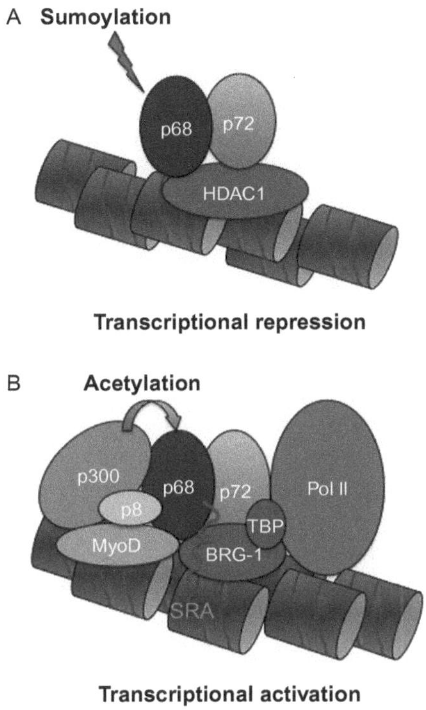

Figure 3.2 (A) Sumoylation of the RNA helicase p68 (forming heterodimers with the RNA helicase p72) can favor repression by promoting p68 interaction with the histone deacetylase HDAC1. (B) At transcribed genes, the RNA helicases p68/p72, the noncoding RNA SRA, form a transcriptional complex with the HAT p300, the high-mobility group-related p8, MyoD, the nucleosome remodeling SWI/SNF complex, the TATA-binding protein (TBP), and PolII. In addition to histones, p300 also acetylates p68. (See Color Insert.)

was affected by p68 require BRG-1 (de la Serna *et al.*, 2005). More recently, the estrogen receptor ERα has been demonstrated to form a protein complex with p300, p68/p72, and MyoD to regulate BRCA2 transcription (Jin *et al.*, 2008). The HMG-related small chromatin protein p8 also associates and is recruited at muscle-specific genes with p300, p68, and MyoD (Fig. 3.2B). p8 knockdown compromises chromatin recruitment of p300, p68, and MyoD, thus impairing muscle transcription (Sambasivan *et al.*, 2009). In addition to a direct role on the formation and stabilization of chromatin-modifying protein complexes, p68/p72 influence also other processes related to gene expression. P68/p72 are part of the nuclear RNase III Drosha complex, which cleaves primary transcripts of miRNA genes (pri-miRNAs) into hairpin precursor intermediates (pre-miRNAs), further processed to mature miRNAs by the cytosolic RNase III Dicer.

Association of p68/72 with the TFs p53 or SMAD regulates the processing of primary miRNAs into precurors miRNAs (Davis *et al.*, 2008; Suzuki *et al.*, 2009). Gene disruption of either p68 or p72 results in early lethality, with specific defects in maturation of selected miRNAs. P72 deletion negatively impacts maturation of miR-214 (Fukuda *et al.*, 2007), a microRNA regulating Ezh2 availability and involved in muscle cell differentiation (Juan *et al.*, 2009). Thus, p72 may regulate muscle gene expression by two independent mechanisms, by coactivating MyoD-dependent transcription and favoring miR-214 processing (Juan and Sartorelli, 2010).

4.8. Switching basal players: TRF3 and TAF3

It might have been naively predicted that a family of tissue-specific TFs (activators), such as the myogenic bHLH, would suffice to ensure cell-lineage restricted gene expression, with the constant core basal transcriptional machinery serving many masters. It turned out that activators and core machinery members display mutual preferences (D'Alessio *et al.*, 2009). Upon muscle cell differentiation, TBP and several TAFs—expressed in undifferentiated cells—are degraded and replaced by TRF3 and TAF3. Indeed, MyoD can directly target TAF3, and such interaction support *in vitro* transcription. Moreover, TRF3/TAF3 are recruited at the myogenin promoter and interference with their accumulation prevents myotube formation (Deato and Tjian, 2007; Deato *et al.*, 2008). As MyoD is transcriptionally active in undifferentiated myoblasts, these findings imply that MyoD can communicate and direct transcription with two alternative core basal machineries. As reported for SWI/SNF (Tamkun *et al.*, 1992), also tissue-specific TAFs counteract PcG proteins to promote terminal differentiation (Chen *et al.*, 2005).

4.9. Conveying signals to the nucleus by the MAPK p38

In skeletal muscle cells, extracellular cues mediated by morphogen gradients, ligand–receptor, and soluble growth factor–receptor interactions are decoded and ultimately conveyed to the genome by chromatin-modifying complexes to regulate transcriptional output (reviewed in Guasconi and Puri, 2009). Protein modifications mediated by kinases and phosphates rapidly and effectively mediate cellular responses to environmental signals and internal processes by regulating protein interactions, enzymes activity, and protein localization (Pawson, 2007).

Within this context, signaling mediated by the MAPK p38 has received a great deal of experimental attention. The important role exerted by p38a in regulating the satellite quiescent state (Jones *et al.*, 2005) is evidenced by the analysis of p38a mutant mice, which shows delayed-cell-cycle exit and altered expression of cell cycle regulators in cultured myoblasts (Perdiguero et al., 2007). Activation of p38, initiated by the transmembrane Ig-fibronectin—type III repeat CDO protein—promotes formation of MyoD-E47 dimers, which in turn activate CDO transcription (Cole *et al.*, 2004; Takaesu *et al.*, 2006). E47 is a direct target of p38-mediated phosphorylation promoting MyoD-E47 heterodimer formation (Lluis *et al.*, 2005). Different members of the MAPK p38 family exert opposing effects on muscle gene expression. While p38a/b favor muscle transcription by promoting MyoD-E47 heterodimer formation (Lluis *et al.*, 2005) and MEF2D binding (Penn *et al.*, 2004), engagement of the SWI/SNF chromatin remodeling complex via phosphorylation of the BAF60c subunit (Simone *et al.*, 2004), and recruitment of the Ash2L-containing MLL protein complex through MEF2D phosphorylation (Rampalli *et al.*, 2007), p38g antagonizes muscle gene expression by promoting recruitment of the H3K9 methyltransferase Suv39h1/KMT1 at the myogenin promoter through direct phosphorylation of Ser199 and Ser200 of MyoD (Gillespie *et al.*, 2009).

An unexpected role of p38a in linking tumor necrosis factor (TNF) signaling to PcG proteins during muscle regeneration has been recently uncovered. Inflammatory cells populating sites of damaged muscles are an essential component of the regenerative phase, characterized by SC expansion and differentiation. Locally released cytokines, including interleukin 1,4,6, and TNFa promote muscle regeneration (Charge and Rudnicki, 2004; Dhawan and Rando, 2005; Horsley *et al.*, 2003; Serrano *et al.*, 2008). TNFa regulates muscle regeneration by activating MAPK p38 (Chen *et al.*, 2007). In proliferating, undifferentiated muscle cells, where p38a signaling is ineffective (Wu *et al.*, 2000), the PRC2 complex prevents unscheduled gene expression by repressing transcription of muscle-specific myosins and muscle creatine kinase genes (Caretti *et al.*, 2004). When proliferating SCs undergo differentiation, PRC2 is released from muscle structural genes—where is replaced by an activation complex (Caretti *et al.*, 2004)—and relocated to Pax7 regulatory regions to repress gene expression (Palacios *et al.*, 2010). Such PRC2 redistribution from muscle structural genes to Pax7 is regulated by p38a, which directly phosphorylates the human PRC2 catalytic subunit Ezh2 at Thr372 (corresponding to mouse Ezh2 Thr367), influencing recruitment of the Polycomb-related transcriptional repressor YY1 (Palacios *et al.*, 2010). Indeed, p38a, Ezh2, YY1, and H3K27me3 are codetected at the Pax7 regulatory regions of differentiating primary myoblasts. Genetically or pharmacologically interfering with p38 or PRC2, results in continued Pax7 expression and expansion of SCs, which retain their ability to differentiate once p38 or PRC2 blockade is removed (Palacios *et al.*, 2010). While chromatin engagement of MAPKs p38 has been previously documented (Chow and Davis, 2006; de Nadal and Posas, 2010; Pokholok *et al.*, 2006), the contribution of p38α to redistribute PcG proteins at specific genomic loci in response to inflammatory signals introduces an additional layer of regulatory refinement.

5. Conclusions

As a result of the concerted action of acetyltransferases and deacetylases, methyltransferases and demethylases, kinases and phosphatases, epigenetic modifications are dynamic and reversible. This property of the epigenome, associated with the availability and development of new small molecules with the ability of regulating the activities of specific chromatin-modifying protein complexes, allows, in principle, a regulatable manipulation of cell fate decisions.

In the context of muscle stem cell biology, the combined investigation of molecules and mechanisms regulating the epigenome and small molecule screening may prove useful toward the development of effective therapies for degenerative diseases.

Acknowledgment

The authors are supported by the Intramural Research Program of the National Institute of Arthritis, Musculoskeletal, and Skin Diseases of the National Institutes of Health.

References

Affar el, B., Gay, F., Shi, Y., Liu, H., Huarte, M., Wu, S., Collins, T., and Li, E. (2006). Essential dosage-dependent functions of the transcription factor yin yang 1 in late embryonic development and cell cycle progression. *Mol. Cell. Biol.* **26,** 3565–3581.

Albini, S., and Puri, P. L. (2010). SWI/SNF complexes, chromatin remodeling and skeletal myogenesis: It's time to exchange!. *Exp. Cell Res.* **316,** 3073–3080.

Atchison, L., Ghias, A., Wilkinson, F., Bonini, N., and Atchison, M. L. (2003). Transcription factor YY1 functions as a PcG protein in vivo. *EMBO J.* **22,** 1347–1358.

Avantaggiati, M. L., Ogryzko, V., Gardner, K., Giordano, A., Levine, A. S., and Kelly, K. (1997). Recruitment of p300/CBP in p53-dependent signal pathways. *Cell* **89,** 1175–1184.

Barski, A., Cuddapah, S., Cui, K., Roh, T. Y., Schones, D. E., Wang, Z., Wei, G., Chepelev, I., and Zhao, K. (2007). High-resolution profiling of histone methylations in the human genome. *Cell* **129,** 823–837.

Bates, G. J., Nicol, S. M., Wilson, B. J., Jacobs, A. M., Bourdon, J. C., Wardrop, J., Gregory, D. J., Lane, D. P., Perkins, N. D., and Fuller-Pace, F. V. (2005). The DEAD box protein p68: A novel transcriptional coactivator of the p53 tumour suppressor. *EMBO J.* **24,** 543–553.

Beauchamp, J. R., Heslop, L., Yu, D. S., Tajbakhsh, S., Kelly, R. G., Wernig, A., Buckingham, M. E., Partridge, T. A., and Zammit, P. S. (2000). Expression of CD34 and Myf5 defines the majority of quiescent adult skeletal muscle satellite cells. *J. Cell Biol.* **151,** 1221–1234.

Benchaouir, R., Meregalli, M., Farini, A., D'Antona, G., Belicchi, M., Goyenvalle, A., Battistelli, M., Bresolin, N., Bottinelli, R., Garcia, L., and Torrente, Y. (2007). Restoration of human dystrophin following transplantation of exon-skipping-engineered DMD patient stem cells into dystrophic mice. *Cell Stem Cell* **1,** 646–657.

Bergstrom, D. A., and Tapscott, S. J. (2001). Molecular distinction between specification and differentiation in the myogenic basic helix-loop-helix transcription factor family. *Mol. Cell. Biol.* **21,** 2404–2412.

Bergstrom, D. A., Penn, B. H., Strand, A., Perry, R. L., Rudnicki, M. A., and Tapscott, S. J. (2002). Promoter-specific regulation of MyoD binding and signal transduction cooperate to pattern gene expression. *Mol. Cell* **9,** 587–600.

Berkes, C. A., Bergstrom, D. A., Penn, B. H., Seaver, K. J., Knoepfler, P. S., and Tapscott, S. J. (2004). Pbx marks genes for activation by MyoD indicating a role for a homeodomain protein in establishing myogenic potential. *Mol. Cell* **14,** 465–477.

Bernstein, B. E., Mikkelsen, T. S., Xie, X., Kamal, M., Huebert, D. J., Cuff, J., Fry, B., Meissner, A., Wernig, M., Plath, K., Jaenisch, R., Wagschal, A., *et al.* (2006). A bivalent chromatin structure marks key developmental genes in embryonic stem cells. *Cell* **125,** 315–326.

Bianchi-Frias, D., Orian, A., Delrow, J. J., Vazquez, J., Rosales-Nieves, A. E., and Parkhurst, S. M. (2004). Hairy transcriptional repression targets and cofactor recruitment in *Drosophila*. *PLoS Biol.* **2,** E178.

Bird, A. (2007). Perceptions of epigenetics. *Nature* **447,** 396–398.

Blais, A., Tsikitis, M., Acosta-Alvear, D., Sharan, R., Kluger, Y., and Dynlacht, B. D. (2005). An initial blueprint for myogenic differentiation. *Genes Dev.* **19,** 553–569.

Boyer, L. A., Plath, K., Zeitlinger, J., Brambrink, T., Medeiros, L. A., Lee, T. I., Levine, S. S., Wernig, M., Tajonar, A., Ray, M. K., Bell, G. W., Otte, A. P., *et al.* (2006). Polycomb complexes repress developmental regulators in murine embryonic stem cells. *Nature* **441,** 349–353.

Brack, A. S., Conboy, I. M., Conboy, M. J., Shen, J., and Rando, T. A. (2008). A temporal switch from notch to Wnt signaling in muscle stem cells is necessary for normal adult myogenesis. *Cell Stem Cell* **2,** 50–59.

Bracken, A. P., Dietrich, N., Pasini, D., Hansen, K. H., and Helin, K. (2006). Genome-wide mapping of Polycomb target genes unravels their roles in cell fate transitions. *Genes Dev.* **20,** 1123–1136.

Brunet, A., Sweeney, L. B., Sturgill, J. F., Chua, K. F., Greer, P. L., Lin, Y., Tran, H., Ross, S. E., Mostoslavsky, R., Cohen, H. Y., Hu, L. S., Cheng, H. L., *et al.* (2004). Stress-dependent regulation of FOXO transcription factors by the SIRT1 deacetylase. *Science* **303,** 2011–2015.

Buckingham, M. (2007). Skeletal muscle progenitor cells and the role of Pax genes. *C R Biol.* **330,** 530–533.

Cao, R., and Zhang, Y. (2004). SUZ12 is required for both the histone methyltransferase activity and the silencing function of the EED–EZH2 complex. *Mol. Cell* **15,** 57–67.

Cao, R., Wang, L., Wang, H., Xia, L., Erdjument-Bromage, H., Tempst, P., Jones, R. S., and Zhang, Y. (2002). Role of histone H3 lysine 27 methylation in Polycomb-group silencing. *Science* **298,** 1039–1043.

Cao, Y., Kumar, R. M., Penn, B. H., Berkes, C. A., Kooperberg, C., Boyer, L. A., Young, R. A., and Tapscott, S. J. (2006). Global and gene-specific analyses show distinct roles for Myod and Myog at a common set of promoters. *EMBO J.* **25,** 502–511.

Cao, Y., Yao, Z., Sarkar, D., Lawrence, M., Sanchez, G. J., Parker, M. H., MacQuarrie, K. L., Davison, J., Morgan, M. T., Ruzzo, W. L., Gentleman, R. C., and Tapscott, S. J. (2010). Genome-wide MyoD binding in skeletal muscle cells: A potential for broad cellular reprogramming. *Dev. Cell* **18,** 662–674.

Caretti, G., Di Padova, M., Micales, B., Lyons, G. E., and Sartorelli, V. (2004). The Polycomb Ezh2 methyltransferase regulates muscle gene expression and skeletal muscle differentiation. *Genes Dev.* **18,** 2627–2638.

Caretti, G., Schiltz, R. L., Dilworth, F. J., Di Padova, M., Zhao, P., Ogryzko, V., Fuller-Pace, F. V., Hoffman, E. P., Tapscott, S. J., and Sartorelli, V. (2006). The RNA helicases p68/p72 and the noncoding RNA SRA are coregulators of MyoD and skeletal muscle differentiation. *Dev. Cell* **11,** 547–560.

Carmon, Y., Czosnek, H., Nudel, U., Shani, M., and Yaffe, D. (1982). DNAase I sensitivity of genes expressed during myogenesis. *Nucleic Acids Res.* **10,** 3085–3098.

Chambers, I., Colby, D., Robertson, M., Nichols, J., Lee, S., Tweedie, S., and Smith, A. (2003). Functional expression cloning of Nanog, a pluripotency sustaining factor in embryonic stem cells. *Cell* **113,** 643–655.

Chambers, I., Silva, J., Colby, D., Nichols, J., Nijmeijer, B., Robertson, M., Vrana, J., Jones, K., Grotewold, L., and Smith, A. (2007). Nanog safeguards pluripotency and mediates germline development. *Nature* **450,** 1230–1234.

Charge, S. B., and Rudnicki, M. A. (2004). Cellular and molecular regulation of muscle regeneration. *Physiol. Rev.* **84,** 209–238.

Chen, X., Hiller, M., Sancak, Y., and Fuller, M. T. (2005). Tissue-specific TAFs counteract Polycomb to turn on terminal differentiation. *Science* **310,** 869–872.

Chen, S. E., Jin, B., and Li, Y. P. (2007). TNF-alpha regulates myogenesis and muscle regeneration by activating p38 MAPK. *Am. J. Physiol. Cell Physiol.* **292,** C1660–C1671.

Chow, C. W., and Davis, R. J. (2006). Proteins kinases: Chromatin-associated enzymes? *Cell* **127,** 887–890.

Chuikov, S., Kurash, J. K., Wilson, J. R., Xiao, B., Justin, N., Ivanov, G. S., McKinney, K., Tempst, P., Prives, C., Gamblin, S. J., Barlev, N. A., and Reinberg, D. (2004). Regulation of p53 activity through lysine methylation. *Nature* **432,** 353–360.

Cole, F., Zhang, W., Geyra, A., Kang, J. S., and Krauss, R. S. (2004). Positive regulation of myogenic bHLH factors and skeletal muscle development by the cell surface receptor CDO. *Dev. Cell* **7,** 843–854.

Collins, C. A., Olsen, I., Zammit, P. S., Heslop, L., Petrie, A., Partridge, T. A., and Morgan, J. E. (2005). Stem cell function, self-renewal, and behavioral heterogeneity of cells from the adult muscle satellite cell niche. *Cell* **122,** 289–301.

Conboy, I. M., and Rando, T. A. (2002). The regulation of Notch signaling controls satellite cell activation and cell fate determination in postnatal myogenesis. *Dev. Cell* **3,** 397–409.

Conboy, I. M., Conboy, M. J., Smythe, G. M., and Rando, T. A. (2003). Notch-mediated restoration of regenerative potential to aged muscle. *Science* **302,** 1575–1577.

Conboy, M. J., Karasov, A. O., and Rando, T. A. (2007). High incidence of non-random template strand segregation and asymmetric fate determination in dividing stem cells and their progeny. *PLoS Biol.* **5,** e102.

Corbel, S. Y., Lee, A., Yi, L., Duenas, J., Brazelton, T. R., Blau, H. M., and Rossi, F. M. (2003). Contribution of hematopoietic stem cells to skeletal muscle. *Nat. Med.* **9,** 1528–1532.

Cossu, G., and Bianco, P. (2003). Mesoangioblasts–vascular progenitors for extravascular mesodermal tissues. *Curr. Opin. Genet. Dev.* **13,** 537–542.

Czermin, B., Melfi, R., McCabe, D., Seitz, V., Imhof, A., and Pirrotta, V. (2002). *Drosophila* enhancer of Zeste/ESC complexes have a histone H3 methyltransferase activity that marks chromosomal Polycomb sites. *Cell* **111,** 185–196.

Dacwag, C. S., Bedford, M. T., Sif, S., and Imbalzano, A. N. (2009). Distinct protein arginine methyltransferases promote ATP-dependent chromatin remodeling function at different stages of skeletal muscle differentiation. *Mol. Cell. Biol.* **29,** 1909–1921.

D'Alessio, J. A., Wright, K. J., and Tjian, R. (2009). Shifting players and paradigms in cell-specific transcription. *Mol. Cell* **36,** 924–931.

Davis, R. L., Weintraub, H., and Lassar, A. B. (1987). Expression of a single transfected cDNA converts fibroblasts to myoblasts. *Cell* **51,** 987–1000.

Davis, B. N., Hilyard, A. C., Lagna, G., and Hata, A. (2008). SMAD proteins control DROSHA-mediated microRNA maturation. *Nature* **454,** 56–61.

de la Serna, I. L., Carlson, K. A., and Imbalzano, A. N. (2001). Mammalian SWI/SNF complexes promote MyoD-mediated muscle differentiation. *Nat. Genet.* **27,** 187–190.

de la Serna, I. L., Ohkawa, Y., Berkes, C. A., Bergstrom, D. A., Dacwag, C. S., Tapscott, S. J., and Imbalzano, A. N. (2005). MyoD targets chromatin remodeling complexes to the myogenin locus prior to forming a stable DNA-bound complex. *Mol. Cell. Biol.* **25,** 3997–4009.

de la Serna, I. L., Ohkawa, Y., and Imbalzano, A. N. (2006). Chromatin remodelling in mammalian differentiation: Lessons from ATP-dependent remodellers. *Nat. Rev. Genet.* **7,** 461–473.

de Nadal, E., and Posas, F. (2010). Multilayered control of gene expression by stress-activated protein kinases. *EMBO J.* **29,** 4–13.

Deato, M. D., and Tjian, R. (2007). Switching of the core transcription machinery during myogenesis. *Genes Dev.* **21,** 2137–2149.

Deato, M. D., Marr, M. T., Sottero, T., Inouye, C., Hu, P., and Tjian, R. (2008). MyoD targets TAF3/TRF3 to activate myogenin transcription. *Mol. Cell* **32,** 96–105.

Dellavalle, A., Sampaolesi, M., Tonlorenzi, R., Tagliafico, E., Sacchetti, B., Perani, L., Innocenzi, A., Galvez, B. G., Messina, G., Morosetti, R., Li, S., Belicchi, M., *et al.* (2007). Pericytes of human skeletal muscle are myogenic precursors distinct from satellite cells. *Nat. Cell Biol.* **9,** 255–267.

Dhawan, J., and Rando, T. A. (2005). Stem cells in postnatal myogenesis: Molecular mechanisms of satellite cell quiescence, activation and replenishment. *Trends Cell Biol.* **15,** 666–673.

Dilworth, F. J., Seaver, K. J., Fishburn, A. L., Htet, S. L., and Tapscott, S. J. (2004). In vitro transcription system delineates the distinct roles of the coactivators pCAF and p300 during MyoD/E47-dependent transactivation. *Proc. Natl. Acad. Sci. USA* **101,** 11593–11598.

Dioum, E. M., Chen, R., Alexander, M. S., Zhang, Q., Hogg, R. T., Gerard, R. D., and Garcia, J. A. (2009). Regulation of hypoxia-inducible factor 2alpha signaling by the stress-responsive deacetylase sirtuin 1. *Science* **324,** 1289–1293.

Duquet, A., Polesskaya, A., Cuvellier, S., Ait-Si-Ali, S., Hery, P., Pritchard, L. L., Gerard, M., and Harel-Bellan, A. (2006). Acetylation is important for MyoD function in adult mice. *EMBO Rep.* **7,** 1140–1146.

Eckner, R., Yao, T. P., Oldread, E., and Livingston, D. M. (1996). Interaction and functional collaboration of p300/CBP and bHLH proteins in muscle and B-cell differentiation. *Genes Dev.* **10,** 2478–2490.

Ezhkova, E., Pasolli, H. A., Parker, J. S., Stokes, N., Su, I. H., Hannon, G., Tarakhovsky, A., and Fuchs, E. (2009). Ezh2 orchestrates gene expression for the stepwise differentiation of tissue-specific stem cells. *Cell* **136,** 1122–1135.

Francis, N. J., Kingston, R. E., and Woodcock, C. L. (2004). Chromatin compaction by a polycomb group protein complex. *Science* **306,** 1574–1577.

Fukuda, T., Yamagata, K., Fujiyama, S., Matsumoto, T., Koshida, I., Yoshimura, K., Mihara, M., Naitou, M., Endoh, H., Nakamura, T., Akimoto, C., Yamamoto, Y., *et al.* (2007). DEAD-box RNA helicase subunits of the Drosha complex are required for processing of rRNA and a subset of microRNAs. *Nat. Cell Biol.* **9,** 604–611.

Fulco, M., Schiltz, R. L., Iezzi, S., King, M. T., Zhao, P., Kashiwaya, Y., Hoffman, E., Veech, R. L., and Sartorelli, V. (2003). Sir2 regulates skeletal muscle differentiation as a potential sensor of the redox state. *Mol. Cell* **12,** 51–62.

Fulco, M., Cen, Y., Zhao, P., Hoffman, E. P., McBurney, M. W., Sauve, A. A., and Sartorelli, V. (2008). Glucose restriction inhibits skeletal myoblast differentiation by activating SIRT1 through AMPK-mediated regulation of Nampt. *Dev. Cell* **14,** 661–673.

Gerber, A. N., Klesert, T. R., Bergstrom, D. A., and Tapscott, S. J. (1997). Two domains of MyoD mediate transcriptional activation of genes in repressive chromatin: A mechanism for lineage determination in myogenesis. *Genes Dev.* **11,** 436–450.

Gillespie, M. A., Le Grand, F., Scime, A., Kuang, S., von Maltzahn, J., Seale, V., Cuenda, A., Ranish, J. A., and Rudnicki, M. A. (2009). p38-{gamma}-dependent gene silencing restricts entry into the myogenic differentiation program. *J. Cell Biol.* **187,** 991–1005.

Gros, J., Manceau, M., Thome, V., and Marcelle, C. (2005). A common somitic origin for embryonic muscle progenitors and satellite cells. *Nature* **435,** 954–958.

Groudine, M., and Weintraub, H. (1981). Activation of globin genes during chicken development. *Cell* **24,** 393–401.

Gu, W., Shi, X. L., and Roeder, R. G. (1997). Synergistic activation of transcription by CBP and p53. *Nature* **387,** 819–823.

Guasconi, V., and Puri, P. L. (2009). Chromatin: The interface between extrinsic cues and the epigenetic regulation of muscle regeneration. *Trends Cell Biol.* **19,** 286–294.

Hansen, K. H., Bracken, A. P., Pasini, D., Dietrich, N., Gehani, S. S., Monrad, A., Rappsilber, J., Lerdrup, M., and Helin, K. (2008). A model for transmission of the H3K27me3 epigenetic mark. *Nat. Cell Biol.* **10,** 1291–1300.

Horsley, V., Jansen, K. M., Mills, S. T., and Pavlath, G. K. (2003). IL-4 acts as a myoblast recruitment factor during mammalian muscle growth. *Cell* **113,** 483–494.

Huang, J., Perez-Burgos, L., Placek, B. J., Sengupta, R., Richter, M., Dorsey, J. A., Kubicek, S., Opravil, S., Jenuwein, T., and Berger, S. L. (2006). Repression of p53 activity by Smyd2-mediated methylation. *Nature* **444,** 629–632.

Huang, J., Dorsey, J., Chuikov, S., Perez-Burgos, L., Zhang, X., Jenuwein, T., Reinberg, D., and Berger, S. L. (2010). G9a and Glp methylate lysine 373 in the tumor suppressor p53. *J. Biol. Chem.* **285,** 9636–9641.

Iezzi, S., Cossu, G., Nervi, C., Sartorelli, V., and Puri, P. L. (2002). Stage-specific modulation of skeletal myogenesis by inhibitors of nuclear deacetylases. *Proc. Natl. Acad. Sci. USA* **99,** 7757–7762.

Iezzi, S., Di Padova, M., Serra, C., Caretti, G., Simone, C., Maklan, E., Minetti, G., Zhao, P., Hoffman, E. P., Puri, P. L., and Sartorelli, V. (2004). Deacetylase inhibitors increase muscle cell size by promoting myoblast recruitment and fusion through induction of follistatin. *Dev. Cell* **6,** 673–684.

Imai, S., and Guarente, L. (2010). Ten years of NAD-dependent SIR2 family deacetylases: Implications for metabolic diseases. *Trends Pharmacol. Sci.* **31,** 212–220.

Imai, S., Armstrong, C. M., Kaeberlein, M., and Guarente, L. (2000). Transcriptional silencing and longevity protein Sir2 is an NAD-dependent histone deacetylase. *Nature* **403,** 795–800.

Jacobs, A. M., Nicol, S. M., Hislop, R. G., Jaffray, E. G., Hay, R. T., and Fuller-Pace, F. V. (2007). SUMO modification of the DEAD box protein p68 modulates its transcriptional activity and promotes its interaction with HDAC1. *Oncogene* **26,** 5866–5876.

Jin, W., Chen, Y., Di, G. H., Miron, P., Hou, Y. F., Gao, H., and Shao, Z. M. (2008). Estrogen receptor (ER) beta or p53 attenuates ERalpha-mediated transcriptional activation on the BRCA2 promoter. *J. Biol. Chem.* **283,** 29671–29680.

Joe, A. W., Yi, L., Natarajan, A., Le Grand, F., So, L., Wang, J., Rudnicki, M. A., and Rossi, F. M. (2010). Muscle injury activates resident fibro/adipogenic progenitors that facilitate myogenesis. *Nat. Cell Biol.* **12,** 153–163.

Jones, N. C., Tyner, K. J., Nibarger, L., Stanley, H. M., Cornelison, D. D., Fedorov, Y. V., and Olwin, B. B. (2005). The p38alpha/beta MAPK functions as a molecular switch to activate the quiescent satellite cell. *J. Cell Biol.* **169,** 105–116.

Juan, A. H., and Sartorelli, V. (2010). MicroRNA-214 and polycomb group proteins: A regulatory circuit controlling differentiation and cell fate decisions. *Cell Cycle* **9,** 1445–1446.

Juan, A. H., Kumar, R. M., Marx, J. G., Young, R. A., and Sartorelli, V. (2009). Mir-214-dependent regulation of the polycomb protein Ezh2 in skeletal muscle and embryonic stem cells. *Mol. Cell* **36,** 61–74.

Kang, J. S., and Krauss, R. S. (2010). Muscle stem cells in developmental and regenerative myogenesis. *Curr. Opin. Clin. Nutr. Metab. Care* **13,** 243–248.

Kassar-Duchossoy, L., Giacone, E., Gayraud-Morel, B., Jory, A., Gomes, D., and Tajbakhsh, S. (2005). Pax3/Pax7 mark a novel population of primitive myogenic cells during development. *Genes Dev.* **19,** 1426–1431.

Kirmizis, A., Bartley, S. M., Kuzmichev, A., Margueron, R., Reinberg, D., Green, R., and Farnham, P. J. (2004). Silencing of human polycomb target genes is associated with methylation of histone H3 Lys 27. *Genes Dev.* **18,** 1592–1605.

Knoepfler, P. S., Bergstrom, D. A., Uetsuki, T., Dac-Korytko, I., Sun, Y. H., Wright, W. E., Tapscott, S. J., and Kamps, M. P. (1999). A conserved motif N-terminal to the DNA-binding domains of myogenic bHLH transcription factors mediates cooperative DNA binding with pbx-Meis1/Prep1. *Nucleic Acids Res.* **27,** 3752–3761.

Kuang, S., Kuroda, K., Le Grand, F., and Rudnicki, M. A. (2007). Asymmetric self-renewal and commitment of satellite stem cells in muscle. *Cell* **129,** 999–1010.

Kuninger, D., Wright, A., and Rotwein, P. (2006). Muscle cell survival mediated by the transcriptional coactivators p300 and PCAF displays different requirements for acetyltransferase activity. *Am. J. Physiol. Cell Physiol.* **291,** C699–C709.

Kuzmichev, A., Nishioka, K., Erdjument-Bromage, H., Tempst, P., and Reinberg, D. (2002). Histone methyltransferase activity associated with a human multiprotein complex containing the Enhancer of Zeste protein. *Genes Dev.* **16,** 2893–2905.

Kuzmichev, A., Margueron, R., Vaquero, A., Preissner, T. S., Scher, M., Kirmizis, A., Ouyang, X., Brockdorff, N., Abate-Shen, C., Farnham, P., and Reinberg, D. (2005). Composition and histone substrates of polycomb repressive group complexes change during cellular differentiation. *Proc. Natl. Acad. Sci. USA* **102,** 1859–1864.

Lee, H., Habas, R., and Abate-Shen, C. (2004). MSX1 cooperates with histone H1b for inhibition of transcription and myogenesis. *Science* **304,** 1675–1678.

Lee, T. I., Jenner, R. G., Boyer, L. A., Guenther, M. G., Levine, S. S., Kumar, R. M., Chevalier, B., Johnstone, S. E., Cole, M. F., Isono, K., Koseki, H., Fuchikami, T., *et al.* (2006). Control of developmental regulators by Polycomb in human embryonic stem cells. *Cell* **125,** 301–313.

Lickert, H., Takeuchi, J. K., Von Both, I., Walls, J. R., McAuliffe, F., Adamson, S. L., Henkelman, R. M., Wrana, J. L., Rossant, J., and Bruneau, B. G. (2004). Baf60c is essential for function of BAF chromatin remodelling complexes in heart development. *Nature* **432,** 107–112.

Lill, N. L., Grossman, S. R., Ginsberg, D., DeCaprio, J., and Livingston, D. M. (1997). Binding and modulation of p53 by p300/CBP coactivators. *Nature* **387,** 823–827.

Lluis, F., Ballestar, E., Suelves, M., Esteller, M., and Munoz-Canoves, P. (2005). E47 phosphorylation by p38 MAPK promotes MyoD/E47 association and muscle-specific gene transcription. *EMBO J.* **24,** 974–984.

Lu, J., McKinsey, T. A., Nicol, R. L., and Olson, E. N. (2000a). Signal-dependent activation of the MEF2 transcription factor by dissociation from histone deacetylases. *Proc. Natl. Acad. Sci. USA* **97,** 4070–4075.

Lu, J., McKinsey, T. A., Zhang, C. L., and Olson, E. N. (2000b). Regulation of skeletal myogenesis by association of the MEF2 transcription factor with class II histone deacetylases. *Mol. Cell* **6,** 233–244.

Luo, D., Renault, V. M., and Rando, T. A. (2005). The regulation of Notch signaling in muscle stem cell activation and postnatal myogenesis. *Semin. Cell Dev. Biol.* **16,** 612–622.
Mal, A. K. (2006). Histone methyltransferase Suv39h1 represses MyoD-stimulated myogenic differentiation. *EMBO J.* **25,** 3323–3334.
Mal, A., Sturniolo, M., Schiltz, R. L., Ghosh, M. K., and Harter, M. L. (2001). A role for histone deacetylase HDAC1 in modulating the transcriptional activity of MyoD: Inhibition of the myogenic program. *EMBO J.* **20,** 1739–1753.
Mallappa, C., Nasipak, B. T., Etheridge, L., Androphy, E. J., Jones, S. N., Sagerstrom, C. G., Ohkawa, Y., and Imbalzano, A. N. (2010). Myogenic microRNA expression requires ATP-dependent chromatin remodeling enzyme function. *Mol. Cell. Biol.* **30,** 3176–3186.
Margueron, R., and Reinberg, D. (2010). Chromatin structure and the inheritance of epigenetic information. *Nat. Rev. Genet.* **11,** 285–296.
Margueron, R., Justin, N., Ohno, K., Sharpe, M. L., Son, J., Drury, W. J., 3rd, Voigt, P., Martin, S. R., Taylor, W. R., De Marco, V., Pirrotta, V., Reinberg, D., *et al.* (2009). Role of the polycomb protein EED in the propagation of repressive histone marks. *Nature* **461,** 762–767.
Maroto, M., Reshef, R., Münsterberg, A., Koester, S., Goulding, M., and Lassar, A. (1997). Ectopic Pax-3 activates MyoD and Myf-5 expression in embryonic mesoderm and neural tissue. *Cell* **89,** 139–148.
Masui, S., Nakatake, Y., Toyooka, Y., Shimosato, D., Yagi, R., Takahashi, K., Okochi, H., Okuda, A., Matoba, R., Sharov, A. A., Ko, M. S., and Niwa, H. (2007). Pluripotency governed by Sox2 via regulation of Oct3/4 expression in mouse embryonic stem cells. *Nat. Cell Biol.* **9,** 625–635.
Mauro, A. (1961). Satellite cell of skeletal muscle fibers. *J. Biophys. Biochem. Cytol.* **9,** 493–495.
McKinnell, I. W., Ishibashi, J., Le Grand, F., Punch, V. G., Addicks, G. C., Greenblatt, J. F., Dilworth, F. J., and Rudnicki, M. A. (2008). Pax7 activates myogenic genes by recruitment of a histone methyltransferase complex. *Nat. Cell Biol.* **10,** 77–84.
Megeney, L. A., Kablar, B., Garrett, K., Anderson, J. E., and Rudnicki, M. A. (1996). MyoD is required for myogenic stem cell function in adult skeletal muscle. *Genes Dev.* **10,** 1173–1183.
Minetti, G. C., Colussi, C., Adami, R., Serra, C., Mozzetta, C., Parente, V., Fortuni, S., Straino, S., Sampaolesi, M., Di Padova, M., Illi, B., Gallinari, P., *et al.* (2006). Functional and morphological recovery of dystrophic muscles in mice treated with deacetylase inhibitors. *Nat. Med.* **12,** 1147–1150.
Mitchell, K. J., Pannerec, A., Cadot, B., Parlakian, A., Besson, V., Gomes, E. R., Marazzi, G., and Sassoon, D. A. (2010). Identification and characterization of a non-satellite cell muscle resident progenitor during postnatal development. *Nat. Cell Biol.* **12,** 257–266.
Mohammad, F., Pandey, R. R., Nagano, T., Chakalova, L., Mondal, T., Fraser, P., and Kanduri, C. (2008). Kcnq1ot1/Lit1 noncoding RNA mediates transcriptional silencing by targeting to the perinucleolar region. *Mol. Cell. Biol.* **28,** 3713–3728.
Mooney, S. M., Goel, A., D'Assoro, A. B., Salisbury, J. L., and Janknecht, R. (2010). Pleiotropic effects of p300-mediated acetylation on p68 and p72 RNA helicase. *J. Biol. Chem.* **285,** 30443–30452.
Motta, M. C., Divecha, N., Lemieux, M., Kamel, C., Chen, D., Gu, W., Bultsma, Y., McBurney, M., and Guarente, L. (2004). Mammalian SIRT1 represses forkhead transcription factors. *Cell* **116,** 551–563.
Mozzetta, C., Minetti, G., and Puri, P. L. (2009). Regenerative pharmacology in the treatment of genetic diseases: The paradigm of muscular dystrophy. *Int. J. Biochem. Cell Biol.* **41,** 701–710.

Muller, J., Hart, C. M., Francis, N. J., Vargas, M. L., Sengupta, A., Wild, B., Miller, E. L., O'Connor, M. B., Kingston, R. E., and Simon, J. A. (2002). Histone methyltransferase activity of a *Drosophila* Polycomb group repressor complex. *Cell* **111,** 197–208.

Nichols, J., Zevnik, B., Anastassiadis, K., Niwa, H., Klewe-Nebenius, D., Chambers, I., Scholer, H., and Smith, A. (1998). Formation of pluripotent stem cells in the mammalian embryo depends on the POU transcription factor Oct4. *Cell* **95,** 379–391.

Oberdoerffer, P., Michan, S., McVay, M., Mostoslavsky, R., Vann, J., Park, S. K., Hartlerode, A., Stegmuller, J., Hafner, A., Loerch, P., Wright, S. M., Mills, K. D., *et al.* (2008). SIRT1 redistribution on chromatin promotes genomic stability but alters gene expression during aging. *Cell* **135,** 907–918.

Ohkawa, Y., Marfella, C. G., and Imbalzano, A. N. (2006). Skeletal muscle specification by myogenin and Mef2D via the SWI/SNF ATPase Brg1. *EMBO J.* **25,** 490–501.

Ohkawa, Y., Yoshimura, S., Higashi, C., Marfella, C. G., Dacwag, C. S., Tachibana, T., and Imbalzano, A. N. (2007). Myogenin and the SWI/SNF ATPase Brg1 maintain myogenic gene expression at different stages of skeletal myogenesis. *J. Biol. Chem.* **282,** 6564–6570.

Oustanina, S., Hause, G., and Braun, T. (2004). Pax7 directs postnatal renewal and propagation of myogenic satellite cells but not their specification. *EMBO J.* **23,** 3430–3439.

Palacios, D., Mozzetta, C., Consalvi, S., Caretti, G., Saccone, V., Proserpio, V., Marquez, V. E., Valente, S., Mai, A., Forcales, S. V., Sartorelli, V., and Puri, P. L. (2010). TNF/p38alpha/Polycomb signaling to Pax7 locus in satellite cells links inflammation to the epigenetic control of muscle regeneration. *Cell Stem Cell* **7,** 455–469.

Pallafacchina, G., Francois, S., Regnault, B., Czarny, B., Dive, V., Cumano, A., Montarras, D., and Buckingham, M. (2010). An adult tissue-specific stem cell in its niche: A gene profiling analysis of in vivo quiescent and activated muscle satellite cells. *Stem Cell Res.* **4,** 77–91.

Partridge, T. (2004). Reenthronement of the muscle satellite cell. *Cell* **119,** 447–448.

Pasini, D., Bracken, A. P., Jensen, M. R., Lazzerini Denchi, E., and Helin, K. (2004). Suz12 is essential for mouse development and for EZH2 histone methyltransferase activity. *EMBO J.* **23,** 4061–4071.

Pawson, T. (2007). Dynamic control of signaling by modular adaptor proteins. *Curr. Opin. Cell Biol.* **19,** 112–116.

Penn, B. H., Bergstrom, D. A., Dilworth, F. J., Bengal, E., and Tapscott, S. J. (2004). A MyoD-generated feed-forward circuit temporally patterns gene expression during skeletal muscle differentiation. *Genes Dev.* **18,** 2348–2353.

Perdiguero, E., Ruiz-Bonilla, V., Gresh, L., Hui, L., Ballestar, E., Sousa-Victor, P., Baeza-Raja, B., Jardi, M., Bosch-Comas, A., Esteller, M., Caelles, C., Serrano, A. L., *et al.* (2007). Genetic analysis of p38 MAP kinases in myogenesis: Fundamental role of p38alpha in abrogating myoblast proliferation. *EMBO J.* **26,** 1245–1256.

Perdiguero, E., Sousa-Victor, P., Ballestar, E., and Munoz-Canoves, P. (2009). Epigenetic regulation of myogenesis. *Epigenetics* **4,** 541–550.

Pokholok, D. K., Zeitlinger, J., Hannett, N. M., Reynolds, D. B., and Young, R. A. (2006). Activated signal transduction kinases frequently occupy target genes. *Science* **313,** 533–536.

Polach, K. J., and Widom, J. (1996). A model for the cooperative binding of eukaryotic regulatory proteins to nucleosomal target sites. *J. Mol. Biol.* **258,** 800–812.

Polesskaya, A., Duquet, A., Naguibneva, I., Weise, C., Vervisch, A., Bengal, E., Hucho, F., Robin, P., and Harel-Bellan, A. (2000). CREB-binding protein/p300 activates MyoD by acetylation. *J. Biol. Chem.* **275,** 34359–34364.

Puri, P. L., Avantaggiati, M. L., Balsano, C., Sang, N., Graessmann, A., Giordano, A., and Levrero, M. (1997a). p300 is required for MyoD-dependent cell cycle arrest and muscle-specific gene transcription. *EMBO J.* **16,** 369–383.

Puri, P. L., Sartorelli, V., Yang, X. J., Hamamori, Y., Ogryzko, V. V., Howard, B. H., Kedes, L., Wang, J. Y., Graessmann, A., Nakatani, Y., and Levrero, M. (1997b). Differential roles of p300 and PCAF acetyltransferases in muscle differentiation. *Mol. Cell* **1,** 35–45.

Puri, P. L., Iezzi, S., Stiegler, P., Chen, T. T., Schiltz, R. L., Muscat, G. E., Giordano, A., Kedes, L., Wang, J. Y., and Sartorelli, V. (2001). Class I histone deacetylases sequentially interact with MyoD and pRb during skeletal myogenesis. *Mol. Cell* **8,** 885–897.

Radman-Livaja, M., and Rando, O. J. (2010). Nucleosome positioning: How is it established and why does it matter? *Dev. Biol.* **339,** 258–266.

Rampalli, S., Li, L., Mak, E., Ge, K., Brand, M., Tapscott, S. J., and Dilworth, F. J. (2007). p38 MAPK signaling regulates recruitment of Ash2L-containing methyltransferase complexes to specific genes during differentiation. *Nat. Struct. Mol. Biol.* **14,** 1150–1156.

Rathbone, C. R., Booth, F. W., and Lees, S. J. (2008). Sirt1 increases skeletal muscle precursor cell proliferation. *Eur. J. Cell Biol.* **88,** 35–44.

Relaix, F., Polimeni, M., Rocancourt, D., Ponzetto, C., Schafer, B. W., and Buckingham, M. (2003). The transcriptional activator PAX3-FKHR rescues the defects of Pax3 mutant mice but induces a myogenic gain-of-function phenotype with ligand-independent activation of Met signaling in vivo. *Genes Dev.* **17,** 2950–2965.

Relaix, F., Rocancourt, D., Mansouri, A., and Buckingham, M. (2005). A Pax3/Pax7-dependent population of skeletal muscle progenitor cells. *Nature* **435,** 948–953.

Rinn, J. L., Kertesz, M., Wang, J. K., Squazzo, S. L., Xu, X., Brugmann, S. A., Goodnough, L. H., Helms, J. A., Farnham, P. J., Segal, E., and Chang, H. Y. (2007). Functional demarcation of active and silent chromatin domains in human HOX loci by noncoding RNAs. *Cell* **129,** 1311–1323.

Rosenberg, M. I., and Parkhurst, S. M. (2002). *Drosophila* Sir2 is required for heterochromatic silencing and by euchromatic Hairy/E(Spl) bHLH repressors in segmentation and sex determination. *Cell* **109,** 447–458.

Rudnicki, M. A., Schnegelsberg, P. N., Stead, R. H., Braun, T., Arnold, H. H., and Jaenisch, R. (1993). MyoD or Myf-5 is required for the formation of skeletal muscle. *Cell* **75,** 1351–1359.

Russo, V. E. A., Martinssen, R. A., and Riggs, A. D. (eds.) (1996). Epigenetic Mechanisms of Gene Regulation, Cold Spring Harbor Laboratory Press, Woodbury.

Sagerstrom, C. G. (2004). PbX marks the spot. *Dev. Cell* **6,** 737–738.

Sambasivan, R., Cheedipudi, S., Pasupuleti, N., Saleh, A., Pavlath, G. K., and Dhawan, J. (2009). The small chromatin-binding protein p8 coordinates the association of anti-proliferative and pro-myogenic proteins at the myogenin promoter. *J. Cell Sci.* **122,** 3481–3491.

Sampaolesi, M., Torrente, Y., Innocenzi, A., Tonlorenzi, R., D'Antona, G., Pellegrino, M. A., Barresi, R., Bresolin, N., De Angelis, M. G., Campbell, K. P., Bottinelli, R., and Cossu, G. (2003). Cell therapy of alpha-sarcoglycan null dystrophic mice through intra-arterial delivery of mesoangioblasts. *Science* **301,** 487–492.

Sampaolesi, M., Blot, S., D'Antona, G., Granger, N., Tonlorenzi, R., Innocenzi, A., Mognol, P., Thibaud, J. L., Galvez, B. G., Barthelemy, I., Perani, L., Mantero, S., *et al.* (2006). Mesoangioblast stem cells ameliorate muscle function in dystrophic dogs. *Nature* **444,** 574–579.

Sartorelli, V., and Caretti, G. (2005). Mechanisms underlying the transcriptional regulation of skeletal myogenesis. *Curr. Opin. Genet. Dev.* **15,** 528–535.

Sartorelli, V., Huang, J., Hamamori, Y., and Kedes, L. (1997). Molecular mechanisms of myogenic coactivation by p300: Direct interaction with the activation domain of MyoD and with the MADS box of MEF2C. *Mol. Cell. Biol.* **17,** 1010–1026.

Sartorelli, V., Puri, P. L., Hamamori, Y., Ogryzko, V., Chung, G., Nakatani, Y., Wang, J. Y., and Kedes, L. (1999). Acetylation of MyoD directed by PCAF is necessary for the execution of the muscle program. *Mol. Cell* **4,** 725–734.

Satijn, D. P., Hamer, K. M., den Blaauwen, J., and Otte, A. P. (2001). The polycomb group protein EED interacts with YY1, and both proteins induce neural tissue in *Xenopus* embryos. *Mol. Cell. Biol.* **21,** 1360–1369.

Sato, T., Rocancourt, D., Marques, L., Thorsteinsdottir, S., and Buckingham, M. (2010). A Pax3/Dmrt2/Myf5 regulatory cascade functions at the onset of myogenesis. *PLoS Genet.* **6,** e1000897.

Seale, P., Sabourin, L. A., Girgis-Gabardo, A., Mansouri, A., Gruss, P., and Rudnicki, M. A. (2000). Pax7 is required for the specification of myogenic satellite cells. *Cell* **102,** 777–786.

Seenundun, S., Rampalli, S., Liu, Q. C., Aziz, A., Palii, C., Hong, S., Blais, A., Brand, M., Ge, K., and Dilworth, F. J. (2010). UTX mediates demethylation of H3K27me3 at muscle-specific genes during myogenesis. *EMBO J.* **29,** 1401–1411.

Serra, C., Palacios, D., Mozzetta, C., Forcales, S. V., Morantte, I., Ripani, M., Jones, D. R., Du, K., Jhala, U. S., Simone, C., and Puri, P. L. (2007). Functional interdependence at the chromatin level between the MKK6/p38 and IGF1/PI3K/AKT pathways during muscle differentiation. *Mol. Cell* **28,** 200–213.

Serrano, A. L., Baeza-Raja, B., Perdiguero, E., Jardi, M., and Munoz-Canoves, P. (2008). Interleukin-6 is an essential regulator of satellite cell-mediated skeletal muscle hypertrophy. *Cell Metab.* **7,** 33–44.

Sherwood, R. I., Christensen, J. L., Weissman, I. L., and Wagers, A. J. (2004). Determinants of skeletal muscle contributions from circulating cells, bone marrow cells, and hematopoietic stem cells. *Stem Cells* **22,** 1292–1304.

Shinin, V., Gayraud-Morel, B., Gomes, D., and Tajbakhsh, S. (2006). Asymmetric division and cosegregation of template DNA strands in adult muscle satellite cells. *Nat. Cell Biol.* **8,** 677–687.

Simon, J. A., and Kingston, R. E. (2009). Mechanisms of polycomb gene silencing: Knowns and unknowns. *Nat. Rev. Mol. Cell Biol.* **10,** 697–708.

Simone, C., Forcales, S. V., Hill, D. A., Imbalzano, A. N., Latella, L., and Puri, P. L. (2004). p38 pathway targets SWI–SNF chromatin-remodeling complex to muscle-specific loci. *Nat. Genet.* **36,** 738–743.

Singhal, N., Graumann, J., Wu, G., Arauzo-Bravo, M. J., Han, D. W., Greber, B., Gentile, L., Mann, M., and Scholer, H. R. (2010). Chromatin-remodeling components of the BAF complex facilitate reprogramming. *Cell* **141,** 943–955.

Squazzo, S. L., O'Geen, H., Komashko, V. M., Krig, S. R., Jin, V. X., Jang, S. W., Margueron, R., Reinberg, D., Green, R., and Farnham, P. J. (2006). Suz12 binds to silenced regions of the genome in a cell-type-specific manner. *Genome Res.* **16,** 890–900.

Srinivasan, L., and Atchison, M. L. (2004). YY1 DNA binding and PcG recruitment requires CtBP. *Genes Dev.* **18,** 2596–2601.

Suzuki, H. I., Yamagata, K., Sugimoto, K., Iwamoto, T., Kato, S., and Miyazono, K. (2009). Modulation of microRNA processing by p53. *Nature* **460,** 529–533.

Tajbakhsh, S., Rocancourt, D., Cossu, G., and Buckingham, M. (1997). Redefining the genetic hierarchies controlling skeletal myogenesis: Pax-3 and Myf-5 act upstream of MyoD. *Cell* **89,** 127–138.

Takaesu, G., Kang, J. S., Bae, G. U., Yi, M. J., Lee, C. M., Reddy, E. P., and Krauss, R. S. (2006). Activation of p38alpha/beta MAPK in myogenesis via binding of the scaffold protein JLP to the cell surface protein Cdo. *J. Cell Biol.* **175,** 383–388.

Takata, T., and Ishikawa, F. (2003). Human Sir2-related protein SIRT1 associates with the bHLH repressors HES1 and HEY2 and is involved in HES1- and HEY2-mediated transcriptional repression. *Biochem. Biophys. Res. Commun.* **301,** 250–257.

Tamkun, J. W., Deuring, R., Scott, M. P., Kissinger, M., Pattatucci, A. M., Kaufman, T. C., and Kennison, J. A. (1992). brahma: A regulator of *Drosophila* homeotic genes structurally related to the yeast transcriptional activator SNF2/SWI2. *Cell* **68,** 561–572.
Tapscott, S. J. (2005). The circuitry of a master switch: Myod and the regulation of skeletal muscle gene transcription. *Development* **132,** 2685–2695.
Tedesco, F. S., Dellavalle, A., Diaz-Manera, J., Messina, G., and Cossu, G. (2010). Repairing skeletal muscle: Regenerative potential of skeletal muscle stem cells. *J. Clin. Invest.* **120,** 11–19.
van der Vlag, J., and Otte, A. P. (1999). Transcriptional repression mediated by the human polycomb-group protein EED involves histone deacetylation. *Nat. Genet.* **23,** 474–478.
van Essen, D., Engist, B., Natoli, G., and Saccani, S. (2009). Two modes of transcriptional activation at native promoters by NF-kappaB p65. *PLoS Biol.* **7,** e73.
Vaquero, A., Scher, M., Erdjument-Bromage, H., Tempst, P., Serrano, L., and Reinberg, D. (2007). SIRT1 regulates the histone methyl-transferase SUV39H1 during heterochromatin formation. *Nature* **450,** 440–444.
Vaute, O., Nicolas, E., Vandel, L., and Trouche, D. (2002). Functional and physical interaction between the histone methyl transferase Suv39H1 and histone deacetylases. *Nucleic Acids Res.* **30,** 475–481.
Vire, E., Brenner, C., Deplus, R., Blanchon, L., Fraga, M., Didelot, C., Morey, L., Van Eynde, A., Bernard, D., Vanderwinden, J. M., Bollen, M., Esteller, M., *et al.* (2006). The Polycomb group protein EZH2 directly controls DNA methylation. *Nature* **439,** 871–874.
Wang, H., Hertlein, E., Bakkar, N., Sun, H., Acharyya, S., Wang, J., Carathers, M., Davuluri, R., and Guttridge, D. C. (2007). NF-kappaB regulation of YY1 inhibits skeletal myogenesis through transcriptional silencing of myofibrillar genes. *Mol. Cell. Biol.* **27,** 4374–4387.
Wang, Z., Zang, C., Cui, K., Schones, D. E., Barski, A., Peng, W., and Zhao, K. (2009). Genome-wide mapping of HATs and HDACs reveals distinct functions in active and inactive genes. *Cell* **138,** 1019–1031.
Weintraub, H., and Groudine, M. (1976). Chromosomal subunits in active genes have an altered conformation. *Science* **193,** 848–856.
Wilson, B. J., Bates, G. J., Nicol, S. M., Gregory, D. J., Perkins, N. D., and Fuller-Pace, F. V. (2004). The p68 and p72 DEAD box RNA helicases interact with HDAC1 and repress transcription in a promoter-specific manner. *BMC Mol. Biol.* **5,** 11.
Wong, C. F., and Tellam, R. L. (2008). microRNA-26a targets the histone methyltransferase enhancer of Zeste homolog 2 during myogenesis. *J. Biol. Chem.* **283,** 9836–9843.
Woo, C. J., Kharchenko, P. V., Daheron, L., Park, P. J., and Kingston, R. E. (2010). A region of the human HOXD cluster that confers polycomb-group responsiveness. *Cell* **140,** 99–110.
Wu, Z., Woodring, P. J., Bhakta, K. S., Tamura, K., Wen, F., Feramisco, J. R., Karin, M., Wang, J. Y., and Puri, P. L. (2000). p38 and extracellular signal-regulated kinases regulate the myogenic program at multiple steps. *Mol. Cell. Biol.* **20,** 3951–3964.
Yamane, K., Toumazou, C., Tsukada, Y., Erdjument-Bromage, H., Tempst, P., Wong, J., and Zhang, Y. (2006). JHDM2A, a JmjC-containing H3K9 demethylase, facilitates transcription activation by androgen receptor. *Cell* **125,** 483–495.
Yuan, W., Condorelli, G., Caruso, M., Felsani, A., and Giordano, A. (1996). Human p300 protein is a coactivator for the transcription factor MyoD. *J. Biol. Chem.* **271,** 9009–9013.
Yue, R., Kang, J., Zhao, C., Hu, W., Tang, Y., Liu, X., and Pei, G. (2009). Beta-arrestin1 regulates zebrafish hematopoiesis through binding to YY1 and relieving polycomb group repression. *Cell* **139,** 535–546.

Zhang, C. L., McKinsey, T. A., and Olson, E. N. (2002). Association of class II histone deacetylases with heterochromatin protein 1: Potential role for histone methylation in control of muscle differentiation. *Mol. Cell. Biol.* **22,** 7302–7312.

Zhao, X., Sternsdorf, T., Bolger, T. A., Evans, R. M., and Yao, T. P. (2005). Regulation of MEF2 by histone deacetylase 4- and SIRT1 deacetylase-mediated lysine modifications. *Mol. Cell. Biol.* **25,** 8456–8464.

Zhao, X. D., Han, X., Chew, J. L., Liu, J., Chiu, K. P., Choo, A., Orlov, Y. L., Sung, W. K., Shahab, A., Kuznetsov, V. A., Bourque, G., Oh, S., *et al.* (2007). Whole-genome mapping of histone H3 Lys4 and 27 trimethylations reveals distinct genomic compartments in human embryonic stem cells. *Cell Stem Cell* **1,** 286–298.

Zhao, J., Sun, B. K., Erwin, J. A., Song, J. J., and Lee, J. T. (2008). Polycomb proteins targeted by a short repeat RNA to the mouse X chromosome. *Science* **322,** 750–756.

CHAPTER FOUR

NF-κB Signaling in Skeletal Muscle Health and Disease

Jennifer M. Peterson, Nadine Bakkar, *and* Denis C. Guttridge

Contents

Abstract

Muscle development, growth, and maintenance require an intricate and timely series of events initiated through a multitude of signaling pathways. The very nature of skeletal muscle requires tremendous plasticity to accommodate the need for anabolism or catabolism, and deregulation of these processes may be a tipping point in the development or progression of various skeletal muscle disorders. Among the relevant signaling pathways, NF-κB has emerged as a critical factor involved in various facets of muscle homeostasis. In this review, we summarize the NF-κB signaling pathway and provide a fresh perspective into

Department of Molecular Virology, Immunology, and Medical Genetics, Human Cancer Genetics Program, Arthur G. James Comprehensive Cancer Center, The Ohio State University, Columbus, Ohio, USA

Current Topics in Developmental Biology, Volume 96
ISSN 0070-2153, DOI: 10.1016/B978-0-12-385940-2.00004-8

the regulation and function of this transcription factor, underlying both the physiological and pathophysiological states of skeletal muscle.

1. Introduction

From the initial discovery of a *n*uclear *f*actor that bound to the *k*appa light chain enhancer in *B* cells (Sen and Baltimore, 1986), the NF-κB transcription factor has become a keen interest of investigation for many laboratories. This interest draws from the appreciation that NF-κB is not only present in B cells, but ubiquitously expressed. Additionally, in most cells, NF-κB does not reside in the nucleus as is typical with transcription factors, but rather is held in an inactive state in the cytoplasm under tight control by a factor appropriately termed *i*nhibitor of NF-*κB* (IκB; Haskill *et al.*, 1991). Release from inhibition by IκB degradation essentially activates NF-κB, allowing its nuclear translocation and subsequent binding to a selective DNA sequence that in turn signals to the basal transcription factor assembly complex to stimulate gene expression. The profile of genes regulated by NF-κB is diverse and numerous, potentially numbering in the hundreds (Pahl, 1999; Sharif *et al.*, 2007). It is through this diversity that NF-κB is considered a vital regulator of multiple cellular processes. The most recognized of these processes is the regulation of the immune system, which is controlled by NF-κB dependent stimulation of cytokine and chemokine genes as well as other genes that influence the proliferation, differentiation, migration, adhesion, and survival of immune cells (Pahl, 1999). Because of its ubiquitous expression pattern, we have come to understand that regulation of such cellular properties by NF-κB is not unique to immune cells, but in fact is shared by other cell lineages that orchestrate the development and homeostasis of multiple tissues from drosophila to humans.

Skeletal muscle cells are included in this lineage repertoire, and since the first documented report of NF-κB activity in myoblast nuclei (Lehtinen *et al.*, 1996), there has been increasing interest in learning more about this signaling pathway and its relevance to muscle differentiation. This interest has been bolstered by observations in both patient samples and animal models associating NF-κB with several skeletal muscle disorders such as Duchenne muscular dystrophy (DMD), inflammatory myopathies, rhabdomyosarcoma, cancer cachexia, denervation, and disuse atrophy (Peterson and Guttridge, 2008). We are only beginning to understand the significance of this association as well as the mechanisms regulated by NF-κB that may underlie some of the causal effects in various muscle diseases. Insight into how NF-κB signaling is regulated and functions in skeletal muscle differentiation is likely to enhance our knowledge on how this signaling pathway contributes to both the physiological and pathophysiological states of skeletal muscle.

The intent of this review is to provide a comprehensive overview of the current state of our understanding of NF-κB in skeletal muscle. For complementary materials, we refer the reader to additional recent reviews (Bakkar and Guttridge, 2010; Creus *et al.*, 2009; Li *et al.*, 2008; Mourkioti and Rosenthal, 2008; Peterson and Guttridge, 2008). Although some overlap with past literature is unavoidable, we have tried to the best of our abilities to expand on topics in this chapter not previously discussed in past reviews. Following a general description of the NF-κB family and the main regulatory signaling pathways, we review how NF-κB activity is regulated and functions in skeletal myogenesis, both in cultured cells and *in vivo*. We then provide a section on NF-κB adaptation to stress signals and exercise, followed by discussion of NF-κB association with skeletal muscle disorders. In line with a relation to muscle disease, a model is presented describing how we currently view NF-κB function in DMD. We conclude with discussion of current therapeutic interventions targeting the pathway, indicating their potential advantages and limitations.

2. The NF-κB Family and Signaling Pathway Activation

In mammalian cells, the NF-κB family is composed of five genes that code for protein subunits, RelA/p65, RelB, c-Rel, p50, and p52 (the latter two being synthesized as the precursor proteins, p105 and p100, respectively). NF-κB functions as a dimer, with each monomeric subunit containing a conserved Rel homology domain (RHD) in the amino half of the protein. The RHD is essential for subunit dimerization, DNA binding, nuclear localization, and interactions with IκB proteins. RelA/p65 (from here on referred to only as p65), RelB, and c-Rel possess additional transcriptional activation domains (TAD) in their carboxyl ends allowing for positive regulation of gene expression (Fig. 4.1A). The most commonly described forms of NF-κB are the p50/p65 heterodimer and p50/p50 homodimer complexes, followed by p50/c-Rel and p52/RelB, depending on cell type and activating stimulus. The absence of TAD domains in p50 and p52 subunits generates p50 and p52 homodimers that are considered to function as transcriptional repressor complexes (Tong *et al.*, 2004; Fig. 4.1B). NF-κB dimers bind to DNA sequences with the consensus site, GGGRNNYYCC, in the promoter and enhancer elements of NF-κ B target genes that are activated by recruitment of transcriptional coactivators and chromatin remodeling complexes (Hayden and Ghosh, 2008).

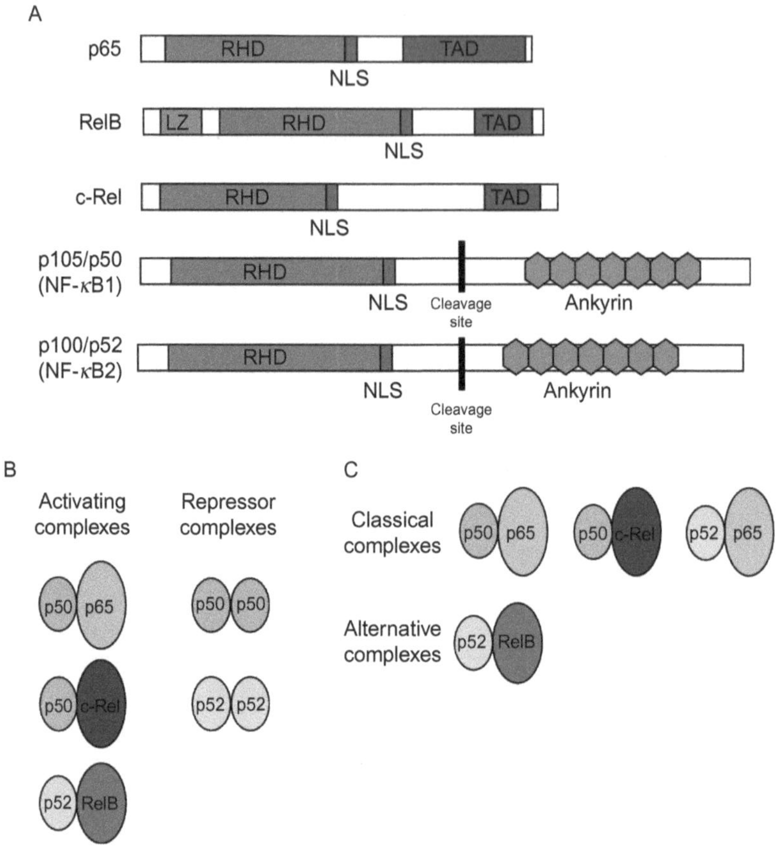

Figure 4.1 The NF-κB family members. (A) NF-κB family members are represented with their specific domains. TAD, transactivation domain; RHD, Rel-homolgy domain; NLS, nuclear localization signal; LZ, leucine zipper; Ankyrin, ankyrin repeats. (B) Activating and repressor NF-κB dimers. (C) Classical and alternative NF-κB complexes.

Activation of NF-κB signaling pathways occurs through the IκB kinase (IKK) complex, composed predominantly of two catalytic subunits, IKKα/IKK1 and IKKβ/IKK2 and an oligomerized regulatory subunit, IKKγ/NEMO (*N*F-κB *e*ssential *mo*difier; Tegethoff *et al.*, 2003; Zandi *et al.*, 1997). Complex formation with NEMO is dependent on the NEMO binding domain (NBD), a short amino acid sequence that maps to the carboxyl terminal end of IKKα and IKKβ (May *et al.*, 2000), whose structure was recently solved (Madge and May, 2009). Although IKKα and IKKβ maintain over 51% sequence identity (Mercurio *et al.*, 1997), they differ in their target specificity. IKKβ is more efficient than IKKα in phosphorylating IκB proteins, leading to their polyubiquitin-dependent degradation (Hacker and Karin, 2006). In addition, while both kinases phosphorylate p65 at Ser 536 to enhance NF-κB transactivation potential, IKKβ can also target Ser 468 of p65 to further stimulate NF-κB activity (Schwabe and Sakurai, 2005). IKKα on the other hand, phosphorylates p100 to induce partial proteolysis and activation of the alternative NF-κB pathway (see below), but can also function independently of NF-κB by acting in the nucleus to phosphorylate transcriptional cofactors CBP and SMRT, as well as DNA histone protein H3 (Anest *et al.*, 2003; Hoberg *et al.*, 2004; Huang *et al.*, 2007).

3. Classical NF-κB Signaling

In response to inflammatory factors, bacterial and viral products, oxidative stress, and DNA damaging agents, NF-κB is activated by what is now referred to as the canonical or classical pathway. Signaling by cytokines, TNFα and IL-1β, or through LPS, triggers receptor-mediated assembly of a multiprotein signaling complex leading to IKK activation. Common to this assembly process are the TNF-receptor associated factor family (TRAF) proteins, TRAF1-7, that with the exception of TRAF1, contain a RING domain and function as E3 ubiquitin ligases. For TNFα, activation of IKK and NF-κB initiates with oligomerization of the TNF receptor (TNFR1) that causes the recruitment of the TNFR1-associated death domain protein (TRADD), TRAF2, and the receptor-interacting protein kinase (RIP1; Chen and Goeddel, 2002). TRAF2 activation of NF-κB depends on its oligomerization and polyubiquitination, which in association with the cellular inhibitor of apoptosis protein (c-IAP) induce K-63 linked polyubiquitination of RIP1. This signal serves to further recruit and activate TGF-β-activated kinase 1 (TAK1). This kinase, in complex with associated TAK1 proteins called TABs directly phosphorylate IKK resulting in phosphorylation of the IκB inhibitory proteins and activation of the predominant classical p50/p65 complex (Fig. 4.1C). In addition to IKK, the TAK1 complex is also responsible for phosphorylating and activating the mitogen activated protein kinase (MAPK) pathway MEKK6, which is upstream of p38, as well as c-Jun N-terminal kinase protein (JNK; Fig. 4.2).

IL-1β and LPS-induced (Toll-like receptor, TLR) signaling functions analogously to TNFα in the receptor-mediated recruitment of a multisubunit complex that in this case operates via the E3 ubiquitin ligase activity of TRAF6. Oligomerization and autoubiquitination of TRAF6 leads to TAK1, TAB2/3 recruitment, and TAK1-induced phosphorylation and activation of IKK (Ea *et al.*, 2006; Kanayama *et al.*, 2004; Wu *et al.*, 2006; Fig. 4.2). For additional details of TNF/IL-1/TLR signaling to IKK, we

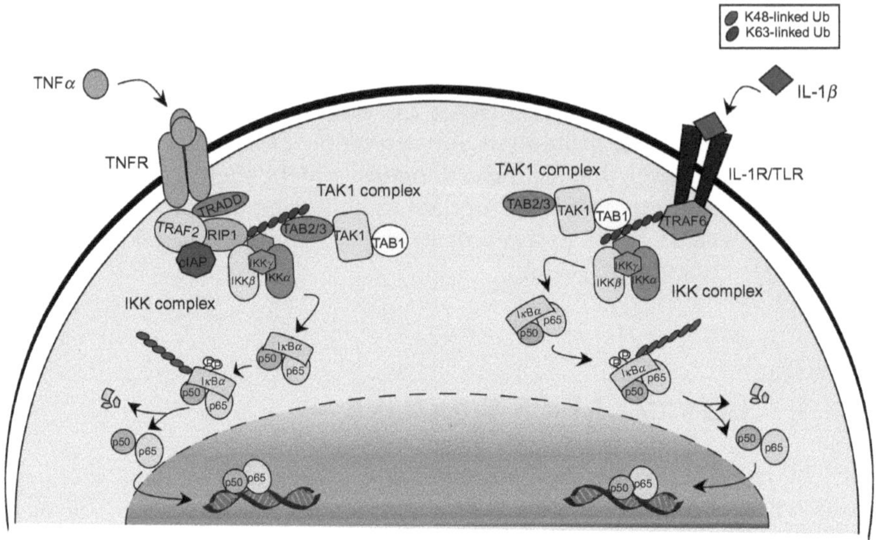

Figure 4.2 The classical NF-κB signaling pathway. In the TNF signaling pathway (left), stimulation of the TNF receptor (TNFR) leads to the recruitment of TRADD (TNF receptor associated death domain), TRAF2, c-IAP, and RIP1. RIP1 is activated via a K63 polyubiquitin tail, resulting in recruitment of IKK (through IKKγ) and TAK1 complexes. TAK1 in turn activates IKKβ, which phosphorylates IkBα, targeting it for K48-linked ubiquitination and proteasomal degradation. Following IκBα degradation, p50/p65 translocates to the nucleus to activate gene transcription. In the IL1-receptor (IL1-R) and Toll-like receptor (TLR) signaling (right), ligand binding leads to receptor dimerization, TRAF6 is recruited to the membrane where it becomes polyubiquitinated at K63. This process again leads to recruitment of TAK1 and IKK complexes, and the subsequent activation of NF-κB dimers. (See Color Insert.)

recommend the reading of a recent comprehensive review from the Chen laboratory (Skaug *et al.*, 2009).

The resulting activation of IKK leads to phosphorylation of IκB proteins. IκBα and IκBβ are the prototypical members of the IκB family that also include the less-described IκBε, IκBγ, BCL-3, p100, and p105 inhibitors. The IκB family of proteins contains ankyrin repeats at their carboxy terminus that regulate interaction with NF-κB subunits, sequestering them in the cytoplasm by masking their nuclear localization site (Hayden and Ghosh, 2004). Phosphorylation of IκBα on residue, Ser 32 and Ser 36, or Ser 19 and Ser 23 for IκBβ, induces K48-linked polyubiquitination by the SCF-βTrCP ubiquitin ligase complex on neighboring lysine residues, thus marking the proteins for 26S-dependent proteasomal degradation. This proteolysis unmasks the nuclear localization signal of p65 (or c-Rel), leading to the translocation of the p50/p65 or p50/c-Rel complex into the nucleus to bind DNA at NF-κB consensus sites and subsequently activate gene transcription (Perkins, 2006). Other signaling inducers including viral proteins, oxidative stress, and DNA damage activate the classical NF-κB pathway in a manner analogous to cytokines and inflammatory factors downstream of IKK.

In response to TNFα or other mediators of classical NF-κB, high-density gene profiling studies have identified in the realm of hundreds of NF-κB dependent genes, many of which remain to be formally validated (Feuerhake *et al.*, 2005; Sharif *et al.*, 2007). Regardless of the accuracy of the profiling data, such findings make the point that NF-κB is responsible for regulating an extensive list of genes that code for proteins mostly involved in inflammatory, cell adhesion, migration, proliferation, extracellular matrix (ECM) remodeling, and cell survival processes. Of these genes, a few have been shown to be absolutely critical for controlling the transient nature of NF-κB activation, which in most cells, depending on the activating signal, lasts only a few hours. Regulating this turnover process is IκBα, which is one of the early immediate genes induced by NF-κB. Once translated, IκBα is imported into the nucleus to bind and remove NF-κB off the DNA, causing the transcription factor to be sequestered back to the cytoplasm (Huang *et al.*, 2000). Although IκBβ also sequesters NF-κB in the cytoplasm, unlike IκBα, this direct inhibitor is not a transcriptional target of NF-κB. Instead it is involved in dynamic interactions with NF-κB that on the one hand causes stabilization of IκBβ via direct binding to the p65 subunit (Hertlein *et al.*, 2005), and on the other allows IκBβ to actually function as a transcriptional coactivator by interacting in a ternary complex with p65 and c-Rel to stimulate TNFα expression in response to LPS (Rao *et al.*, 2010).

Another rapidly transcribed NF-κB-dependent gene that functions analogously to IκBα in feedback inhibition of NF-κB is A20 (TNFAIP3; Krikos *et al.*, 1992). A20 is unique in the sense that the protein contains both ubiquitin and deubiquitinase activities, the latter of which is critical for editing K63-linked polyubiquitination of RIP1 and TRAF6. Deubiquitination attenuates the recruitment activities of these proteins thereby inhibiting IKK activation. In addition, A20 utilizes its ubiquitin ligase activity to induce K48 polyubiquitination of TRAF6 and RIP1, thereby causing their degradation and further silencing IKK and classical NF-κB activation (Boone *et al.*, 2004; Wertz *et al.*, 2004).

3.1. Alternative NF-κB signaling

NF-κB can also be activated in a noncanonical, alternative pathway, mediated by an IKKα homodimer complex. This alternative pathway has been best described in B cells, and in fact is essential for development of lymphoid tissues (Senftleben *et al.*, 2001). B cell activators such as B cell activating factor (BAFF), lymphotoxin β, or CD40 ligands bind to their respective receptors, causing stabilization of the NF-κB inducing kinase

(NIK; Pomerantz and Baltimore, 2002; Senftleben *et al.*, 2001). Recent findings support that NIK stabilization is controlled by a c-IAP1/2 and TRAF3 containing complex. In resting cells, the E3 ubiquitinase activity of c-IAP causes TRAF3 and NIK degradation, whereas upon B cell and lymphotoxin signaling, c-IAP is itself degraded leading to NIK stability (Razani *et al.*, 2010; Sanjo *et al.*, 2010). NIK in turn phosphorylates IKKα, resulting in activation and subsequent phosphorylation of the IκB member p100 on sites, Ser 866 and 870 (Xiao *et al.*, 2004). Unlike the complete proteolysis that occurs with other IκB proteins, p100 is only partially proteolyzed at the site of its C-terminal ankyrin repeats, resulting in the production of a functional p52 subunit. The p52 subunit is then released and shuttled to the nucleus in complex with RelB (Fig. 4.1C). Although classical and alternative NF-κB pathways are considered to possess distinct functions via their unique transcriptional targets, to date little mechanistic insight has been provided to elucidate the regulatory processes that underlie the distinctness of these pathways. One possibility is related to the subtle difference in DNA consensus binding sites that has been proposed to exist between classical and alternative NF-κB dimers. For example, investigators revealed that the B cell specific genes, ELC and BLC, contained NF-κB binding sites in their promoters that diverged slightly from the consensus sequence, 5′ GGGRCTTTCC3′, where a guanine base was substituted for cytosine in the most 3′ position (underlined). This one base modification was sufficient to promote a switch in binding favoring a p52/RelB complex (Bonizzi *et al.*, 2004). Although such data support the uniqueness of both NF-κB pathways, evidence also exists that cross talk between pathways is possible, whereby in response to lymphotoxin, processed p52 is able to complex with p65 to activate downstream classical signaling genes (Basak *et al.*, 2007).

4. Regulation of NF-κB in Skeletal Myogenesis

Skeletal muscle development during embryogenesis is marked by the commitment of myoblasts in the dermomyotome that proliferate and migrate to proximal limbs and the surrounding body wall (Borycki and Emerson, 2000; Buckingham, 2001; Perry and Rudnicki, 2000). Primary myogenesis ensues with the growth arrest of myoblasts that subsequently align and fuse to form multinucleated myotubes. The maturation process extends into secondary myogenesis during fetal stage, where additional myotubes are formed and surround primary fibers. Muscle maturation continues in early postnatal development characterized by muscle growth resulting from a hypertrophic response.

Primary myogenesis can be replicated in culture with primary or immortalized myoblast cells such as the traditional C2C12 cell line, which when switched from a proliferating to differentiating medium adopts a similar myogenic differentiation program as that seen *in vivo*. Thus, cultured myoblasts receiving their differentiation cues progress through stages of cell cycle arrest, myoblast fusion, and myofiber maturation (Perry and Rudnicki, 2000). At the molecular level, this myogenic program is controlled by master switch transcription factors such as MyoD, which along with Myf-5, are part of the bHLH family of transcription factors specifically expressed in skeletal muscle (Murre *et al.*, 1989). MyoD is expressed in myoblasts, but by mechanisms that are still not well defined remains inactive. Upon differentiation, MyoD is activated and in complex with E box transcription factors (E12/E47). These complexes regulate many of the downstream events that coordinate exit from the cell cycle, migration and fusion, and expression of the contractile apparatus (Bergstrom *et al.*, 2002; Cao *et al.*, 2010).

As opposed to most cells at rest where NF-κB activity is present at a low basal state, the relatively abundant constitutive activation of NF-κB in cultured myoblasts has captured the attention of our laboratory as well as others. Attempts have been made since those initial observations (Lehtinen *et al.*, 1996) to elucidate the functional relevance of this nuclear and transactivation activity. Classical NF-κB signaling has been shown to play an integral role in regulating differentiation in cultured myoblasts by several mechanisms. NF-κB binds to the cyclin D1 promoter to direct its transcription and thereby maintain myoblasts in a proliferative state by stimulating the progression of cells into S phase (Guttridge *et al.*, 1999). For myogenesis, this process is regulated by classical p50/p65 heterodimers (Guttridge *et al.*, 1999). An additional method by which NF-κB controls cyclin D1 expression is through protein stabilization. Specifically, p65 binds indirectly to cyclin D1 to increase its stability as well as enhance the kinase activity of CDK4, which exists in complex with cyclin D1 (Dahlman *et al.*, 2009). This regulation is thought to contribute to the molecular program of differentiating muscle cells by coordinating the timing of myoblasts exit from cell cycle. Regulation of cyclin D1 by NF-κB is not unique to myogenesis, as transcriptional control by p65 and other subunits has been seen in other cell types (Guo *et al.*, 2009; Hinz *et al.*, 1999; Westerheide *et al.*, 2001), and implicated as a mechanism underlying cell proliferation in breast cancer development (Cao *et al.*, 2001).

NF-κB also maintains myoblasts in an undifferentiated state by repressing muscle specific gene expression. One way this occurs is by limiting the expression of MyoD at the level of the transcript and protein (Dogra *et al.*, 2006; Guttridge *et al.*, 2000; Langen *et al.*, 2004). This activity of NF-κB is specific to p65, which causes MyoD mRNA destabilization through a region mapping to 130 base pairs within the coding sequence (Guttridge *et al.*, 2000; Sitcheran *et al.*, 2003). Interestingly, this same region was identified in the transcript of Sox10, a transcription factor that analogously to MyoD promotes chondrocyte differentiation. Thus, in both myoblasts and chondroblasts, activation of NF-κB through p65 causes the destabilization of MyoD and Sox10 mRNA that leads to the inhibition of their respective differentiation processes (Guttridge *et al.*, 2000; Sitcheran *et al.*, 2003). Classical NF-κB signaling is also capable of inducing the degradation of MyoD protein, which again results in limiting myogenesis (Dogra *et al.*, 2006; Langen *et al.*, 2004). Interestingly, TRAF7, which activates NF-κB, was recently found to be under the transcriptional control of MyoD in myoblasts (Tsikitis *et al.*, 2010) potentially forming a feedback loop to prevent excessive NF-κB signaling. Moreover, NF-κB indirectly represses the expression of muscle specific genes, such as isoforms of troponin and myosin heavy chain, by stimulating the transcription of the Polycomb repressive complex associated member, Yin Yang1 (YY1; Wang *et al.*, 2007). YY1, under control of NF-κB, binds to enhancer elements of muscle genes such as troponin I2, MyHCIIb, and α-actin and recruits members of the Polycomb group, as well as the histone deacetylase, HDAC-1, to repress transcription (Caretti *et al.*, 2004; Wang *et al.*, 2007). In addition to being a repressor of muscle gene expression, YY1 acts in a similar manner to epigenetically silence microRNA-29 (miR-29) expression in proliferating myoblasts (Wang *et al.*, 2008). Once differentiation ensues, YY1 levels decrease due to a concomitant loss of NF-κB nuclear activity, which in turn causes the derepression of miR-29 that functions in a feedback loop to target its own inhibitor, YY1. The combined decline of YY1, from decreased NF-κB activity and induced miR-29 expression, leads to the upregulation of myofibrillar genes and downregulation of other described miR-29 targets, such as collagen type IV that may contribute to skeletal muscle remodeling during myogenesis (van Rooij *et al.*, 2008).

Although an increasing number of papers are becoming aligned in their conclusions that NF-κB acts in cultured myoblasts to prevent differentiation (Acharyya *et al.*, 2010; Tubaro *et al.*, 2010), there remains a body of literature, some even recently cited (Wang *et al.*, 2010), that point to a promyogenic role for this transcription factor. We have analyzed in length in a recent review (Bakkar and Guttridge, 2010) the experimental variations that may account for these differing interpretations, and do not intend to repeat that discussion in this chapter. However, as an additional point on this topic, our own approach to dissecting how NF-κB is regulated and functions during skeletal myogenesis has been to examine its upstream signaling pathway. Since there is general acceptance that p65 is active in cultured myoblasts, and is part of the classical signaling pathway, we analyzed whether other components of this pathway would behave in a similar manner to the activities of p65. As discussed above, immediately upstream regulating NF-κB is the IKK complex. Results show that mouse embryonic fibroblasts (MEFs) null for classical pathway subunits, IKKβ and IKKγ, and reconstituted with MyoD, display a similar antimyogenic activity as that seen with *p65*$^{-/-}$ MEFs (Bakkar *et al.*, 2008). Similar findings have also been recapitulated in primary *p65*$^{-/-}$ and *IKKβ*$^{-/-}$ myoblasts (Bakkar and Guttridge, unpublished observations; Bakkar *et al.*, 2008). These data support the conclusion that NF-κB classical signaling functions to repress skeletal muscle differentiation, a finding that is also consistent with *in vivo* deletion of these NF-κB pathway subunits (Bakkar *et al.*, 2008; Dahlman *et al.*, 2010).

Recently, a similar experimental approach was described to examine the myogenic activity of TAK1, which as described above is the immediate upstream regulator of IKK in the classical pathway. Biochemical analysis of TAK1 expression and associated activity during C2C12 myogenesis exhibited a comparable downregulated pattern as that described for NF-κB (Bakkar *et al.*, 2008; Catani *et al.*, 2004; Lehtinen *et al.*, 1996). However, unlike p65 and IKKβ/IKKγ (Bakkar *et al.*, 2008), *TAK1*$^{-/-}$ MEFs expressing MyoD were impaired in their myogenic potential (Bhatnagar *et al.*, 2010). Besides IKK, TAK1, in association with TAB proteins, can signal to other pathways such as MAPK p38, JNK (Frazier *et al.*, 2007; Skaug *et al.*, 2009), and AMPK (Momcilovic *et al.*, 2006). Investigators therefore explored the consequences of these downstream signaling modulators in *TAK1*$^{-/-}$ cells undergoing differentiation. Findings showed that only p38 activity was significantly affected under these conditions. Whereas p38α activity is known to be elevated during skeletal myogenesis (Lluis *et al.*, 2005; Wu *et al.*, 2000), the phosphorylation of p38α was pronouncedly diminished in differentiating *TAK1*$^{-/-}$ MyoD expressing fibroblasts (Bhatnagar *et al.*, 2010). No changes were described in NF-κB and AMPK, suggesting that TAK activation in myoblasts predominantly signals through the MAPK/MKK6/p38 pathway. Activation of p38 by TAK1 was further shown to be required for growth arrest through expression of the cyclin dependent kinase inhibitor, p21$^{CIP/WAF1}$ (Bhatnagar *et al.*, 2010). Although it remains to be seen whether such differences in TAK1 signaling in myoblasts is regulated by distinct TAK1/TAB complexes, these findings are nevertheless provocative as they suggest that constitutive IKK and NF-κB classical signaling in proliferating muscle cells may be TAK1 independent.

5. Alternative NF-κB Signaling During Muscle Differentiation

In contrast to classical NF-κB signaling, evidence to date supports that the alternative pathway is activated during myotube formation, as characterized by coordinated increases in IKK activity and p100 to p52 processing

(Bakkar *et al.*, 2008). In support of the activation of this pathway, similar processing of p100 to p52 was observed in postnatal murine muscle development, at a time when classical signaling becomes downregulated (Dahlman *et al.*, 2010). Thus, it appears that during muscle development a signaling switch occurs between the classical and alternative NF-κB signaling pathways. Factors controlling this switch point are unknown and it is not well understood whether downregulation of the classical pathway is a prerequisite for the activation of alternative signaling.

Further, unlike the classical pathway, which inhibits myotube formation, activation of the alternative pathway in differentiating cells may regulate mitochondrial biogenesis and maintain the homeostasis of mature myotubes (Bakkar *et al.*, 2008). Such observations were made in a C2C12 system and will need to be recapitulated *in vivo* (Bakkar and Guttridge, unpublished observations).

6. NF-κB Function in Postnatal Skeletal Muscle Development

Following birth, primary and secondary stages of myogenesis have already been completed, yet a tremendous amount of growth and remodeling occurs in early postnatal muscle to enable the organism to gain full muscle function and achieve weight-bearing capacity. During this time, myoblasts continue to fuse into existing muscle fibers, thereby donating their nuclei to accommodate this growth phase. Although NF-κB activity had been described in postnatal muscle (Acharyya *et al.*, 2007), the cellular signaling source(s) and functional relevance of this activation is just beginning to be explored.

Classical NF-κB signaling is highly activated in limb muscle for approximately 1.5 weeks after birth, then declines rather rapidly, and is maintained at low levels thereafter (Acharyya *et al.*, 2007; Dahlman *et al.*, 2010). Using transgenic NF-κB-EGFP reporter mice, postnatal muscle sections were analyzed to visualize the cellular location of NF-κB activity (Dahlman *et al.*, 2010). On postnatal day 0 (P0), myofibers were determined to be a major cellular source of NF-κB signaling. However, by P8 when muscle structure and organization is more complete, NF-κB–EGFP expression declines in myofibers and becomes highly expressed in the interstitial compartment of skeletal muscles (Dahlman *et al.*, 2010). Stromal fibroblasts were revealed as the likely cellular source of this expression, and NF-κB was shown to function in these cells by secreting inducible nitric oxide (iNOS), which was further proposed to promote myoblast fusion into preexisting myofibers. These data revealed a beneficial role for NF-κB activation in fibroblasts during postnatal muscle development.

In addition to the interstitial compartmentalization of NF-κB in skeletal muscle, observations made 10 years earlier had localized NF-κB at the neuromuscular junctions (NMJ) in normal human adult muscle, but the functional relevance of this localization was undetermined (Yang *et al.*, 1998). Recent work has begun to explore this area of study, elucidating a specialized role for NF-κB in regulating the clustering of acetylcholine receptors (AChR) at the NMJ (Wang *et al.*, 2010). The p65 subunit of NF-κB was demonstrated to regulate expression of rapsyn, a protein involved in the clustering of AChR at the NMJ. These data therefore extend our understanding of NF-κB function in postnatal muscle by suggesting a novel role for this transcription factor in the specialized postsynaptic nuclei that reside at NMJs.

7. Regulation of NF-κB in Skeletal Muscle Adaptations

Skeletal muscle is a plastic organ capable of efficiently responding to changes in homeostasis. When mechanical stress and caloric intake are constant, protein synthesis and degradation rates remain balanced and no net changes in protein content occur. However, skeletal muscle is also proficient at responding to homeostatic perturbations, readily adapting to increased mechanical load and energy availability, which tip the balance to favor hypertrophy. Classical NF-κB becomes activated in this process and plays a central role in aiding muscle adaptations.

Cellular stress is a known activator of the classical NF-κB signaling pathway. Passive maintenance of muscle in a stretched position is sufficient to transiently activate NF-κB in a manner that is dependent on reactive oxygen species (ROS) production, but independent from calcium signaling (Kumar and Boriek, 2003). However, the relevance of NF-κB activation in response to acute injury goes beyond that of a basic cellular stress response. When skeletal muscle is subjected to freeze or chemical (cardiotoxin) injury, inhibiting NF-κB activation has been shown to reduce inflammation, enhance regeneration, and limit fibrosis (Acharyya *et al.*, 2007; Mittal *et al.*, 2010b; Mourkioti *et al.*, 2006; Thaloor *et al.*, 1999; Wang *et al.*, 2007), indicating that NF-κB signaling is instrumental in retarding muscle repair after severe injury. Multiple mechanisms have been elucidated in muscle disease models to explain how NF-κB-dependent signaling from individual cell types within the injured muscle environment contributes to muscle injury and impair regeneration. These mechanisms are discussed more comprehensively in the disease section of this review.

NF-κB has also been implicated in skeletal muscle ischemia reperfusion (I/R) injury. Reperfusion of skeletal muscle after a period of ischemia induces an acute inflammatory response, edema, tissue necrosis, and microvascular damage. Administration of agents that inhibit NF-κB activation have been reported to improve overall histological appearance of muscle, reduce edema, iNOS expression, apoptosis, and increase muscle blood flow in rodent I/R models (Andrade-Silva *et al.*, 2009; Lille *et al.*, 2001; Park *et al.*, 2007; Qi *et al.*, 2004). Additionally, reduced intercellular adhesion molecule-1 expression, inflammatory cell infiltration, neutrophil activation, and oxidative damage were also reported (Andrade-Silva *et al.*, 2009; Lille *et al.*, 2001; Park *et al.*, 2007). These studies clearly demonstrate that NF-κB promotes muscle damage after I/R, but do not provide mechanistic understanding of how the injury is mediated. Interestingly, biphasic activation of NF-κB was observed in rat muscles after I/R (Lille *et al.*, 2001). Activated NF-κB was reportedly diminished 3 h after I/R, but then elevated again 6 post I/R. The functional significance of this second phase of activation will need to be further explored to more fully understand the implications of activated NF-κB signaling in response to I/R.

Although acute injury models provide insight into understanding how skeletal muscle is capable of rebuilding itself, adaptations in response to exercise add an extra layer of complexity to the remodeling paradigm of injury and repair. Exercise induces both mechanical and metabolic stresses on skeletal muscles and the specific adaptations required to accommodate these stresses depends on the type and intensity of exercise performed. Strictly speaking, adaptations to strength training focus on the addition of contractile machinery within the muscles resulting in hypertrophy to accommodate lifting heavy loads, whereas adaptations to endurance exercise are focused on metabolic adaptations to accommodate the increased energy demands of prolonged exercise (Hawley, 2009). Besides adaptations that take place within myofibers, remodeling of the stroma occurs, which can include connective tissue remodeling (Chiquet *et al.*, 2009) and increased vascularization (Laughlin and Roseguini, 2008) to accommodate increased mechanical stress and metabolic demands, respectively. During novel bouts of damaging exercise, muscle injury along with an ensuing inflammatory response also occur, adding complexity to the microenvironment of exercised muscle. How NF-κB signaling regulates these adaptations is still being elucidated.

Increased NF-κB activation has been reported in rodent skeletal muscle after a novel bout of exercise entailing treadmill running (Brooks *et al.*, 2008; Gomez-Cabrera *et al.*, 2005; Ho *et al.*, 2005; Hollander *et al.*, 1999; Ji *et al.*, 2004; Liao *et al.*, 2010; Lima-Cabello *et al.*, 2010; Song *et al.*, 2006; Spangenburg *et al.*, 2006), *in situ* isometric contractions (Vasilaki *et al.*, 2006), and eccentric contractions of isolated muscle (Ho *et al.*, 2005), and after a novel bout of cycling in humans (Tantiwong *et al.*, 2010). ROS and reactive nitrogen species (NOS) produced by contracting muscle have been implicated in the exercise-induced activation of NF-κB (Brooks *et al.*, 2008;

Gomez-Cabrera *et al.*, 2005), and various investigations have suggested a beneficial role for this response in skeletal muscles. Regular exercise training results in less ROS production concurrent with diminished NF-κB activation in mouse muscles (Brooks *et al.*, 2008). It has been hypothesized that NF-κB regulates the expression of antioxidants, potentially accounting for this observation. Increased expression of antioxidant enzymes such as copper zinc superoxide dismutase (CuZn SOD; Hollander *et al.*, 1999), manganese (Mn) SOD (Gomez-Cabrera *et al.*, 2005; Hollander *et al.*, 1999), iNOS (Gomez-Cabrera *et al.*, 2005), and endothelial NOS (eNOS; Gomez-Cabrera *et al.*, 2005) have been reported after exercise in skeletal muscles in parallel with increased NF-κB activity. A recent study using chromatin immunoprecipitation assays reported that all three isoforms of NOS (iNOS, eNOS, and neuronal NOS (nNOS)) are direct transcriptional targets for p65 after acute exercise (Lima-Cabello *et al.*, 2010). In that same study, prior exercise training or administration of the NF-κB inhibitor, pyrrolidine dithiocarbamate (PDTC), reduced the binding of p65 to NOS genes. Further, administration of the antioxidant Allopurinol (Gomez-Cabrera *et al.*, 2005) or PDTC (Ji *et al.*, 2004) have been shown to reduce expression of antioxidants and NF-κB activation after exercise. Together, these studies suggest that one role for exercise-induced NF-κB activation may be to facilitate a protective response against future bouts of exercise-induced oxidative stress.

Beyond the role of oxidative stress responses, the relevance of NF-κB signaling in exercise-induced injury, repair, and adaptation is largely unexplored. Several studies have reported activation or lack thereof of NF-κB in different populations and in response to exercise training (Brooks *et al.*, 2008; Lima-Cabello *et al.*, 2010; Song *et al.*, 2006; Spangenburg *et al.*, 2006; Tantiwong *et al.*, 2010; Vasilaki *et al.*, 2006), but the mechanisms responsible for these variable responses are currently unknown. Interestingly, several investigations reported chronic NF-κB activation in the muscles of exercise-trained rodents (Brooks *et al.*, 2008; Song *et al.*, 2006; Spangenburg *et al.*, 2006). Although this elevation could be indicative of adaptations to protect against oxidative stress, chronic classical NF-κB activation induces muscle atrophy making this adaptation seem counterintuitive.

With so much still not understood regarding the conditions in which exercise activates NF-κB and the implications of that activation, studies aimed at determining the role of NF-κB in response to different types of exercise is vital to understanding when this signaling pathway plays beneficial and deleterious roles in the adaptation process. Because NF-κB activity is observed in various cell types during postnatal development and in muscle disease, its role in adaptation to exercise likely extends beyond protecting the muscle against oxidative stress. In fact, NF-κB activation has been reported in skeletal muscles of sedentary obese (Tantiwong *et al.*, 2010), aged patients (Buford *et al.*, 2010), and patients with chronic disease such as diabetes (Tantiwong *et al.*, 2010). Understanding how NF-κB regulates exercise adaptations in skeletal muscles of healthy individuals could serve as a guide for determining the appropriate approach when prescribing exercise for less healthy population.

8. NF-κB Involvement in Skeletal Muscle Disorders

Skeletal muscles are susceptible to a variety of different diseases. Although the etiology and pathological progression can vary greatly between each disease state, a commonality they all share is that the primary cause and/or secondary effects lead to the deregulation of normal signaling pathways. Chronic NF-κB activation has been reported in various states of atrophy/cachexia (Acharyya *et al.*, 2005; Hunter and Kandarian, 2004; Li *et al.*, 2008; Mourkioti *et al.*, 2006), in idiopathic inflammatory myopathies (Monici *et al.*, 2003; Yang *et al.*, 1998), rhabdomyosarcomas (RMS; Wang *et al.*, 2008), and muscular dystrophies (Acharyya *et al.*, 2007; Haslbeck *et al.*, 2005; Kumar and Boriek, 2003; Messina *et al.*, 2006a; Monici *et al.*, 2003). We will focus this portion of the chapter on states of atrophy, RMS, and DMD to provide an update to a previous review pertaining to NF-κB in muscle disease (Peterson and Guttridge, 2008).

8.1. Muscle atrophy

Muscle mass, although metabolically expensive, must be maintained to ensure the health and mobility of an organism. During periods of reduced loading or disuse, pathways that favor protein breakdown are upregulated while those that regulate synthesis are maintained or reduced resulting in net protein breakdown or catabolism (Glass, 2010; McCarthy and Esser, 2010). The discovery that chronic classical NF-κB activation caused atrophy in skeletal muscle (Cai *et al.*, 2004), whereas inhibition of this pathway prevented atrophy (Cai *et al.*, 2004; Hunter and Kandarian, 2004; Mourkioti *et al.*, 2006) provided insight into the normal regulatory role that NF-κB signaling plays in reducing physiologically unnecessary muscle mass as well as the pathological role that chronic NF-κB activation plays in disease-associated muscle atrophy or cachexia.

Two E3 ubiquitin ligases have been found to play an integral role in regulating muscle catabolism, atrogin-1/MAFbx, and MuRF1 (Bodine *et al.*, 2001; Gomes *et al.*, 2001). Classical NF-κB promotes atrophy through the regulation of MuRF1 (Cai *et al.*, 2004), and data support that cachectic factors such as TNFα, TNF-like weak inducer of apoptosis (TWEAK), proteolysis-inducing factor, and angiotensin II signal through NF-κB to mediate myofiber atrophy and activation of the ubiquitin proteasome pathway (Li and Reid, 2000; Li *et al.*, 2003; Mittal *et al.*, 2010a; Russell *et al.*, 2006; Whitehouse and Tisdale, 2003). Consistently, direct activation of the IKK complex through transient overexpression of a constitutively active form of either IKKβ or IKKα leads to fiber atrophy (Van Gammeren *et al.*, 2009).

In addition to NF-κB, the FoxO3/4 transcription factors have also been shown to promote atrophy in various states of catabolism or conditions resulting in decreased IGF-1 signaling (Glass, 2003). FoxO-induced muscle atrophy is mediated through binding of the E3 ubiquitin ligase genes (Sandri *et al.*, 2004; Stitt *et al.*, 2004) as well as genes promoting autophagosome formation (Mammucari *et al.*, 2007; Zhao *et al.*, 2007). Currently, it is unclear whether NF-κB and FoxO can synergize in the catabolic response, and if so whether such activity occurs via MuRF1 regulation or possibly by one or more genes regulating formation of the autophagosome. Nevertheless, solid advancements have been made in identifying catabolic factors that regulate NF-κB and FoxO. In one study, ROS were shown to increase during disuse atrophy and promote muscle mass loss through induced activities of NF-κB and FoxO3 (Dodd *et al.*, 2010). In contrast to these findings, direct treatment of hydrogen peroxide to C2C12 myotubes induced atrophy through p38 MAPK without affecting NF-κB and FoxO3 (McClung *et al.*, 2010), suggesting that other factors in addition to ROS are required in disuse atrophy to stimulate these transcription factors and their downstream effects to regulate loss of muscle mass. In addition, several recently identified factors such as heat shock proteins Hsp27 (Dodd *et al.*, 2009) and Hsp70 (Senf *et al.*, 2008) as well as the coactivator peroxisome proliferator-activated receptor gamma coactivator 1alpha (PGC-1α; Brault *et al.*, 2010), have been shown to restrict NF-κB and FoxO3 activities under physiological conditions, as loss of Hsp27/70 and PGC-1α expression during disuse or denervation-induced atrophy, respectively, led to concomitant upregulation of NF-κB and FoxO3.

8.2. Rhabdomyosarcoma

RMS is the most common soft tissue sarcoma of childhood and adolescence. It is thought to arise from poorly differentiated highly proliferative muscle progenitors cells. Categorization of RMS is based on histological and genetic signatures. Embryonal type of RMS (eRMS) is characterized by a loss of heterozygosity on the short arm of chromosome 11 (Gallego Melcon and Sanchez de Toledo Codina, 2007), and recapitulates phenotypical features of embryonic muscle (Qualman *et al.*, 1998). The more aggressive type of RMS, alveolar RMS (aRMS), is associated with loosely organized tumors with poor muscle differentiation and results in high mortality rate. One of the key features of aRMS is a reciprocal translocation between chromosome 2 and 13 or 1 and 13, resulting in PAX3-FOXO1, or PAX7-FOXO1 fusion genes, respectively (Linardic, 2008). Fusion of these genes can induce cellular proliferation and act as potent repressors of myogenic differentiation. Surgery and radiation followed by chemotherapy are currently the only therapeutic options for RMS, but characterization of the deregulated signaling pathways involved in RMS may be instrumental in providing the knowledge necessary to develop more targeted therapies. Gene expression profiling of RMS samples has thus been conducted, implicating many new culprits in the disease process (Huh and Skapek, 2010).

NF-κB was one of the signaling pathways found deregulated in RMS. Consistent with undifferentiated myoblast-like properties, primary murine and human RMS tissues display high levels of p65 and its active phosphorylated form as compared to adjacent normal tissue (De Bortoli *et al.*, 2007; Wang *et al.*, 2008). Such activity potentially mediates proliferation of RMS cells, since inhibiting p65 through MDM2 overexpression suppresses NF-κ B activity, reduces cellular growth rate, and increases apoptosis (Cheney *et al.*, 2008). In addition, the RMS differentiation process mirrors that of C2C12 myoblasts with regard to NF-κB signaling, namely greatly diminished DNA binding as cells became more differentiated (Lehtinen *et al.*, 1996). Mechanistically, NF-κB was recently reported to be required for regulation of the chemokine receptor CXCR7 (Tarnowski *et al.*, 2010), which earlier had been shown to regulate the metastatic potential of RMS cells (Grymula *et al.*, 2010). In our own studies, we found that in undifferentiated myoblasts, NF-κB induces YY1 expression. This transcriptional repressor in turn binds a conserved regulatory region upstream of the miR-29b/c cluster, recruiting the Polycomb group Ezh2 and the histone deacetylase HDAC1, and epigenetically silencing expression levels of this promyogenic miRNA (Wang *et al.*, 2008). Downregulated miR-29 levels were similarly observed in RMS samples following a miRNA profiling study (Subramanian *et al.*, 2008), further supporting the tumor suppressor function of this miRNA in RMS. Importantly, increasing miR-29 levels in RMS cells blocked YY1 translation, allowing myofibrillar gene expression and myogenic differentiation to ensue. This feedback loop between YY1 and miR-29 is under NF-κB control and functions to maintain an efficient balance between proliferation and differentiation, which if dysregulated may facilitate RMS development or progression (Wang *et al.*, 2008).

8.3. Duchenne muscular dystrophy

DMD is an X-linked recessive disease caused by a mutation in the dystrophin gene. Characterized by severe muscle weakness and dysfunction, this disorder leads to premature death in affected boys. In normal muscle, dystrophin assembles into a complex known as the dystrophin–glycoprotein complex (DGC), which provides a link between the cytoskeleton of muscle fibers and the ECM. This link functions to maintain muscle membrane stability (Lapidos *et al.*, 2004). In DMD, the loss of dystrophin protein expression destabilizes the DGC, rendering muscle membranes vulnerable to contraction-induced damage. Ultimately, membrane instability leads to a cascade of downstream effects including myofiber degeneration, chronic inflammation, impaired regeneration, and fibrosis (Blake *et al.*, 2002). To gain a better understanding into the mechanisms involved in DMD disease progression, the *mdx* mouse model is commonly used. Similar to DMD patients, the mutation in *mdx* mice results in a complete loss of dystrophin protein expression; however, the severity of disease is much milder in *mdx* mice than DMD patients. A more severe mouse model of DMD has also been developed in which the gene for dystrophin and its homolog utrophin have both been mutated. Disease progression in this mouse, referred to as a double knockout (*dko*), is greatly exacerbated compared to *mdx* mice (Deconinck *et al.*, 1997). However, the *dko* model is not as commonly used as *mdx*, likely due to the challenges that accelerated disease progression presents.

Following the first reports that activated NF-κB was detectable in muscles from DMD patients (Monici *et al.*, 2003) and *mdx* mice (Kumar and Boriek, 2003), several studies followed shortly thereafter indicating that NF-κB activation was indeed relevant to dystrophic disease progression (Acharyya *et al.*, 2007; Carlson *et al.*, 2005; Messina *et al.*, 2006a,b). Several mechanisms were also proposed for NF-κB-dependent disease progression (Acharyya *et al.*, 2007). Since those initial investigations, numerous studies confirming the requirement for NF-κB in dystrophic disease development have been published (see below) and chronic NF-κB activation is now recognized as a central component in the progression of dystrophic pathology.

NF-κB activation has proved to be a multifactorial signaling pathway in dystrophinopathy, functioning to promote inflammation and myofiber necrosis and impair muscle regeneration. Inhibition of the classical NF-κB pathway, either through general antioxidant administration or more targeted NF-κB inhibitors, ameliorates the pathology in *mdx* mice as measured by improved histological and/or functional parameters (Acharyya *et al.*, 2007; Carlson *et al.*, 2005; Messina *et al.*, 2006a,b, 2009; Pan *et al.*, 2008; Siegel *et al.*, 2009; Whitehead *et al.*, 2008). In addition to improvements in *mdx* mice, we recently showed that NF-κB inhibition also results in functional improvements in the diaphragm of *dko* mice (Peterson *et al.*, 2011).

Through immunolabeling of *mdx* limb muscles with an antibody specific for activated p65, NF-κB activation was observed localized to the nuclei of infiltrating immune cells and regenerating myofibers (Acharyya *et al.*, 2007). To dissect the individual contribution of NF-κB activation to *mdx* pathology in myeloid cells and myofibers, the cre-lox system was used to genetically ablate IKKβ signaling conditionally from each of these cell types. *Mdx* mice floxed for IKKβ were bred either to mice expressing cre recombinase on a myosin light chain (MLC) or lysozyme (Lys) specific promoter to conditionally delete IKKβ in muscle or myeloid cells, respectively (Acharyya *et al.*, 2007). Inhibiting IKKβ specifically from myeloid cells was found to have a protective effect on *mdx* muscles. Histopathology revealed greatly reduced myofiber necrosis, and gene expression analysis showed an attenuation of inflammatory cytokine gene expression. Consistent with these findings, inflammatory cytokine expression from *mdx* macrophages has since been found to be NF-κB-dependent (Chen *et al.*, 2005; Vidal *et al.*, 2008). Through Mac-1 receptor binding, *mdx* macrophages express cytokines in response to fibrinogen. Deposition of fibrinogen was detected in muscle sections of DMD patients and *mdx* mice, and macrophage activation by fibrinogen through an NF-κB dependent mechanism provides further understanding for how macrophages contribute to promoting a proinflammatory environment within dystrophic muscle (Vidal *et al.*, 2008). Together these studies demonstrate that myeloid-dependent NF-κB signaling contributes to dystrophic pathology through myofiber necrosis and inflammatory cytokine production.

While myeloid-specific IKKβ ablation prevented the activation of NF-κB in inflammatory cells, it did not influence their accumulation in dystrophic muscles (Acharyya *et al.*, 2007). Muscle-specific IKKβ was also insufficient to reduce macrophage accumulation in *mdx* muscles. This is in contrast to what occurs following more global NF-κB inhibition. Reduced NF-κB signaling accomplished genetically, using *mdx* mice heterozygous for p65 ($p65^{+/-}$) or pharmacologically, through administration of a peptide inhibitor of NF-κB (NBD peptide) effectively reduced muscle macrophage accumulation. These findings suggest that macrophage recruitment to dystrophic muscles is independent of NF-κB signaling in muscle and inflammatory cells, but dependent on an undetermined cellular source of NF-κB. The mechanism by which NF-κB becomes activated in myofibers also remains to be determined. Because myeloid-specific IKKβ depletion did not influence NF-κB activation in myofibers, this suggests that NF-κB activation in muscle cells is not dependent on NF-κB signaling from infiltrating inflammatory cells. However, because inflammatory cell infiltration was not blunted as a result of myeloid-specific IKKβ depletion, the possibility still remains that inflammatory cells could be signaling the activation of NF-κ B in myofibers through some, yet to be determined mechanism. We have observed increased macrophage concentrations in *mdx* muscle sections prior to the onset of necrosis and increased NF-κB DNA binding that occurs at 3 weeks of age in these mice (Peterson and Guttridge, unpublished observations) indicating that they are present within muscle when NF-κB becomes activated; however, macrophage depletion studies will need to be conducted to definitively determine whether they are activating NF-κB in myofibers.

Another potential mechanism by which NF-κB activation may promote myofiber necrosis has been suggested. Matrix metalloproteinase-9 (MMP-9) is a protease found in the ECM of skeletal muscle and is a known transcriptional target of NF-κB. In muscle, dynamic remodeling of the ECM is directed by MMP activity and excessive MMP activation can result in deleterious tissue remodeling and fibrosis. MMP activation has been demonstrated to contribute to myofiber necrosis in *mdx* mice by further destabilizing the myofiber membrane (Kumar and Bhatnagar, 2010; Li *et al.*, 2009a). Specifically, MMP-9 has been implicated in cleaving the DGC complex member β-dystroglycan (Hnia *et al.*, 2007). Downregulation of β-dystroglycan is evident in *mdx* muscles, and ablation of MMP-9 (Li *et al.*, 2009a) or administration of antioxidants that attenuate NF-κB activation (Hnia *et al.*, 2007, 2008; Whitehead *et al.*, 2008) result in increased expression of β-dystroglycan. Stabilization of β-dystroglycan has also been shown to coincide with increased expression of the DGC members utrophin (Hnia *et al.*, 2008; Whitehead *et al.*, 2008) and n-NOS (Li *et al.*, 2009a), while decreasing the levels of the dystrophin/utrophin competitor protein caveolin-3 (Li *et al.*, 2009a; Whitehead *et al.*, 2008) that results in stabilization of the myofiber membrane. Although NF-κB has been shown to regulate MMP-9 expression in muscle cells (Li *et al.*, 2009b), and inhibition of MMP-9 is capable of stabilizing β-dystroglycan expression (Li *et al.*, 2009a), whether NF-κB is solely regulating this process in *mdx* muscle has yet to be deciphered. Thus far, only treatments that nonspecifically inhibit the NF-κB signaling pathway have been shown to stabilize the DGC and thus directly prevent myofiber necrosis (Hnia *et al.*, 2007, 2008; Whitehead *et al.*, 2008). Since this effect has yet to be reported in studies using genetic models or specific inhibitors of NF-κB, this may indicate that although NF-κB may be involved in MMP-9 inhibition, it may not be the only regulator of this process and perhaps addition of antioxidants provide cumulative protection against membrane destruction independent of NF-κB signaling.

NF-κB activation in myofibers functions to impair regeneration in dystrophic muscles. When IKKβ signaling was ablated specifically in muscle cells, enhanced regeneration as measured by increased number of embryonic myosin heavy expressing myofibers, increased number of central nucleated myofibers (Acharyya *et al.*, 2007), and larger cross-sectional area of central nucleated fibers was reported (Acharyya *et al.*, 2007; Mourkioti *et al.*, 2006). These data are consistent with our understanding of NF-κB as a negative regulator of muscle differentiation. A recent report revealed that miR-29, located on the miR-29b2/c loci was downregulated in *mdx* muscles and as well as cultured human DMD myotubes (Cacchiarelli *et al.*, 2010; Eisenberg *et al.*, 2007). We previously reported in cultured myoblasts and RMS cells that activated NF-κB-YY1 epigenetically silenced miR-29 via the same b2/c loci, thereby contributing to a block in muscle differentiation (Wang *et al.*, 2008). The repression of miR-29 in dystrophic muscle suggests that this mechanism may also be relevant in limiting regeneration in DMD. Interestingly, overexpression of miR-29 from the miR-29a/b1 loci ameliorated *mdx* pathology, not by rescuing regeneration, but instead by reducing collagen expression and overall muscle fibrosis (Cacchiarelli *et al.*, 2010). Although these miRNAs contain identical seed sequences, and in theory are capable of targeting common gene targets, their expression is mediated via separate chromosomes. While the miR-29a/b1 locus, located on human chromosome 7, was recently described to be repressed in acute myeloid leukemia by an Sp1/NF-κB DNA binding complex that recruits HDAC-1 to silence miR-29 expression (Liu *et al.*, 2010; Wang *et al.*, 2008), a second miR-29b2/c locus localized to human chromosome 1 is negatively regulated, at least in undifferentiated myoblasts and RMS cells, by a YY1 and Ezh2 containing Polycomb complex (Liu *et al.*, 2010; Wang *et al.*, 2008). Since NF-κB is common to both repressor mechanisms, it is possible that chronic activation of NF-κB in DMD contributes to the downregulation of miR-29 and presumably fibrosis due to increasing levels of collagen protein.

Recent findings also revealed that NF-κB inhibition in *mdx* muscle effectively reduced the proinflammatory cytokine, TNFα, which corresponded to a concomitant enhanced population of muscle satellite cells (Acharyya *et al.*, 2007). This suggested that chronic TNF signaling might impact DMD pathology by inhibiting muscle regeneration through the negative regulation of satellite cells. This regulation was shown to be due to inhibition of Notch-1 receptor transcription as a result of chronic TNF signaling (Acharyya *et al.*, 2010). Repression of Notch-1 occurred by a TNF-dependent recruitment of the Polycomb group protein Ezh2 resulting in K27 trimethylation on histone H3. Ezh2 binding was localized to a small CpG island within the Notch-1 promoter resulting in sequential recruitment of the DNA methyltransferase, Dnmt-3b, which in turn caused site-specific DNA methylation contributing to further transcriptional repression of Notch-1 (Acharyya *et al.*, 2010).

Taken together, we envision NF-κB functioning in DMD by the following model (Fig. 4.3). As a result of dystrophin mutations leading to the loss of dystrophin protein and DGC integrity, sarcolemma damage leads to an innate immune response dominated by the presence of macrophages. NF-κB is activated in myeloid cells promoting a proinflammatory environment through stimulated expression and secretion of a number of cytokines and chemokines. This proinflammatory condition is not responsible for recruitment of macrophages, but rather for contributing to myofiber necrosis. By a mechanism that again is independent of NF-κB signaling in myeloid cells, NF-κB is activated in myofibers, and contributes to the production of TNFα. Accumulating levels of TNFα signal to satellite cells and cause the epigenetic silencing of the Notch-1 receptor through sequential recruitment of the Polycomb and DNA methyltransferase complexes.

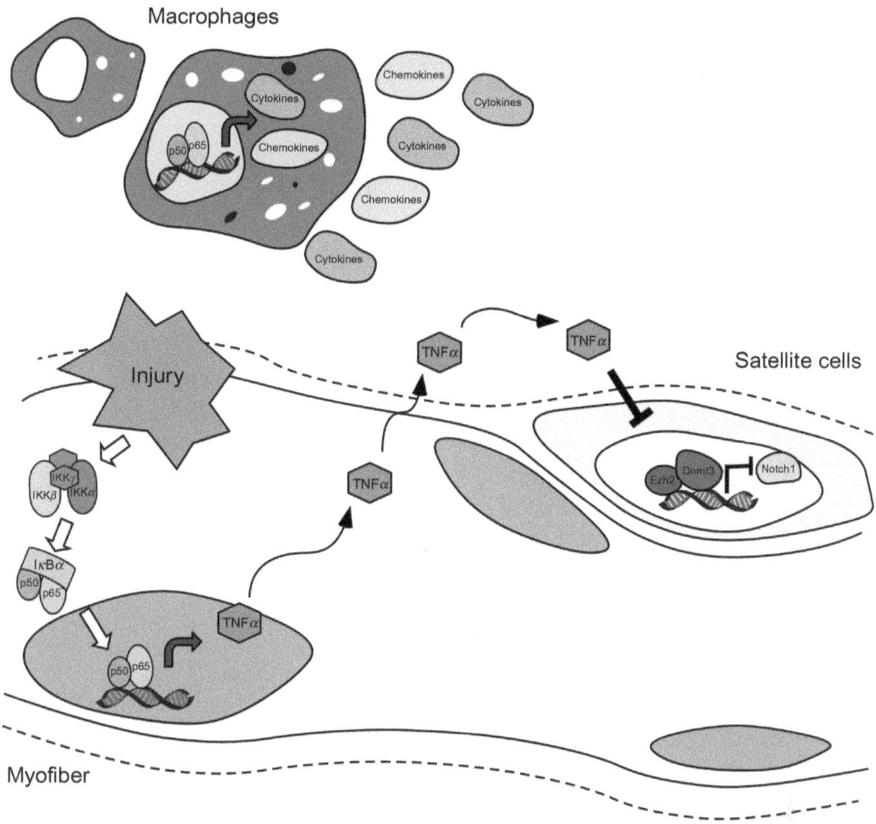

Figure 4.3 Modeling how NF-κB functions in DMD. The figure depicts macrophages recruited to dystrophic muscles that contain activated classical NF-κB signaling. This signaling functions to stimulate the expression and secretion of various cytokines and chemokines that participate in myofiber necrosis. By an unknown mechanism, classical NF-κB signaling is also activated in the myofiber compartment leading to the accumulated production of TNFα. The inflammatory cytokine in turn possesses antidifferentiation activity on proximal satellite cells by suppressing production of the Notch-1 receptor, resulting in repressed satellite cell activation. TNF-mediated repression of Notch-1 occurs through an epigenetic mechanism whereby Ezh2 and Dnmt3b are sequentially recruited to the Notch-1 promoter to induce histone and DNA methylation, respectively. This regulation results in limiting Notch-1 transcription as well as skeletal muscle regeneration. (See Color Insert.)

Given that Notch-1 is considered a key determinant of satellite cell activation (Conboy *et al.*, 2003), its absence would be predicted to lead to an NF-κB-dependent inhibition of skeletal muscle regeneration. This model proposes that NF-κB functions in at least two cellular compartments, immune cells and myofibers. As discussed further below, these cellular associated functions of NF-κB make the signaling pathway an attractive therapeutic target for DMD treatment, since dampening the immune response and boosting muscle regeneration would be expected to have a beneficial effect on disease severity and progression.

9. NF-κB Therapeutics

NF-κB signaling plays an integral role in regulating skeletal muscle homeostasis, but the deleterious effects that result from its chronic activation make NF-κB inhibition an attractive candidate for treatment of skeletal muscle disorders. Genetic and pharmacological studies have provided proof of concept that inhibiting classical NF-κB activation has the potential to ameliorate DMD pathology and muscle atrophy. The challenge now becomes to translate the knowledge we have gained into therapies that are clinically beneficial. These emerging therapies include gene- and cell-based strategies and pharmacological inhibitors, all possessing their own set of strengths and limitations.

9.1. Gene therapy

Gene therapy strategies are currently being pursued as a corrective measure to replace dystrophin in DMD. This same system may also be useful for dampening muscle NF-κB activation, potentially making the local muscle environment more hospitable for dystrophin gene delivery and combating inflammation and fibrosis that severely afflicts DMD patients. Recently, a study was conducted to explore the efficacy of using an adeno-associated viral (AAV) vector to inhibit NF-κB activation and improve dystrophic pathology in *mdx* mice (Tang *et al.*, 2010). Delivery of a dominant negative form of IKKβ (AAV-IKKβ-dn) was sufficient to blunt NF-κB activation in 11-month-old *mdx* mice. One month after AAV-IKKβ-dn was delivered to these older mice, reduced myofiber necrosis and enhanced muscle regeneration was observed. Because muscle regeneration is insufficient in DMD patients and loss of skeletal muscle is concurrent with fibrosis, it is encouraging that NF-κB was able to stimulate regeneration in older mice, lending credence to the possibility that NF-κB inhibition may be capable of stimulating muscle regeneration in DMD patients.

9.2. Stem cell therapy

Stem cell therapy is an attractive candidate for treating DMD. If successful, delivery of muscle stem cells that contain a functional dystrophin gene to DMD patients would effectively repair damaged muscle. However, effective transplantation of these cells has proved difficult, mainly attributed to poor engraftment (Fan *et al.*, 1996; Skuk *et al.*, 2007). Because NF-κB inhibition enhances muscle regeneration (Acharyya *et al.*, 2007; Wang *et al.*, 2007), the possibility of dampening NF-κB activation in muscle stem cells for transplantation is being explored. One approach has been to use transplantation of $p65^{+/-}$ muscle-derived stem cells (MDSC) into *mdx* muscles. Conclusions from these experiments have resulted in more than double the number of dystrophin positive muscle fibers when compared to transplantation with $p65^{+/+}$ MDSCs (Huard *et al.*, unpublished observations). Additionally, reduced myonecrosis and inflammatory cell infiltration was observed in muscles transplanted from $p65^{+/-}$ MDSCs. The idea that inhibition of NF-κB in a small number of transplanted cells could have such a profound effect on a whole muscle suggests that pursuit of this therapy or perhaps administering an NF-κB inhibitor to patients along with stem cell transplantation may be a way to increase the success of cell-based therapies.

9.3. Pharmacological therapy

Pharmacological inhibitors of NF-κB are an appealing option for treating muscle disease because they do not require administration of cells or viruses that are subject to host rejection and are not targeted to one particular muscle. This makes them capable of providing benefit to the limb muscles, diaphragm, and heart at the same time. In the case of DMD, specific inhibitors of the NF-κB pathway have not been developed yet for patient trials. However, there are indications derived from nonspecific NF-κB inhibitors that targeting this pathway will provide benefit to DMD patients.

Glucocorticoids, such as prednisone, are currently the most commonly prescribed treatment for DMD patients and these drugs exert their beneficial effects partially by inhibiting NF-κB activation (Du *et al.*, 2000). Variable results have been reported on the effectiveness of glucocorticoid treatment in DMD patients, but overall the consensus is that patients able to tolerate the side effects benefit from this treatment. Reported benefits include improved muscle strength, delayed time until wheelchair and ventilator use is required, reduced scoliosis, and delayed cardiomyopathy (Biggar *et al.*, 2006; Manzur *et al.*, 2008; Markham *et al.*, 2008). However, side effects for this class of steroids are also common, which can limit the treatment benefit and reduce patient quality of life. Thus, additional targeted pharmacological agents that provide all the benefits of NF-κB inhibition without the side effects of steroid use may provide a more effective means for treating DMD patients.

Curcumin is a natural agent derived from the herb turmeric (Goel *et al.*, 2008). Curcumin possesses NF-κB inhibiting properties (Jobin *et al.*, 1999), but the literature is mixed concerning its beneficial effects on skeletal muscle. Several studies have shown that curcumin is capable of blunting muscle loss in atrophy conditions (Jin and Li, 2007), and also promoting regeneration after acute (Thaloor *et al.*, 1999) or chronic (Pan *et al.*, 2008) muscle injury. Interestingly, the ability of curcumin to prevent atrogin-1/MAFbx induced muscle loss was independent of NF-κB (Jin and Li, 2007), suggesting that curcumin may improve muscle pathology through NF-κB-dependent and independent mechanisms. Although the efficacious nature of curcumin may be unclear, if muscle preservation can be achieved in atrophy and/or DMD patients through curcumin administration, the favorable safety profile and availability of this agent make it a viable therapeutic candidate.

PDTC has been shown to specifically inhibit NF-κB activation by stabilizing IκBα, but also acts on other pathways as well. Although this agent has been instrumental in determining the efficacy of NF-κB inhibition in *mdx* mice (Carlson *et al.*, 2005; Messina *et al.*, 2006b), and has demonstrated a role for NF-κB in I/R injury (Qi *et al.*, 2004), PDTC, in its current formulation and efficacious dose may not be suitable for clinical use due to toxicity concerns (Calviello *et al.*, 2005; Kim *et al.*, 1999).

The NBD peptide has also demonstrated therapeutic potential for treating not only dystrophic muscles but other diseased tissues as well where chronic NF-κB activity is thought to be an underlying mechanism of the pathology (Baima *et al.*, 2010; Chapoval *et al.*, 2007; Dasgupta *et al.*, 2004; di Meglio *et al.*, 2005; Ghosh *et al.*, 2007; May *et al.*, 2000; Shibata *et al.*, 2007; Tas *et al.*, 2006). Use of this peptide is in the early stages of exploration, but studies conducted thus far have indicated that NBD has the potential to be translated to clinical use. To date, toxicity has not been associated with administration in any rodent model (Baima *et al.*, 2010; Chapoval *et al.*, 2007; Dasgupta *et al.*, 2004; di Meglio *et al.*, 2005; Ghosh *et al.*, 2007; May *et al.*, 2000; Shibata *et al.*, 2007; Tas *et al.*, 2006). Additionally, NBD has been shown to circumvent one of the biggest concerns with NF-κB inhibition, which is its ability to specifically inhibit inflammatory signaling through IKKβ without influencing basal NF-κB functions (Baima *et al.*, 2010; May *et al.*, 2000). Because NF-κB is important for normal cellular activities such as survival, inhibiting all signaling through this pathway would not likely be a feasible option therapeutically. In *mdx* mice, NBD administration improves histopathology as measured by reduced necrosis and inflammatory cell infiltration and increased regeneration (Acharyya *et al.*, 2007). Histological improvements translated into significant functional rescue as well (Acharyya *et al.*, 2007). In addition, NBD treatment of *mdx* mice was shown to improve whole body function as well as function of isolated diaphragm muscles (Peterson *et al.*, 2011). Current advances in the formulation of NBD have now also proven efficacious with the development of a GLP grade compound soluble in water (Peterson *et al.*, 2011). Although these results are encouraging, development of the NBD peptide into a therapeutic presents several challenges. In general, peptide-based therapies can be problematic due to limited stability and expensive production. Additionally, whether NBD peptide administration will prove effective in treating DMD patients remains to be determined.

10. Concluding Remarks

Several key points were discussed in this review pertaining to NF-κB as an integral signaling pathway involved in many facets of skeletal muscle regulation. During myogenesis, classical NF-κB activation promotes myoblast proliferation, effectively preventing premature myogenic differentiation, whereas alternative signaling becomes activated upon myotube differentiation and functions to support myotube homeostasis. During postnatal growth, classical NF-κB activation in fibroblasts aids the process of myoblast fusion, thus encouraging the maturation of muscles during this early stage of growth development. Additionally, classical NF-κB signaling in postsynaptic nuclei may be directing the clustering of AChRs at NMJs. When additional muscle is needed or excess muscle must be removed, classical NF-κB signaling is responsive to these perturbations and aids in the hypertrophy or atrophy process as required. Chronic classical NF-κB activation induces skeletal muscle atrophy, exacerbates I/R-associated muscle damage, supports tumor development in RMS, and perpetuates dystrophic pathology by promoting inflammation, myonecrosis, and fibrosis, and impairing muscle regeneration. Altogether, these functions lie in stark contrast to one another, but sum up both the beneficial and deleterious ways we understand NF-κB activation to dichotomously support normal skeletal muscle function and promote muscle disease. Our understanding of how NF-κB regulates skeletal muscle processes has increased greatly over the last few years, but much has yet to be learned. As more mechanisms are uncovered, we will more fully appreciate the context in which NF-κB signaling is beneficial or deleterious to muscle function, which should make us better equipped to combat skeletal muscle diseases.

ACKNOWLEDGMENTS

Support provided by NIH grants R01 AR052787 and U01 NS058451 to D. C. G. and an MDA award to J. M. P.

REFERENCES

Acharyya, S., *et al.* (2005). Dystrophin glycoprotein complex dysfunction: A regulatory link between muscular dystrophy and cancer cachexia. *Cancer Cell* **8,** 421–432.

Acharyya, S., *et al.* (2007). Interplay of IKK/NF-kappaB signaling in macrophages and myofibers promotes muscle degeneration in Duchenne muscular dystrophy. *J. Clin. Invest.* **117,** 889–901.

Acharyya, S., *et al.* (2010). TNF inhibits notch-1 in skeletal muscle cells by Ezh2 and DNA methylation mediated repression: Implications in Duchenne muscular dystrophy. *PLoS One* **5,** e12479.

Andrade-Silva, A. R., *et al.* (2009). Effect of NFkappaB inhibition by CAPE on skeletal muscle ischemia-reperfusion injury. *J. Surg. Res.* **153,** 254–262.

Anest, V., *et al.* (2003). A nucleosomal function for IkappaB kinase-alpha in NF-kappaB-dependent gene expression. *Nature* **423,** 659–663.

Baima, E. T., *et al.* (2010). Novel insights into the cellular mechanisms of the anti-inflammatory effects of NF-kappaB essential modulator binding domain peptides. *J. Biol. Chem.* **285,** 13498–13506.

Bakkar, N., and Guttridge, D. C. (2010). NF-kappaB signaling: A tale of two pathways in skeletal myogenesis. *Physiol. Rev.* **90,** 495–511.

Bakkar, N., *et al.* (2008). IKK/NF-kappaB regulates skeletal myogenesis via a signaling switch to inhibit differentiation and promote mitochondrial biogenesis. *J. Cell Biol.* **180,** 787–802.

Basak, S., *et al.* (2007). A fourth IkappaB protein within the NF-kappaB signaling module. *Cell* **128,** 369–381.

Bergstrom, D. A., *et al.* (2002). Promoter-specific regulation of MyoD binding and signal transduction cooperate to pattern gene expression. *Mol. Cell* **9,** 587–600.

Bhatnagar, S., *et al.* (2010). Transforming growth factor-beta-activated kinase 1 is an essential regulator of myogenic differentiation. *J. Biol. Chem.* **285,** 6401–6411.

Biggar, W. D., *et al.* (2006). Long-term benefits of deflazacort treatment for boys with Duchenne muscular dystrophy in their second decade. *Neuromuscul. Disord.* **16,** 249–255.

Blake, D. J., *et al.* (2002). Function and genetics of dystrophin and dystrophin-related proteins in muscle. *Physiol. Rev.* **82,** 291–329.

Bodine, S. C., *et al.* (2001). Identification of ubiquitin ligases required for skeletal muscle atrophy. *Science* **294,** 1704–1708.

Bonizzi, G., *et al.* (2004). Activation of IKKalpha target genes depends on recognition of specific kappaB binding sites by RelB:p52 dimers. *EMBO J.* **23,** 4202–4210.

Boone, D. L., *et al.* (2004). The ubiquitin-modifying enzyme A20 is required for termination of Toll-like receptor responses. *Nat. Immunol.* **5,** 1052–1060.

Borycki, A. G., and Emerson, C. P., Jr. (2000). Multiple tissue interactions and signal transduction pathways control somite myogenesis. *Curr. Top. Dev. Biol.* **48,** 165–224.

Brault, J. J., *et al.* (2010). Peroxisome proliferator-activated receptor gamma coactivator 1alpha or 1beta overexpression inhibits muscle protein degradation, induction of ubiquitin ligases, and disuse atrophy. *J. Biol. Chem.* **285,** 19460–19471.

Brooks, S. V., *et al.* (2008). Repeated bouts of aerobic exercise lead to reductions in skeletal muscle free radical generation and nuclear factor kappaB activation. *J. Physiol.* **586,** 3979–3990.

Buckingham, M. (2001). Skeletal muscle formation in vertebrates. *Curr. Opin. Genet. Dev.* **11,** 440–448.

Buford, T. W., *et al.* (2010). Effects of age and sedentary lifestyle on skeletal muscle NF-kappaB signaling in men. *J. Gerontol. A Biol. Sci. Med. Sci.* **65,** 532–537.

Cacchiarelli, D., *et al.* (2010). MicroRNAs involved in molecular circuitries relevant for the Duchenne muscular dystrophy pathogenesis are controlled by the dystrophin/nNOS pathway. *Cell Metab.* **12,** 341–351.

Cai, D., *et al.* (2004). IKKbeta/NF-kB activation causes severe muscle wasting in mice. *Cell* **119,** 285–298.

Calviello, G., *et al.* (2005). Repeated exposure to pyrrolidine-dithiocarbamate induces peripheral nerve alterations in rats. *Toxicol. Lett.* **158,** 61–71.

Cao, Y., *et al.* (2001). IKKalpha provides an essential link between RANK signaling and cyclin D1 expression during mammary gland development. *Cell* **107,** 763–775.

Cao, Y., *et al.* (2010). Genome-wide MyoD binding in skeletal muscle cells: A potential for broad cellular reprogramming. *Dev. Cell* **18,** 662–674.

Caretti, G., *et al.* (2004). The Polycomb Ezh2 methyltransferase regulates muscle gene expression and skeletal muscle differentiation. *Genes Dev.* **18,** 2627–2638.

Carlson, C. G., *et al.* (2005). Chronic treatment with agents that stabilize cytosolic IkappaB-alpha enhances survival and improves resting membrane potential in MDX muscle fibers subjected to chronic passive stretch. *Neurobiol. Dis.* **20,** 719–730.

Catani, M. V., *et al.* (2004). Nuclear factor kappaB and activating protein 1 are involved in differentiation-related resistance to oxidative stress in skeletal muscle cells. *Free Radic. Biol. Med.* **37,** 1024–1036.

Chapoval, S. P., *et al.* (2007). Inhibition of NF-kappaB activation reduces the tissue effects of transgenic IL-13. *J. Immunol.* **179,** 7030–7041.

Chen, G., and Goeddel, D. V. (2002). TNF-R1 signaling: A beautiful pathway. *Science* **296,** 1634–1635.

Chen, Y. W., *et al.* (2005). Early onset of inflammation and later involvement of TGFbeta in Duchenne muscular dystrophy. *Neurology* **65,** 826–834.

Cheney, M. D., *et al.* (2008). MDM2 displays differential activities dependent upon the activation status of NFkappaB. *Cancer Biol. Ther.* **7,** 38–44.

Chiquet, M., *et al.* (2009). From mechanotransduction to extracellular matrix gene expression in fibroblasts. *Biochim. Biophys. Acta* **1793,** 911–920.

Conboy, I. M., *et al.* (2003). Notch-mediated restoration of regenerative potential to aged muscle. *Science* **302,** 1575–1577.

Creus, K. K., *et al.* (2009). Idiopathic inflammatory myopathies and the classical NF-kappaB complex: current insights and implications for therapy. *Autoimmun. Rev.* **8,** 627–631.

Dahlman, J. M., *et al.* (2009). The RelA/p65 subunit of NF-kappaB specifically regulates cyclin D1 protein stability: Implications for cell cycle withdrawal and skeletal myogenesis. *J. Cell. Biochem.* **106,** 42–51.

Dahlman, J. M., *et al.* (2010). NF-kappaB functions in stromal fibroblasts to regulate early postnatal muscle development. *J. Biol. Chem.* **285,** 5479–5487.

Dasgupta, S., *et al.* (2004). Antineuroinflammatory effect of NF-kappaB essential modifier-binding domain peptides in the adoptive transfer model of experimental allergic encephalomyelitis. *J. Immunol.* **173,** 1344–1354.

De Bortoli, M., *et al.* (2007). Patched haploinsufficient mouse rhabdomyosarcoma overexpress secreted phosphoprotein 1 and matrix metalloproteinases. *Eur. J. Cancer* **43,** 1308–1317.

Deconinck, A. E., *et al.* (1997). Utrophin-dystrophin-deficient mice as a model for Duchenne muscular dystrophy. *Cell* **90,** 717–727.

di Meglio, P., *et al.* (2005). Amelioration of acute inflammation by systemic administration of a cell-permeable peptide inhibitor of NF-kappaB activation. *Arthritis Rheum.* **52,** 951–958.

Dodd, S. L., *et al.* (2009). Hsp27 inhibits IKKbeta-induced NF-kappaB activity and skeletal muscle atrophy. *FASEB J.* **23,** 3415–3423.

Dodd, S. L., *et al.* (2010). Ros-mediated activation of NF-kappaB and Foxo during muscle disuse. *Muscle Nerve* **41,** 110–113.

Dogra, C., *et al.* (2006). Tumor necrosis factor-like weak inducer of apoptosis inhibits skeletal myogenesis through sustained activation of nuclear factor-kappaB and degradation of MyoD protein. *J. Biol. Chem.* **281,** 10327–10336.

Du, J., *et al.* (2000). Glucocorticoids induce proteasome C3 subunit expression in L6 muscle cells by opposing the suppression of its transcription by NF-kappa B. *J. Biol. Chem.* **275,** 19661–19666.

Ea, C. K., *et al.* (2006). Activation of IKK by TNFalpha requires site-specific ubiquitination of RIP1 and polyubiquitin binding by NEMO. *Mol. Cell* **22,** 245–257.

Eisenberg, I., *et al.* (2007). Distinctive patterns of microRNA expression in primary muscular disorders. *Proc. Natl. Acad. Sci. USA* **104,** 17016–17021.

Fan, Y., *et al.* (1996). Rapid death of injected myoblasts in myoblast transfer therapy. *Muscle Nerve* **19,** 853–860.
Feuerhake, F., *et al.* (2005). NFkappaB activity, function, and target-gene signatures in primary mediastinal large B-cell lymphoma and diffuse large B-cell lymphoma subtypes. *Blood* **106,** 1392–1399.
Frazier, D. P., *et al.* (2007). PKC-alpha and TAK-1 are intermediates in the activation of c-Jun NH2-terminal kinase by hypoxia-reoxygenation. *Am. J. Physiol. Heart Circ. Physiol.* **292,** H1675–H1684.
Gallego Melcon, S., and Sanchez de Toledo Codina, J. (2007). Molecular biology of rhabdomyosarcoma. *Clin. Transl. Oncol.* **9,** 415–419.
Ghosh, A., *et al.* (2007). Selective inhibition of NF-kappaB activation prevents dopaminergic neuronal loss in a mouse model of Parkinson's disease. *Proc. Natl. Acad. Sci. USA* **104,** 18754–18759.
Glass, D. J. (2003). Signalling pathways that mediate skeletal muscle hypertrophy and atrophy. *Nat. Cell Biol.* **5,** 87–90.
Glass, D. J. (2010). Signaling pathways perturbing muscle mass. *Curr. Opin. Clin. Nutr. Metab. Care* **13,** 225–229.
Goel, A., *et al.* (2008). Curcumin as "Curecumin": From kitchen to clinic. *Biochem. Pharmacol.* **75,** 787–809.
Gomes, M. D., *et al.* (2001). Atrogin-1, a muscle-specific F-box protein highly expressed during muscle atrophy. *Proc. Natl. Acad. Sci. USA* **98,** 14440–14445.
Gomez-Cabrera, M. C., *et al.* (2005). Decreasing xanthine oxidase-mediated oxidative stress prevents useful cellular adaptations to exercise in rats. *J. Physiol.* **567,** 113–120.
Grymula, K., *et al.* (2010). Overlapping and distinct role of CXCR7-SDF-1/ITAC and CXCR4-SDF-1 axes in regulating metastatic behavior of human rhabdomyosarcomas. *Int. J. Cancer* **127,** 2554–2568.
Guo, Z., *et al.* (2009). An inhibitory role of p53 via NF-kappaB element on the cyclin D1 gene under heat shock. *Biochim. Biophys. Acta* **1789,** 758–762.
Guttridge, D. C., *et al.* (1999). NF-kappaB controls cell growth and differentiation through transcriptional regulation of cyclin D1. *Mol. Cell. Biol.* **19,** 5785–5799.
Guttridge, D. C., *et al.* (2000). NF-kB-induced loss of MyoD messenger RNA: Possible role in muscle decay and cachexia. *Science* **289,** 2363–2366.
Hacker, H., and Karin, M. (2006). Regulation and function of IKK and IKK-related kinases. *Sci. STKE* **357,** re13.
Haskill, S., *et al.* (1991). Characterization of an immediate-early gene induced in adherent monocytes that encodes I kappa B-like activity. *Cell* **65,** 1281–1289.
Haslbeck, K. M., *et al.* (2005). The RAGE pathway in inflammatory myopathies and limb girdle muscular dystrophy. *Acta. Neuropathol.* **110,** 247–254.
Hawley, J. A. (2009). Molecular responses to strength and endurance training: Are they incompatible? *Appl. Physiol. Nutr. Metab.* **34,** 355–361.
Hayden, M. S., and Ghosh, S. (2004). Signaling to NF-kappaB. *Genes Dev.* **18,** 2195–2224.
Hayden, M. S., and Ghosh, S. (2008). Shared principles in NF-kappaB signaling. *Cell* **132,** 344–362.
Hertlein, E., *et al.* (2005). RelA/p65 regulation of IkappaBbeta. *Mol. Cell. Biol.* **25,** 4956–4968.
Hinz, M., *et al.* (1999). NF-kB function in growth control: Regulation of cyclin D1 expression and G0/G1-to-S-phase transition. *Mol. Cell. Biol.* **19,** 2690–2698.
Hnia, K., *et al.* (2007). Modulation of p38 mitogen-activated protein kinase cascade and metalloproteinase activity in diaphragm muscle in response to free radical scavenger administration in dystrophin-deficient Mdx mice. *Am. J. Pathol.* **170,** 633–643.

Hnia, K., *et al.* (2008). L-arginine decreases inflammation and modulates the nuclear factor-kappaB/matrix metalloproteinase cascade in mdx muscle fibers. *Am. J. Pathol.* **172,** 1509–1519.
Ho, R. C., *et al.* (2005). Regulation of IkappaB kinase and NF-kappaB in contracting adult rat skeletal muscle. *Am. J. Physiol. Cell Physiol.* **289,** C794–C801.
Hoberg, J. E., *et al.* (2004). SMRT derepression by the IkappaB kinase alpha: A prerequisite to NF-kappaB transcription and survival. *Mol. Cell* **16,** 245–255.
Hollander, J., *et al.* (1999). Superoxide dismutase gene expression in skeletal muscle: Fiber-specific adaptation to endurance training. *Am. J. Physiol.* **277,** R856–R862.
Huang, T. T., *et al.* (2000). A nuclear export signal in the N-terminal regulatory domain of IkappaBalpha controls cytoplasmic localization of inactive NF-kappaB/IkappaBalpha complexes. *Proc. Natl. Acad. Sci. USA* **97,** 1014–1019.
Huang, W. C., *et al.* (2007). Phosphorylation of CBP by IKKalpha promotes cell growth by switching the binding preference of CBP from p53 to NF-kappaB. *Mol. Cell* **26,** 75–87.
Huh, W. W., and Skapek, S. X. (2010). Childhood rhabdomyosarcoma: New insight on biology and treatment. *Curr. Oncol. Rep.* **12,** 402–410.
Hunter, R. B., and Kandarian, S. C. (2004). Disruption of either the Nfkb1 or the Bcl3 gene inhibits skeletal muscle atrophy. *J. Clin. Invest.* **114,** 1504–1511.
Ji, L. L., *et al.* (2004). Acute exercise activates nuclear factor (NF)-kappaB signaling pathway in rat skeletal muscle. *FASEB J.* **18,** 1499–1506.
Jin, B., and Li, Y. P. (2007). Curcumin prevents lipopolysaccharide-induced atrogin-1/MAFbx upregulation and muscle mass loss. *J. Cell. Biochem.* **100,** 960–969.
Jobin, C., *et al.* (1999). Curcumin blocks cytokine-mediated NF-kappa B activation and proinflammatory gene expression by inhibiting inhibitory factor I-kappa B kinase activity. *J. Immunol.* **163,** 3474–3483.
Kanayama, A., *et al.* (2004). TAB2 and TAB3 activate the NF-kappaB pathway through binding to polyubiquitin chains. *Mol. Cell* **15,** 535–548.
Kim, C. H., *et al.* (1999). Pyrrolidine dithiocarbamate induces bovine cerebral endothelial cell death by increasing the intracellular zinc level. *J. Neurochem.* **72,** 1586–1592.
Krikos, A., *et al.* (1992). Transcriptional activation of the tumor necrosis factor alpha-inducible zinc finger protein, A20, is mediated by kappa B elements. *J. Biol. Chem.* **267,** 17971–17976.
Kumar, A., and Bhatnagar, S. (2010). Matrix metalloproteinase inhibitor batimastat alleviates pathology and improves skeletal muscle function in dystrophin-deficient mdx mice. *Am. J. Pathol.* **177,** 248–260.
Kumar, A., and Boriek, A. M. (2003). Mechanical stress activates the nuclear factor-kappaB pathway in skeletal muscle fibers: A possible role in Duchenne muscular dystrophy. *FASEB J.* **17,** 386–396.
Langen, R. C., *et al.* (2004). Tumor necrosis factor-alpha inhibits myogenic differentiation through MyoD protein destabilization. *FASEB J.* **18,** 227–237.
Lapidos, K. A., *et al.* (2004). The dystrophin glycoprotein complex: Signaling strength and integrity for the sarcolemma. *Circ. Res.* **94,** 1023–1031.
Laughlin, M. H., and Roseguini, B. (2008). Mechanisms for exercise training-induced increases in skeletal muscle blood flow capacity: Differences with interval sprint training versus aerobic endurance training. *J. Physiol. Pharmacol.* **59**(Suppl. 7), 71–88.
Lehtinen, S. K., *et al.* (1996). Down-regulation of transcription factors AP-1, Sp-1, and NF-kappa B precedes myocyte differentiation. *Biochem. Biophys. Res. Commun.* **229,** 36–43.
Li, Y. P., and Reid, M. B. (2000). NF-kappaB mediates the protein loss induced by TNF-alpha in differentiated skeletal muscle myotubes. *Am. J. Physiol. Regul. Integr. Comp. Physiol.* **279,** R11650–R11670.
Li, Y. P., *et al.* (2003). TNF-alpha increases ubiquitin-conjugating activity in skeletal muscle by up-regulating UbcH2/E220k. *FASEB J.* **17,** 1048–1057.

Li, H., *et al.* (2008). Nuclear factor-kappa B signaling in skeletal muscle atrophy. *J. Mol. Med.* **86,** 1113–1126.

Li, H., *et al.* (2009a). Matrix metalloproteinase-9 inhibition ameliorates pathogenesis and improves skeletal muscle regeneration in muscular dystrophy. *Hum. Mol. Genet.* **18,** 2584–2598.

Li, H., *et al.* (2009b). Tumor necrosis factor-related weak inducer of apoptosis augments matrix metalloproteinase 9 (MMP-9) production in skeletal muscle through the activation of nuclear factor-kappaB-inducing kinase and p38 mitogen-activated protein kinase: A potential role of MMP-9 in myopathy. *J. Biol. Chem.* **284,** 4439–4450.

Liao, P., *et al.* (2010). Eccentric contraction induces inflammatory responses in rat skeletal muscle: Role of tumor necrosis factor-alpha. *Am. J. Physiol. Regul. Integr. Comp. Physiol.* **298,** R599–R607.

Lille, S. T., *et al.* (2001). Inhibition of the initial wave of NF-kappaB activity in rat muscle reduces ischemia/reperfusion injury. *Muscle Nerve* **24,** 534–541.

Lima-Cabello, E., *et al.* (2010). Eccentric exercise induces nitric oxide synthase expression through nuclear factor-kappaB modulation in rat skeletal muscle. *J. Appl. Physiol.* **108,** 575–583.

Linardic, C. M. (2008). PAX3-FOXO1 fusion gene in rhabdomyosarcoma. *Cancer Lett.* **270,** 10–18.

Liu, S., *et al.* (2010). Sp1/NFkappaB/HDAC/miR-29b regulatory network in KIT-driven myeloid leukemia. *Cancer Cell* **17,** 333–347.

Lluis, F., *et al.* (2005). E47 phosphorylation by p38 MAPK promotes MyoD/E47 association and muscle-specific gene transcription. *EMBO J.* **24,** 974–984.

Madge, L. A., and May, M. J. (2009). Inhibiting proinflammatory NF-kappaB signaling using cell-penetrating NEMO binding domain peptides. *Methods Mol. Biol.* **512,** 209–232.

Mammucari, C., *et al.* (2007). FoxO3 controls autophagy in skeletal muscle in vivo. *Cell Metab.* **6,** 458–471.

Manzur, A. Y., *et al.* (2008). Glucocorticoid corticosteroids for Duchenne muscular dystrophy. *Cochrane Database Syst. Rev.* (1), CD003725.

Markham, L. W., *et al.* (2008). Corticosteroid treatment retards development of ventricular dysfunction in Duchenne muscular dystrophy. *Neuromuscul. Disord.* **18,** 365–370.

May, M. J., *et al.* (2000). Selective inhibition of NF-kappaB activation by a peptide that blocks the interaction of NEMO with the IkappaB kinase complex. *Science* **289,** 1550–1554.

McCarthy, J. J., and Esser, K. A. (2010). Anabolic and catabolic pathways regulating skeletal muscle mass. *Curr. Opin. Clin. Nutr. Metab. Care* **13,** 230–235.

McClung, J. M., *et al.* (2010). p38 MAPK links oxidative stress to autophagy-related gene expression in cachectic muscle wasting. *Am. J. Physiol. Cell Physiol.* **298,** C542–C549.

Mercurio, F., *et al.* (1997). IKK-1 and IKK-2: Cytokine-activated IkappaB kinases essential for NF-kappaB activation. *Science* **278,** 860–866.

Messina, S., *et al.* (2006a). Lipid peroxidation inhibition blunts nuclear factor-kappaB activation, reduces skeletal muscle degeneration, and enhances muscle function in mdx mice. *Am. J. Pathol.* **168,** 918–926.

Messina, S., *et al.* (2006b). Nuclear factor kappa-B blockade reduces skeletal muscle degeneration and enhances muscle function in Mdx mice. *Exp. Neurol.* **198,** 234–241.

Messina, S., *et al.* (2009). Flavocoxid counteracts muscle necrosis and improves functional properties in mdx mice: A comparison study with methylprednisolone. *Exp. Neurol.* **220,** 349–358.

Mittal, A., *et al.* (2010a). The TWEAK-Fn14 system is a critical regulator of denervation-induced skeletal muscle atrophy in mice. *J. Cell Biol.* **188,** 833–849.

Mittal, A., *et al.* (2010b). Genetic ablation of TWEAK augments regeneration and post-injury growth of skeletal muscle in mice. *Am. J. Pathol.* **177,** 1732–1742.
Momcilovic, M., *et al.* (2006). Mammalian TAK1 activates Snf1 protein kinase in yeast and phosphorylates AMP-activated protein kinase in vitro. *J. Biol. Chem.* **281,** 25336–25343.
Monici, M. C., *et al.* (2003). Activation of nuclear factor-kappaB in inflammatory myopathies and Duchenne muscular dystrophy. *Neurology* **60,** 993–997.
Mourkioti, F., and Rosenthal, N. (2008). NF-kappaB signaling in skeletal muscle: Prospects for intervention in muscle diseases. *J. Mol. Med.* **86,** 747–759.
Mourkioti, F., *et al.* (2006). Targeted ablation of IKK2 improves skeletal muscle strength, maintains mass, and promotes regeneration. *J. Clin. Invest.* **116,** 2945–2954.
Murre, C., *et al.* (1989). Interactions between heterologous helix-loop-helix proteins generate complexes that bind specifically to a common DNA sequence. *Cell* **58,** 537–544.
Pahl, H. L. (1999). Activators and target genes of Rel/NF-kappaB transcription factors. *Oncogene* **18,** 6853–6866.
Pan, Y., *et al.* (2008). Curcumin alleviates dystrophic muscle pathology in mdx mice. *Mol. Cells* **25,** 531–537.
Park, J. W., *et al.* (2007). Proteasome inhibitor attenuates skeletal muscle reperfusion injury by blocking the pathway of nuclear factor-kappaB activation. *Plast. Reconstr. Surg.* **120,** 1808–1818.
Perkins, N. D. (2006). Post-translational modifications regulating the activity and function of the nuclear factor kappa B pathway. *Oncogene* **25,** 6717–6730.
Perry, R. L., and Rudnicki, M. A. (2000). Molecular mechanisms regulating myogenic determination and differentiation. *Front. Biosci.* **5,** D750–D767.
Peterson, J. M., *et al.* (2011). Peptide-based inhibition of NF-κB rescues diaphragm muscle contractile dysfunction in a murine model of Duchenne muscular dystrophy. *Mol. Med.* (Epub ahead of print).
Peterson, J. M., and Guttridge, D. C. (2008). Skeletal muscle diseases, inflammation, and NF-kappaB signaling: Insights and opportunities for therapeutic intervention. *Int. Rev. Immunol.* **27,** 375–387.
Pomerantz, J. L., and Baltimore, D. (2002). Two pathways to NF-kappaB. *Mol. Cell* **10,** 693–695.
Qi, W. N., *et al.* (2004). NF-kappaB p65 involves in reperfusion injury and iNOS gene regulation in skeletal muscle. *Microsurgery* **24,** 316–323.
Qualman, S. J., *et al.* (1998). Intergroup Rhabdomyosarcoma study: Update for pathologists. *Pediatr. Dev. Pathol.* **1,** 550–561.
Rao, P., *et al.* (2010). IkappaBbeta acts to inhibit and activate gene expression during the inflammatory response. *Nature* **466,** 1115–1119.
Razani, B., *et al.* (2010). Negative feedback in noncanonical NF-kappaB signaling modulates NIK stability through IKKalpha-mediated phosphorylation. *Sci. Signal* **3,** ra41.
Russell, S. T., *et al.* (2006). Mechanism of induction of muscle protein degradation by angiotensin II. *Cell. Signal.* **18,** 1087–1096.
Sandri, M., *et al.* (2004). Foxo transcription factors induce the atrophy-related ubiquitin ligase atrogin-1 and cause skeletal muscle atrophy. *Cell* **117,** 399–412.
Sanjo, H., *et al.* (2010). Allosteric regulation of the ubiquitin:NIK and ubiquitin:TRAF3 E3 ligases by the lymphotoxin-beta receptor. *J. Biol. Chem.* **285,** 17148–17155.
Schwabe, R. F., and Sakurai, H. (2005). IKKbeta phosphorylates p65 at S468 in transactivaton domain 2. *FASEB J.* **19,** 1758–1760.
Sen, R., and Baltimore, D. (1986). Multiple nuclear factors interact with the immunoglobulin enhancer sequences. *Cell* **46,** 705–716.
Senf, S. M., *et al.* (2008). Hsp70 overexpression inhibits NF-kappaB and Foxo3a transcriptional activities and prevents skeletal muscle atrophy. *FASEB J.* **22,** 3836–3845.

Senftleben, U., *et al.* (2001). Activation by IKKalpha of a second, evolutionary conserved, NF-kappa B signaling pathway. *Science* **293,** 1495–1499.

Sharif, O., *et al.* (2007). Transcriptional profiling of the LPS induced NF-kappaB response in macrophages. *BMC Immunol.* **8,** 1.

Shibata, W., *et al.* (2007). Cutting edge: The IkappaB kinase (IKK) inhibitor, NEMO-binding domain peptide, blocks inflammatory injury in murine colitis. *J. Immunol.* **179,** 2681–2685.

Siegel, A. L., *et al.* (2009). Treatment with inhibitors of the NF-kappaB pathway improves whole body tension development in the mdx mouse. *Neuromuscul. Disord.* **19,** 131–139.

Sitcheran, R., *et al.* (2003). NF-kappaB mediates inhibition of mesenchymal cell differentiation through a posttranscriptional gene silencing mechanism. *Genes Dev.* **17,** 2368–2373.

Skaug, B., *et al.* (2009). The role of ubiquitin in NF-kappaB regulatory pathways. *Annu. Rev. Biochem.* **78,** 769–796.

Skuk, D., *et al.* (2007). Ischemic central necrosis in pockets of transplanted myoblasts in nonhuman primates: Implications for cell-transplantation strategies. *Transplantation* **84,** 1307–1315.

Song, W., *et al.* (2006). Exercise training attenuates age-induced changes in apoptotic signaling in rat skeletal muscle. *Antioxid. Redox Signal.* **8,** 517–528.

Spangenburg, E. E., *et al.* (2006). Exercise increases SOCS-3 expression in rat skeletal muscle: Potential relationship to IL-6 expression. *J. Physiol.* **572,** 839–848.

Stitt, T. N., *et al.* (2004). The IGF-1/PI3K/Akt pathway prevents expression of muscle atrophy-induced ubiquitin ligases by inhibiting FOXO transcription factors. *Mol. Cell* **14,** 395–403.

Subramanian, S., *et al.* (2008). MicroRNA expression signature of human sarcomas. *Oncogene* **27,** 2015–2026.

Tang, Y., *et al.* (2010). Inhibition of the IKK/NF-kappaB pathway by AAV gene transfer improves muscle regeneration in older mdx mice. *Gene Ther.* **17,** 1476–1483.

Tantiwong, P., *et al.* (2010). NF-kappaB activity in muscle from obese and type 2 diabetic subjects under basal and exercise-stimulated conditions. *Am. J. Physiol. Endocrinol. Metab.* **299,** E794–E801.

Tarnowski, M., *et al.* (2010). Regulation of expression of stromal-derived factor-1 receptors: CXCR4 and CXCR7 in human rhabdomyosarcomas. *Mol. Cancer Res.* **8,** 1–14.

Tas, S. W., *et al.* (2006). Local treatment with the selective IkappaB kinase beta inhibitor NEMO-binding domain peptide ameliorates synovial inflammation. *Arthritis Res. Ther.* **8,** R86.

Tegethoff, S., *et al.* (2003). Tetrameric oligomerization of IkappaB kinase gamma (IKK-gamma) is obligatory for IKK complex activity and NF-kappaB activation. *Mol. Cell. Biol.* **23,** 2029–2041.

Thaloor, D., *et al.* (1999). Systemic administration of the NF-kappaB inhibitor curcumin stimulates muscle regeneration after traumatic injury. *Am. J. Physiol.* **277,** C320–C329.

Tong, X., *et al.* (2004). The p50-p50 NF-kappaB complex as a stimulus-specific repressor of gene activation. *Mol. Cell. Biochem.* **265,** 171–183.

Tsikitis, M., *et al.* (2010). Traf7, a MyoD1 transcriptional target, regulates nuclear factor-kappaB activity during myogenesis. *EMBO Rep.* **11,** 969–976.

Tubaro, C., *et al.* (2010). S100B protein in myoblasts modulates myogenic differentiation via NF-kappaB-dependent inhibition of MyoD expression. *J. Cell. Physiol.* **223,** 270–282.

Van Gammeren, D., *et al.* (2009). The IkappaB kinases IKKalpha and IKKbeta are necessary and sufficient for skeletal muscle atrophy. *FASEB J.* **23,** 362–370.

van Rooij, E., *et al.* (2008). Dysregulation of microRNAs after myocardial infarction reveals a role of miR-29 in cardiac fibrosis. *Proc. Natl. Acad. Sci. USA* **105,** 13027–13032.

Vasilaki, A., *et al.* (2006). Adaptive responses of mouse skeletal muscle to contractile activity: The effect of age. *Mech. Ageing Dev.* **127,** 830–839.

Vidal, B., *et al.* (2008). Fibrinogen drives dystrophic muscle fibrosis via a TGFbeta/alternative macrophage activation pathway. *Genes Dev.* **22,** 1747–1752.
Wang, H., *et al.* (2007). NF-kappaB regulation of YY1 inhibits skeletal myogenesis through transcriptional silencing of myofibrillar genes. *Mol. Cell. Biol.* **27,** 4374–4387.
Wang, H., *et al.* (2008). NF-kappaB-YY1-miR-29 regulatory circuitry in skeletal myogenesis and rhabdomyosarcoma. *Cancer Cell* **14,** 369–381.
Wang, J., *et al.* (2010). Nuclear factor kappaB controls acetylcholine receptor clustering at the neuromuscular junction. *J. Neurosci.* **30,** 11104–11113.
Wertz, I. E., *et al.* (2004). De-ubiquitination and ubiquitin ligase domains of A20 downregulate NF-kappaB signalling. *Nature* **430,** 694–699.
Westerheide, S. D., *et al.* (2001). The putative oncoprotein Bcl-3 induces cyclin D1 to stimulate G(1) transition. *Mol. Cell. Biol.* **21,** 8428–8436.
Whitehead, N. P., *et al.* (2008). N-Acetylcysteine ameliorates skeletal muscle pathophysiology in mdx mice. *J. Physiol.* **586,** 2003–2014.
Whitehouse, A. S., and Tisdale, M. J. (2003). Increased expression of the ubiquitin-proteasome pathway in murine myotubes by proteolysis-inducing factor (PIF) is associated with activation of the transcription factor NF-kappaB. *Br. J. Can.* **89,** 1116–1122.
Wu, Z., *et al.* (2000). p38 and extracellular signal-regulated kinases regulate the myogenic program at multiple steps. *Mol. Cell. Biol.* **20,** 3951–3964.
Wu, C. J., *et al.* (2006). Sensing of Lys 63-linked polyubiquitination by NEMO is a key event in NF-kappaB activation [corrected]. *Nat. Cell Biol.* **8,** 398–406.
Xiao, G., *et al.* (2004). Induction of p100 processing by NF-kappaB-inducing kinase involves docking IkappaB kinase alpha (IKKalpha) to p100 and IKKalpha-mediated phosphorylation. *J. Biol. Chem.* **279,** 30099–30105.
Yang, C. C., *et al.* (1998). Immunolocalization of transcription factor NF-kappaB in inclusion-body myositis muscle and at normal human neuromuscular junctions. *Neurosci. Lett.* **254,** 77–80.
Zandi, E., *et al.* (1997). The IkappaB kinase complex (IKK) contains two kinase subunits, IKKalpha and IKKbeta, necessary for IkappaB phosphorylation and NF-kappaB activation. *Cell* **91,** 243–252.
Zhao, J., *et al.* (2007). FoxO3 coordinately activates protein degradation by the autophagic/lysosomal and proteasomal pathways in atrophying muscle cells. *Cell Metab.* **6,** 472–483.

CHAPTER FIVE

Blood Vessels and the Satellite Cell Niche

Rémi Mounier,*,‡,§ Fabrice Chrétien,† *and* Bénédicte Chazaud*,‡,§

Contents

Abstract

The fate of stem cell is regulated by cues received from the surrounding area. Recently, the concept of "stem cell zone"—rather than a predefined niche—introduced the notion of dynamic and permanent interactions between stem cells and their microenvironment. In adult skeletal muscle, satellite cells are considered as the main stem cells responsible for muscle repair and maintenance. They are localized close to vessels regardless their state of activation and differentiation. Moreover, the number of satellite cells is positively correlated to the capillarization of the myofiber. Angiogenesis has been known for a long time to be essential for muscle repair. However, relationships between vessel cells and satellite/myogenic cells that govern myogenic cell expansion, myogenesis, and angiogenesis have been only recently investigated. In this chapter, we discuss the possible existence of a vascular amplifying/differentiating niche, in an attempt to reconciliate several recent observations showing that satellite/myogenic cells

* Inserm, U1016, Institut Cochin, 24 Rue du Faubourg Saint Jacques, Paris, France
† Institut Pasteur, Unité Histopathologie humaine et modèles animaux, 28 Rue du Docteur Roux, Paris, France
‡ CNRS, UMR8104, Paris, France
§ Univ Paris Descartes, Paris, France

Current Topics in Developmental Biology, Volume 96
ISSN 0070-2153, DOI: 10.1016/B978-0-12-385940-2.00005-X

interact with various cell types during the time course of muscle regeneration. Indeed, endothelial cells (ECs) stimulate myogenic cell growth and, inversely, differentiating myogenic cells promote angiogenesis. However, stromal cells may also provide some proliferating or differentiating cues to satellite/myogenic cells in this vascular area. Although some molecular effectors have been identified, including growth factors and cytokines, molecular regulations that occur within this vascular amplifying/differentiating niche requires further investigation. At the end of muscle repair, maturation of newly formed vessels takes place. In this context, we discuss the potential quiescence niche of satellite cells and the specific role of periendothelial cells. Indeed, periendothelial cells promote the return to quiescence of a subset of satellite/myogenic cells and maintain their quiescence (through Angiopoietin-1/Tie-2 signaling). We ask to what extent the environment may control the fate choice of satellite/myogenic cells and we also question the "hypoxic niche" in skeletal muscle, such a quiescence niche having being observed in the bone marrow.

1. Introduction

Satellite cells are considered as the most efficient stem cells of the adult skeletal muscle. They provide myogenic progenitors cells for muscle repair and are capable of self-renewal to ensure a pool of stem cells during the life span. Most of the time, they remain in a quiescent G0 state laying along the myofiber under the basal lamina, until a damage signal activates them and makes them reentering into the cell cycle. Although this specific localization makes the sublaminal zone a potential "niche" for satellite cells, it is likely that this sole location does not define a niche *per se*. Moreover, although under the basal lamina, satellite cells are mobile in the steady state adult muscle (F. Chrétien, unpublished observations), suggesting dynamic interactions between satellite cells and their environment. Microenvironmental signals, regulated by neighboring cells, likely determine "stem cell zones" rather than strictly defined niche, as it has been recently proposed in the bone marrow for hematopoietic stem cells (HSCs; Li and Clevers, 2010).

The overwhelming majority of satellite cells are close to vessels. For all that, does a vascular stem cell niche exist in skeletal muscle? Vascular niches have been described for stem cells in several tissues. In the bone marrow, HSC niches have been particularly well studied and form a paradigm for the niche concept. Numerous studies have established that HSCs reside in several places (Kiel and Morrison, 2008): (i) the endosteal niche where HSCs interact with endosteal cells to maintain their quiescent G0 state. Various factors have been involved in this interaction among which Angiopoietin 1 (Ang1), Ca^{2+} ions, SHH, osteopontin, stem cell factor (SCF), thrombopoietin, and CXCL12, secreted by neighboring perivascular reticular cells; (ii) the vascular niche, specifically the specialized blood vessels called sinusoids, where many medullar HSCs reside. Sinusoids may promote extramedullary haematopoiesis, indicating their ability to create an environment capable of sustaining adult HSCs. The concept of two niches, or more preferably, two "stem cell zones," has recently emerged, proposing the following. Long-term dormant HSCs would reside near the endosteum (i.e., surrounded by CXCL12-secreting perivascular reticular cells and sinusoids), while more active HSCs would reside in a perivascular zone lacking osteoblastic cells, rendering possible their cycling. This slow HSC cycling would be the support of the huge daily expansion of bone marrow cells (Li and Clevers, 2010). Further investigations will help to define more precisely the role of each zone in the maintenance of HSC homeostasis but this proposal reconciliates the data obtained until now.

Our knowledge on satellite and muscle stem cells is less significant than on HSCs. However, as satellite cells are close to vessels at various times of their life and as they interact in different ways with various neighboring cells, we can hypothesize the existence of several "stem cell zones." The characteristics of such zones would vary upon the cell types that are present in the zone and that interact with the satellite cells at each step of their "cycle," that is, quiescence, expansion, differentiation, and self-renewal.

2. Satellite Cell Proximity to Blood Vessel

The first study to mention the localization of satellite cells is the electron microscopy analysis performed by Schmalbruch and Hellhammer (1977) which notes that more than half of the satellite cells are closely associated with a capillary in adult hind limb rat muscle. In a more recent study, we further explored this observation (Christov *et al.*, 2007). In normal adult human muscle, satellite cells are not randomly distributed but are preferentially close to capillaries: 88% of them are less than 20 μm from a capillary. Similar results were obtained in several human muscles and in dog, rat, and mice muscle. For example, in $Myf5^{nlacZ/+}$ or $Myf5^{GFP\text{-}P/+}$ mice, 82% of satellite cells were at 5 μm from endothelial cells (ECs). Interestingly, long-term label retaining satellite cells were as close to vessels as unlabeled satellite cells in BrdU pulse-chase experiments. Moreover, bone marrow-derived satellite cells (after whole body irradiation and bone marrow transplantation) were also in close proximity to vessels as 89% were at 5 μm from ECs. In regenerating human muscle (dystrophic muscle), 82% of cycling satellite cells (Ki67+/CD56+) are colocalized with capillaries. These observations indicate that most of satellite cells remain in close proximity to capillaries regardless of their state of quiescence, proliferation, differentiation, and of their origin.

Our study (Christov *et al.*, 2007) extended to the capillarization of the fibers (i.e., the number of capillaries associated with the fiber). The number of satellite cells increases with the capillarization of the myofiber. It is known for a long time that capillarization depends on the fiber type, type II (fast) being less capillarized than type I (slow; Schmalbruch and Hellhammer, 1977). The satellite cell number is correlatively lower for type II than for type I myofibers. The correlation between capillarization and satellite cell number has been particularly evidenced in two paradigmatic situations that dramatically reduce or increase the number of capillaries in skeletal muscle. Amyopathic dermatomyositis is characterized by a capillary loss observed in the absence of both myofiber damage and inflammation (Emslie-Smith and Engel, 1990). A similar loss in the satellite cell number per myofiber is observed. Inversely, workload is known to increase muscle capillary density (Jensen *et al.*, 2004). In heavy-trained athletes muscle analyzed, increase of both numbers of capillaries and satellite cells per myofiber is observed. This careful morphometric analysis indicates that satellite cells are related to vessels and therefore likely receive cues from vascular and/or perivascular cells.

3. Angiogenesis, Myogenesis, and the Regulation of Muscle Homeostasis

Exercise provokes a strong angiogenic stimulus in the active muscle. Thus, training is associated with an increase in capillarity (Andersen and Henriksson, 1977; Hudlicka *et al.*, 1992). The most studied angiogenic factor involved in muscle is vascular endothelial growth factor (VEGF). VEGF has been regularly involved in skeletal muscle angiogenesis induced by acute aerobic or resistance exercise in both human and animals (Breen *et al.*, 1996; Gavin *et al.*, 2004; Gustafsson *et al.*, 2007; Hiscock *et al.*, 2003; Prior *et al.*, 2003, 2004). Moreover, training also induces modifications of VEGF mRNA and protein expression (Amaral *et al.*, 2001; Campos *et al.*, 2002; Gustafsson *et al.*, 2007; Lloyd *et al.*, 2003; Olfert *et al.*, 2001). Chronic pathological conditions, such as chronic obstruction pulmonary disease (COPD) and diabetes (Barreiro *et al.*, 2008; Kivela *et al.*, 2008), induce capillary rarefaction suggesting that VEGF is also important in the maintenance of adult skeletal muscle microvasculature. Indeed, VEGF has been implicated as an essential survival factor (Tang *et al.*, 2004) for basal skeletal muscle capillarization (Wagner *et al.*, 2006). Moreover, skeletal muscle VEGF deficient mice presents a marked decrease of capillary-to-fiber ratio (−48%) and capillary density (−39%); this indicates that VEGF is crucial in the physiological regulation of postnatal muscle capillarity and in the maintenance of adult skeletal muscle microvasculature (Olfert *et al.*, 2009). Interleukin (IL)-8 has been also shown to be a potent angiogenic factor in several tissues and muscle-derived IL-8 has been suggested to act locally to stimulate angiogenesis through its receptor (Akerstrom *et al.*, 2005; Frydelund-Larsen *et al.*, 2007). Effectively, acute exercise induces an increase in IL-8 protein and mRNA quantities and IL-8 receptor mRNA content within skeletal muscle fibers (Akerstrom *et al.*, 2005; Frydelund-Larsen *et al.*, 2007).

Muscle regeneration is associated with an increase in both capillarization and cross-sectional area of individual regenerating myofibers (Luque *et al.*, 1995). Angiogenesis and myogenesis proceed at the same time as formation of new capillaries and formation of new myofibers are observed concomitantly (Scholz *et al.*, 2003). Accordingly, expression of angiogenesis-related factors, VEGF, Angiopoietin (Ang)1 and 2, monocyte chemoattractant protein (MCP)-1, and their receptors strongly increases after injury (Wagatsuma, 2007). The delivery of AAV-VEGF markedly improves muscle fiber reconstitution in ischemic limb (Arsic *et al.*, 2004). Contrarily, in mice deficient for the main receptor for MCP-1, CCR2, a delay and an impairment of muscle regeneration are observed after injury (Ochoa *et al.*, 2007). In these conditions, VEGF level is decreased until day 21, time of restoration of maximal capillary density (Ochoa *et al.*, 2007). The authors showed that maximal capillary density develops concurrent with the restoration of tissue VEGF, and observed an inverse relationship between the size of regenerated muscle fibers and the number of capillaries (Ochoa *et al.*, 2007). These studies indicates the existence of VEGF-dependent mechanisms at work in the dynamic processes of capillary formation and muscle regeneration.

VEGF has been shown to be expressed by satellite cells, myogenic cells, and particularly by differentiating myogenic cells and regenerating myofibers (Bryan *et al.*, 2008; Chazaud *et al.*, 2003; Germani *et al.*, 2003). Importantly, VEGF secreted by myogenic cells confers to them an angiogenic activity (cf. below). As myogenic cells bear the VEGF-R, it also acts by stimulating their migration and protecting them from apoptosis (Arsic *et al.*, 2004; Bryan *et al.*, 2008). VEGF may also participate in the myogenic differentiation program through its regulation by the myogenic transcription factor MyoD. Indeed, VEGF-null embryonic stem cells show reduced differentiation while inversely, VEGF stimulation promotes myotube hypertrophy (Bryan *et al.*, 2008). As hypertrophy induced by Akt is accompanied by an increase of VEGF levels and high capillary density at focal regions of high Akt transgene expression, it has been proposed that VEGF secretion in myogenic cells is regulated through the Akt pathway (Takahashi *et al.*, 2002).

Beside VEGF, other factors are involved in the regulation of angiogenesis in skeletal muscle. For example, β-catenin overexpression induces both angiogenesis (through EC proliferation) and muscle regeneration (increases VEGF expression in myogenic cells; Kim *et al.*, 2006). Fibroblast growth factors (FGFs) seem also involved: myogenic cells express bFGF that leads to angiogenesis in a paracrine fashion to help muscle regeneration in ischemic limb (Walgenbach *et al.*, 1995). Introducing FGF2 or FGF6 transgenes into muscle enhances both the number of vessels (CD31+ structures) and muscle repair (Doukas *et al.*, 2002). Finally, Nerve Growth Factor (NGF) has been shown to enhance angiogenesis and arteriogenesis after ischemia *in vivo* while it protects ECs and myofibers from apoptosis *in vitro* (Emanueli *et al.*, 2002).

These observations suggest the existence of regulatory mechanisms through which blood vessel recruitment can be coupled to muscle tissue growth and repair. Thus, muscle repair requires myogenesis and angiogenesis to be spatiotemporally coordinated.

4. The Amplifying/Differentiating Niche: Role of ECs and Surrounding Stromal Cells

As angiogenesis and myogenesis are processes that occur simultaneously, it is likely that satellite cells—and their progenitors—and ECs interact. *In vitro* three-dimensional coculture models, using either isolated cells or multicellular models such as microvascular fragments (consisting of endothelial, pericyte, and smooth muscle cells) showed that satellite cells promote angiogenesis (Christov *et al.*, 2007; Rhoads *et al.*, 2009). This effect is mediated through the secretion of soluble factors, of which VEGF plays a crucial role (Rhoads *et al.*, 2009). For now, no other factor has been involved. Of interest, the angiogenic property is more pronounced in differentiating myogenic cells (myogenin+) than in undifferentiated satellite cells (Christov *et al.*, 2007). This is agreement with their secretion of VEGF which is increased by 2.5-fold (Chazaud *et al.*, 2003). *In vivo* morphometric analyses confirmed these *in vitro* data. In regenerating conditions (dystrophic human muscle), differentiating myogenic cells (myogenin+) are closer to ECs than nondifferentiated myogenic progenitors (MYF5+; 97.4% vs. 88.5%). Moreover, EC area and capillary network associated with differentiating cells were higher than those associated with non differentiated progenitors, indicative of neoangiogenesis associated with differentiating myogenic cells (Christov *et al.*, 2007). These data reveal strong spatiotemporal correlation between myogenic cell differentiation and neoangiogenesis.

Inversely, do ECs have an effect on satellite cells? When satellite cells are in indirect coculture with other myogenic cells or other cell types including smooth muscle cells, fibroblasts, and ECs, only those latter induce an increase in satellite cell growth. This indicates that EC monolayers specifically release soluble factors that promote myogenic cell growth (Christov *et al.*, 2007). Among them, we identified Insulin growth Factor-1 (IGF-1), hepatocyte growth factor (HGF), bFGF, platelet-derived growth factor-BB (PDGF-BB), and VEGF, that account for 90% of the EC-stimulated satellite cell growth in these *in vitro* conditions (Christov *et al.*, 2007).

Coordination of myogenesis and angiogenesis may therefore be accomplished through the secretion of soluble factors, by satellite cells and ECs vice versa interacting. The juxtavascular position of satellite cells during regeneration and while they differentiate suggests that they benefit *in vivo* from EC supportive cues. Of special note, while angiogenesis is in process, ECs are not surrounded by mural cells, so they can directly interact with other cell types localized at proximity.

This feature of myogenesis/angiogenesis is reminiscent of the neurovascular unit described for neural stem cells (NSCs). First evidence for an angiogenic niche for neurogenesis has been described in the adult hippocampus (Palmer *et al.*, 2000). In the adult songbird striatum, where neurogenesis proceeds through life in a specific area (the higher vocal center) upon testosterone stimulus, it has been demonstrated that angiogenesis and neurogenesis are coupled processes, through the EC-derived BDNF (Louissaint *et al.*, 2002). Bilateral interactions have been evidenced between ECs and NSCs. ECs stimulate NSC self-renewal and expansion (Shen *et al.*, 2004) while NSCs stimulate angiogenesis (NO-derived NSC simulates secretion of BDNF and VEGF by ECs) and promote vessel stabilization (Li *et al.*, 2006).

The similarities of the cross talks between ECs and neural or muscle progenitors suggest the existence of an amplifying vascular niche where cells may expand and further differentiate. However, ECs may not be the sole cell type neighboring the expanding myogenic progenitors at time of muscle regeneration, and other stromal cells likely participate in this environment. We have shown that monocyte-derived macrophages sequentially act on satellite cells to first stimulate their proliferation then their differentiation (Arnold *et al.*, 2007). This is allowed, thanks to a tight regulation of their activation state with time; the activation state of macrophages governs the delivery of specific cues. In their recent paper, Rossi's team has shown that fibro/adipogenic precursors participate, in a very timely regulated manner, to the differentiation of satellite cells at time of muscle regeneration (Joe *et al.*, 2010). They also propose the existence of transient niches that would be required for a short time after acute damage. Thus, what does the presence of satellite cells close to vessels during all the time course of muscle regeneration mean? This suggests that ECs would be an essential component of the niche, that may be altered by some transient and specific environmental cues supplied by stromal cells as fibro/adipogenic cells and/or macrophages. These latter cells may rapidly change their phenotype or quickly move away or die depending on demand, as it has been shown for macrophages and fibro/adipogenic cells, respectively (Arnold *et al.*, 2007; Joe *et al.*, 2010). The stem cell zone concept, that is composed of specialized cell types which can evolve as the tissue is remodeling, has to be assessed by experimental data. However, this model is interesting because, it can explain the proximity and specific interactions of satellite cells/myogenic precursor cells with various cell types during muscle regeneration.

5. The Quiescence Niche: Role of Periendothelial Cells

5.1. What are the perivascular cells?

The end of angiogenesis is characterized by the maturation of the vessels (Jain, 2003). Newly formed capillaries fuse with others, allowing the emerging of blood flow, therefore lowering hypoxia—and its target genes. Maturation is characterized by the recruitment of mural cells to the new vessels. Mural cells include pericytes, smooth muscle cells, and fibroblastic cells. Pericytes form walls of capillaries while large vessel (arteries and veins) walls are formed by several layers of smooth muscle cells. Perivascular cells regulate the capillary environment by: (1) communication with ECs by soluble mediators and cell–cell contact, (2) synthesis, remodeling, and maintenance of the basal membrane, and (3) regulation of microvascular tone. All of these mechanisms involve an overlapping array of biochemical and biomechanical signaling pathways (Kutcher and Herman, 2009).

Thus at the end of muscle regeneration, ECs are not free to contact other cells anymore but start to be covered by perivascular cells that begin the vessel maturation. When homeostasis of the tissue is reached, the quiescent satellite cells are therefore very close to the perivascular cells (Abou-Khalil *et al.*, 2009). These have been shown to be either pericytes/smooth muscle cells or cells of "fibroblastic" phenotype, poorly characterized. Several "fibroblastic" cell types have been identified in skeletal muscle, using different markers, and under various conditions: the fibro/adipogenic precursors that may give rise to fibroblastic cells (Joe *et al.*, 2010), tcf-4 highly positive cells (Kardon *et al.*, 2003), and CD45-CD34+ connective reticular cells (F. Chrétien, unpublished observations). There are some interstitial cells at steady state in the perivascular space that resembles fibroblasts, in a periendothelial localization (Abou-Khalil *et al.*, 2009). The exact localization in steady state and in regenerating muscle as well as full characterization of these (various?) cell types require further investigations.

A subset of these perivascular cells have stemness properties. Cossu's group demonstrated in mouse, dog, and human that some pericytes are stem cells with myogenic potential and with regenerating capacities in dystrophic muscle (Dellavalle *et al.*, 2007). Human perivascular cells with mesenchymal stemness capacities have been prospectively isolated from muscle (Crisan *et al.*, 2008). In mouse, interstitial cells expressing the cell stress mediator PW1 (PICs), located in a perivascular position, also present a myogenic potential and efficiently contribute to skeletal muscle regeneration *in vivo* as well as generating satellite cells and new PICs (Mitchell *et al.*, 2010). Perivascular fibro/adipogenic progenitors may present adipogenic or fibrogenic properties, depending on the cues they encounter (Joe *et al.*, 2010). It has been recently shown in the bone marrow that HSCs and mesenchymal stem cells (MSCs) share the same niche and interdependently control the homeostasis of each stem lineage (Mendez-Ferrer *et al.*, 2010). This indicates that stem cells and progenitors themselves provide environmental cues for other stem/progenitor cells in the same stem cell zone. This may be the case in skeletal muscle between satellite cells and perivascular MSCs.

5.2. Mechanisms of satellite cell quiescence

As muscle regeneration is ending, satellite cells reoccupy their localization, under the basal lamina near capillaries, where they stay in the quiescent state. However, the mechanisms that control the exit from the cell cycle to enter in the quiescence state remain poorly understood. In mice, several markers have been associated with quiescent satellite cells, including M-cadherin, syndecan 3 and 4, CD34, calcitonin receptor, and Myf5 (Tedesco *et al.*, 2010). In human cell cultures, p130 from the Rb family is involved in the constitution of the reserve cell pool (i.e., quiescent undifferentiated cells that can give rise to both differentiated cells and new reserve cells when replated), by blocking cell-cycle progression and differentiation (Carnac *et al.*, 2000). In mice, Pax7 transcription factor is required for satellite cell maintenance and survival in young animals (i.e., before 3 weeks; Lepper *et al.*, 2009; Olguin and Olwin, 2004; Oustanina *et al.*, 2004). Calcium signaling, via calcineurin and NFAT, upregulates Myf5 expression in reserve cells at time of fate choice between self-renewal and myogenic differentiation (Friday and Pavlath, 2001). Wnt and Notch signalings are crucial regulators of myogenic progenitor cell proliferation and differentiation that are finely regulated with time (Brack *et al.*, 2008). Their role in satellite cell self-renewal is not yet deciphered, although Notch activation alters reserve cell recruitment into myotubes (Kitzmann *et al.*, 2006) and β-catenin promotes self-renewal of satellite cells, likely through Wnt pathway (Perez-Ruiz *et al.*, 2008). Receptor tyrosine kinase (RTK) signaling seems also crucial since a downstream negative regulator of RTK, sprouty, has been shown to regulate satellite cell quiescence during muscle growth and repair (Shea *et al.*, 2010). Recently, it has been shown that concentrations of HGF above 20 ng/ml trigger a quiescence signal in satellite cells from adult rats and that this occurs following the synthesis of myostatin

(Yamada *et al.*, 2010). HGF was known to induce, at low concentrations, the activation (i.e., exit from quiescence) of satellite cells (Tatsumi *et al.*, 1998). HGF being mainly produced by satellite cells, it would induce an autocrine amplification loop, thus provoking accumulation of HGF in the extracellular milieu (Yamada *et al.*, 2010). However, macrophages may also contribute to local HGF production (Galimi *et al.*, 2001).

Effectors mentioned above have been studied independently of the satellite cell environment, although the microenvironment is acknowledged to be crucial in the control of satellite cell fate (Dhawan and Rando, 2005). We have addressed this question by analyzing the involvement of the molecular system Ang1/Tie-2 in the quiescence of satellite cells. This main organizer of vascular homeostasis plays pleiotropic roles in cell survival, proliferation, migration, chemotaxis, and quiescence (Fiedler and Augustin, 2006; Shim *et al.*, 2007). It is involved in the quiescence of HSCs in the endosteal niche: osteoblast-secreted Ang1 binds to its tyrosine kinase Tie-2 receptor borne by HSCs to promote their survival and maintenance in G0 (Arai *et al.*, 2004). In the vessel, Ang1—secreted by periendothelial cells—binding to Tie-2-expressed by ECs is required to maintain vascular integrity (Shim *et al.*, 2007).

We have shown that Tie-2 is highly expressed by the most quiescent satellite cells in murine muscle (Abou-Khalil *et al.*, 2009). Ang1 binding to Tie-2 prevents satellite cell apoptosis, decreases their growth, proliferation, and differentiation and triggers their entry into G0. Ang1/Tie-2 exert their effect through ERK1/2 signaling to control upregulation of markers associated with the reserve cell phenotype (p130, Pax7, Myf-5, M-cadherin) and downregulation of markers associated with myogenic differentiation (MyoD, p57, myogenin; Abou-Khalil *et al.*, 2009). Our data suggest that neighboring interstitial cells such as smooth muscle cells and fibroblastic cells could be a source of Ang1 that acts on satellite cells to promote their quiescence. *Ex vivo*, Ang1 increases the number of Pax7+/MyoD- cells, that is, self-renewing cells, in differentiating clusters derived from satellite cells on single myofibers (Abou-Khalil *et al.*, 2009). *In vivo* gain-of-function and loss-of-function assays showed that blocking Tie-2 induces an increase in cycling satellite cells, while conversely, inducing overexpression of Ang1 by myofibers triggers the entry into quiescence of satellite cells in regenerating muscle (Abou-Khalil *et al.*, 2009). Altogether, these observations show that Ang1 binding to Tie-2 is involved in the regulation of the return to a quiescent state of a subset of satellite cells.

How this system is regulated by the microenvironment? Our data indicate that the subset of cells able to return to quiescence is not unlimited. It is tempting to hypothesize that Ang1/Tie-2-dependent signaling affects only cells that have reached a state of competence after a phase of proliferation. Similarly, Brack's group suggests that only a pool of quiescent-competent cells (less committed?) is expanded and depends on its cellular environment to either execute the quiescence program or maintain a state of competency (Abou-Khalil and Brack, 2010). In this scheme, only a subset of cells would become responsive and return to quiescence. Our data suggest that, as myogenic precursor cells express and produce Ang1, an autocrine loop that could operate to generate reserve cells and commit self-renewal. However, the contribution of Ang1 originating from neighboring cell types, such as periendothelial and interstitial cells, cannot be excluded. These cells could participate in the entry into quiescence of the satellite cells, but would be mainly involved in the maintenance of the quiescent status of these responsive Tie-2+/Pax7+ satellite cells that reside at proximity.

5.3. What about hypoxia?

Most of the cellular biology works studying the mechanisms of stem cell quiescence is done *in vitro* in a 20% oxygen atmosphere, conditions non relevant to what is observed in tissues where the partial pressure oxygen is much lower. During the last decade, numerous works demonstrated the crucial role of the oxygen level both for maintaining quiescence or promoting differentiation of stem cells. For example, in the bone marrow, the presence of hypoxic areas is determined by the architecture of medullary sinuses and the pattern of arterial blood flow in the marrow (Lichtman, 1981). It has been demonstrated that HSCs and progenitors are distributed along an O_2 gradient, with stem cells residing in the most hypoxic areas and proliferating progenitors in O_2-rich areas (Parmar *et al.*, 2007). A same oxygen gradient is thought to be present in the skeletal muscle from capillaries to deep parenchyma and depending on the tissue perfusion. Moreover, the heterogeneity in intramuscular oxygenation could be related to the muscle fiber typology (Greenbaum *et al.*, 1997; Richardson *et al.*, 1998). Although most of the satellite cells are very close to vessels, few cells remain far from vasculature and certainly receive less oxygen. The evidence of a hypoxic niche for satellite cells needs to be demonstrated but it has been previously shown *in vitro* that hypoxia (3% oxygen), stimulates the proliferation of satellite cells and promotes their myogenic differentiation with MyoD playing an important role (Chakravarthy *et al.*, 2001). Inversely, deeper hypoxia ($<$2% oxygen) inhibits myogenesis; the strongest inhibition was observed upon 0.01% O_2 (Yun *et al.*, 2005).

6. Conclusions and Few Questions

Altogether (Fig. 5.1), these data suggest that at the time of muscle regeneration, satellite cells remain under a vascular influence where they benefit of cues from "mural-free" ECs and also from other cell types such as

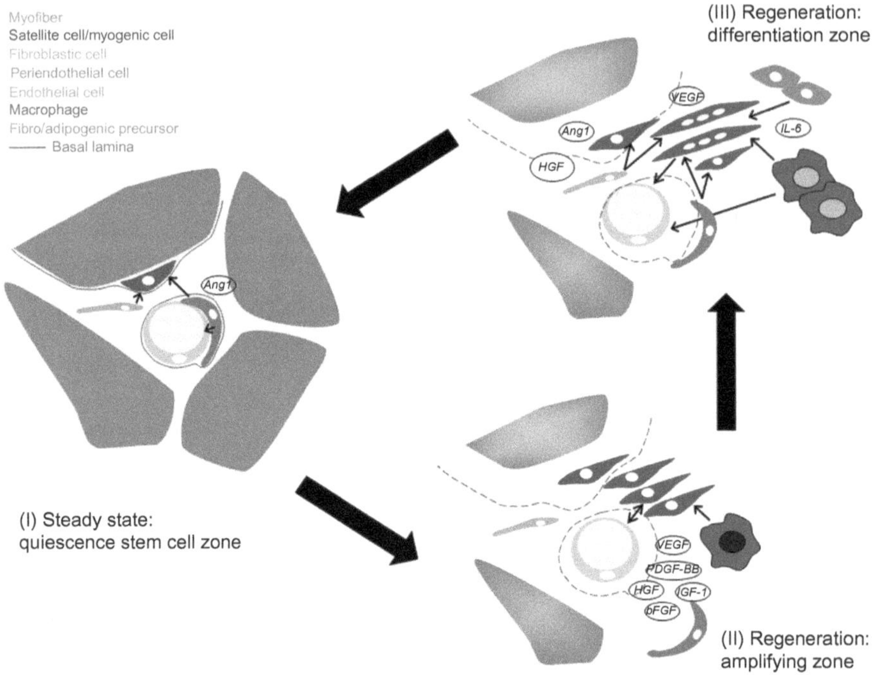

Figure 5.1 Interactions between vessels and satellite cells during muscle regeneration. (I) At steady state, satellite cells lay at quiescent state under basal lamina, under the influence of perivascular Ang1 that helps the maintenance in the G0 state. Other molecular systems and particularly cell adhesion systems are likely involved, although they are still unknown. It is also not known whether a portion of the basal lamina, or the fiber itself, has some particularities to define the quiescence zone for the satellite cells. (II) At the time of regeneration, degenerating/regenerating fibers are still surrounded by basal lamina or at least remnants but cells seem to easily cross it. Vessels are growing, thus ECs are free of vascular mural cells. The behavior of those latter is not clear: maybe they also expand to ensure later stabilization of the newly formed capillaries and vessels. So the status of the basal lamina and mural cells is not well known during muscle regeneration. Expanding myogenic cells and ECs, facing each other, can bidirectionally interact to promote myogenesis and angiogenesis through the secretion of various growth factors. Early infiltrating macrophages, that have an inflammatory phenotype, also stimulate the amplification of myogenic cells. Various growth factors, synthesized by either macrophages or ECs, have been identified to promote myogenic cell growth. (III) Later during the repair process, macrophages have changed their phenotype into antiinflammatory cells that promote differentiation of myogenic cells. Fibro/adipogenic cells also stimulate myogenic differentiation (through IL-6 secretion). Young differentiating myogenic cells promote angiogenesis (through VEGF secretion). Further, macrophages are renowned to have angiogenic properties. When myogenic progenitor expansion is ending, cells have to choose between differentiation and self-renewal. This can be achieved, at least through high concentrations of HGF and through Ang1/Tie-2 signaling that arises from myogenic progenitors themselves, and may be from perivascular cells (fibroblastic or smooth muscle cells) that start to stabilize the newly formed vessels. Homeostasis is then reached after the disappearance of fibro/adipogenic

fibro/adipogenic cells and macrophages to expand and further commit and differentiate. These cues must be tightly and timely regulated to sequentially promote cell proliferation and proper myogenesis and reciprocal angiogenesis. Meantime, as the tissue is getting back to homeostasis, a subset of progenitor cells appear to be responsive to quiescent signals and will eventually find, always in a perivascular localization, cues from periendothelial cells to reinforce the quiescent signal that help to maintain them in the G0 state. Of special note, peri-ECs that secrete Ang1 stabilize satellite cells quiescence on one side and ECs (that also bear Tie-2) on the other side. This scenario requires further experimental investigations, but it is likely that considering the tissue in a dynamic way will help to understand the changing relationships between satellite cells with their neighboring vascular zone.

Previous studies have demonstrated that satellite cells can divide in an asymmetric fashion (Kuang *et al.*, 2007; Shinin *et al.*, 2006) but no relationship has been investigated regarding the environment, excepted that of the basal lamina: the cell that remains in contact with the basal lamina self-renews while the one that is in contact with the myofiber commits into the myogenic lineage (Kuang *et al.*, 2008). Does the vascular environment provide cues for asymmetric division?

The structural and biological roles of the basal lamina remain also to be determined during the various steps of a satellite cell life. Are there some specialized portions of the basal lamina near the vascular zone where satellite cells can find or define their quiescence niche? Previous studies (Christov *et al.*, 2007; Schmalbruch and Hellhammer, 1977) showed that satellite cells and vessels are separated by the basal lamina, but transient cytoplasmic extensions through basal lamina may ensure contacts between satellite cells and vessel cells. At time of regeneration, a minimal basal lamina structure must remain to ensure the muscle reconstruction, and muscle progenitor cells are observed equally under and outside the basal lamina (F. Chrétien, unpublished data). These observations suggest dynamic interactions between basal lamina and satellite cells, that are still now poorly investigated.

Finally, it is not known to what extent environmental cues participate to the entry of a subset of progenitor cells into quiescence/self-renewal by sending signals to a responsive subset of cells present in the expanding progenitor population. In human muscle progenitor culture, supernatants of human smooth muscle cells and fibroblasts trigger both an increase of Pax7 expression and quiescence, in an Ang-1/Tie-2 dependent way, and an

cells, by apoptosis, of macrophages, whose fate is not known, but that are likely drained in lymph nodes. Then, the association of perivascular cells with the vessel and the stabilization of the basal lamina around the vessels and the fibers further achieve the return to the steady-state status of the skeletal muscle. (See Color Insert.)

increase of MyoD expression in a Tie-2 independent way. These results suggest that periendothelial cells would act on both quiescence-responsive cells to induce their self-renewal and on the main myogenic progenitor population (not responsive to Ang1 signaling) to stimulate their further differentiation. This would mean that the niche or "stem cell zone," by sending various cues, acts differently on responsive/unresponsive cells. Depending on their state, stem cells answer—and act—differently to the cues provided by the niche, thus establishing a permanent dialogue between stem cells and their environment.

REFERENCES

Abou-Khalil, R., and Brack, A. S. (2010). Muscle stem cells and reversible quiescence: The role of sprouty. *Cell Cycle* **9,** 2573–2578.

Abou-Khalil, R., Le Grand, F., Pallafacchina, G., Valable, S., Authier, F. J., Rudnicki, M. A., Gherardi, R. K., Germain, S., Chretien, F., Sotiropoulos, A., Lafuste, P., Montarras, D., *et al.* (2009). Autocrine and paracrine angiopoietin 1/Tie-2 signaling promotes muscle satellite cell self-renewal. *Cell Stem Cell* **5,** 298–309.

Akerstrom, T., Steensberg, A., Keller, P., Keller, C., Penkowa, M., and Pedersen, B. K. (2005). Exercise induces interleukin-8 expression in human skeletal muscle. *J. Physiol.* **563,** 507–516.

Amaral, S. L., Papanek, P. E., and Greene, A. S. (2001). Angiotensin II and VEGF are involved in angiogenesis induced by short-term exercise training. *Am. J. Physiol. Heart Circ. Physiol.* **281,** H1163–H1169.

Andersen, P., and Henriksson, J. (1977). Training induced changes in the subgroups of human type II skeletal muscle fibres. *Acta Physiol. Scand.* **99,** 123–125.

Arai, F., Hirao, A., Ohmura, M., Sato, H., Matsuoka, S., Takubo, K., Ito, K., Koh, G. Y., and Suda, T. (2004). Tie2/angiopoietin-1 signaling regulates hematopoietic stem cell quiescence in the bone marrow niche. *Cell* **118,** 149–161.

Arnold, L., Henry, A., Poron, F., Baba-Amer, Y., van Rooijen, N., Plonquet, A., Gherardi, R. K., and Chazaud, B. (2007). Inflammatory monocytes recruited after skeletal muscle injury switch into antiinflammatory macrophages to support myogenesis. *J. Exp. Med.* **204,** 1071–1081.

Arsic, N., Zacchigna, S., Zentilin, L., Ramirez-Correa, G., Pattarini, L., Salvi, A., Sinagra, G., and Giacca, M. (2004). Vascular endothelial growth factor stimulates skeletal muscle regeneration in vivo. *Mol. Ther.* **10,** 844–854.

Barreiro, E., Schols, A. M., Polkey, M. I., Galdiz, J. B., Gosker, H. R., Swallow, E. B., Coronell, C., and Gea, J. (2008). Cytokine profile in quadriceps muscles of patients with severe COPD. *Thorax* **63,** 100–107.

Brack, A. S., Conboy, I. M., Conboy, M. J., Shen, J., and Rando, T. A. (2008). A temporal switch from notch to wnt signaling in muscle stem cells is necessary for normal adult myogenesis. *Cell Stem Cell* **2,** 50–59.

Breen, E. C., Johnson, E. C., Wagner, H., Tseng, H. M., Sung, L. A., and Wagner, P. D. (1996). Angiogenic growth factor mRNA responses in muscle to a single bout of exercise. *J. Appl. Physiol.* **81,** 355–361.

Bryan, B. A., Walshe, T. E., Mitchell, D. C., Havumaki, J. S., Saint-Geniez, M., Maharaj, A. S., Maldonado, A. E., and D'Amore, P. A. (2008). Coordinated vascular endothelial growth factor expression and signaling during skeletal myogenic differentiation. *Mol. Biol. Cell* **19,** 994–1006.

Campos, G. E., Luecke, T. J., Wendeln, H. K., Toma, K., Hagerman, F. C., Murray, T. F., Ragg, K. E., Ratamess, N. A., Kraemer, W. J., and Staron, R. S. (2002). Muscular adaptations in response to three different resistance-training regimens: Specificity of repetition maximum training zones. *Eur. J. Appl. Physiol.* **88,** 50–60.

Carnac, G., Fajas, L., L'Honore, A., Sardet, C., Lamb, N. J., and Fernandez, A. (2000). The retinoblastoma-like protein p130 is involved in the determination of reserve cells in differentiating myoblasts. *Curr. Biol.* **10,** 543–546.

Chakravarthy, M. V., Spangenburg, E. E., and Booth, F. W. (2001). Culture in low levels of oxygen enhances in vitro proliferation potential of satellite cells from old skeletal muscles. *Cell. Mol. Life Sci.* **58,** 1150–1158.

Chazaud, B., Sonnet, C., Lafuste, P., Bassez, G., Rimaniol, A. C., Poron, F., Authier, F. J., Dreyfus, P. A., and Gherardi, R. K. (2003). Satellite cells attract monocytes and use macrophages as a support to escape apoptosis and enhance muscle growth. *J. Cell Biol.* **163,** 1133–1143.

Christov, C., Chretien, F., Abou-Khalil, R., Bassez, G., Vallet, G., Authier, F. J., Bassaglia, Y., Shinin, V., Tajbakhsh, S., Chazaud, B., and Gherardi, R. K. (2007). Muscle satellite cells and endothelial cells: Close neighbors and privileged partners. *Mol. Biol. Cell* **18,** 1397–1409.

Crisan, M., Yap, S., Casteilla, L., Chen, C. W., Corselli, M., Park, T. S., Andriolo, G., Sun, B., Zheng, B., Zhang, L., Norotte, C., Teng, P. N., *et al.* (2008). A perivascular origin for mesenchymal stem cells in multiple human organs. *Cell Stem Cell* **3,** 301–313.

Dellavalle, A., Sampaolesi, M., Tonlorenzi, R., Tagliafico, E., Sacchetti, B., Perani, L., Innocenzi, A., Galvez, B. G., Messina, G., Morosetti, R., Li, S., Belicchi, M., *et al.* (2007). Pericytes of human skeletal muscle are myogenic precursors distinct from satellite cells. *Nat. Cell Biol.* **9,** 255–267.

Dhawan, J., and Rando, T. A. (2005). Stem cells in postnatal myogenesis: Molecular mechanisms of satellite cell quiescence, activation and replenishment. *Trends Cell Biol.* **15,** 666–673.

Doukas, J., Blease, K., Craig, D., Ma, C., Chandler, L. A., Sosnowski, B. A., and Pierce, G. F. (2002). Delivery of FGF genes to wound repair cells enhances arteriogenesis and myogenesis in skeletal muscle. *Mol. Ther.* **5,** 517–527.

Emanueli, C., Salis, M. B., Pinna, A., Graiani, G., Manni, L., and Madeddu, P. (2002). Nerve growth factor promotes angiogenesis and arteriogenesis in ischemic hindlimbs. *Circulation* **106,** 2257–2262.

Emslie-Smith, A. M., and Engel, A. G. (1990). Microvascular changes in early and advanced dermatomyositis: A quantitative study. *Ann. Neurol.* **27,** 343–356.

Fiedler, U., and Augustin, H. G. (2006). Angiopoietins: A link between angiogenesis and inflammation. *Trends Immunol.* **27,** 552–558.

Friday, B. B., and Pavlath, G. K. (2001). A calcineurin- and NFAT-dependent pathway regulates Myf5 gene expression in skeletal muscle reserve cells. *J. Cell Sci.* **114,** 303–310.

Frydelund-Larsen, L., Penkowa, M., Akerstrom, T., Zankari, A., Nielsen, S., and Pedersen, B. K. (2007). Exercise induces interleukin-8 receptor (CXCR2) expression in human skeletal muscle. *Exp. Physiol.* **92,** 233–240.

Galimi, F., Cottone, E., Vigna, E., Arena, N., Boccaccio, C., Giordano, S., Naldini, L., and Comoglio, P. M. (2001). Hepatocyte growth factor is a regulator of monocyte-macrophage function. *J. Immunol.* **166,** 1241–1247.

Gavin, T. P., Robinson, C. B., Yeager, R. C., England, J. A., Nifong, L. W., and Hickner, R. C. (2004). Angiogenic growth factor response to acute systemic exercise in human skeletal muscle. *J. Appl. Physiol.* **96,** 19–24.

Germani, A., Di Carlo, A., Mangoni, A., Straino, S., Giacinti, C., Turrini, P., Biglioli, P., and Capogrossi, M. C. (2003). Vascular endothelial growth factor modulates skeletal myoblast function. *Am. J. Pathol.* **163,** 1417–1428.

Greenbaum, A. R., Etherington, P. J., Manek, S., O'Hare, D., Parker, K. H., Green, C. J., Pepper, J. R., and Winlove, C. P. (1997). Measurements of oxygenation and perfusion in skeletal muscle using multiple microelectrodes. *J. Muscle Res. Cell Motil.* **18,** 149–159.

Gustafsson, T., Rundqvist, H., Norrbom, J., Rullman, E., Jansson, E., and Sundberg, C. J. (2007). The influence of physical training on the angiopoietin and VEGF-A systems in human skeletal muscle. *J. Appl. Physiol.* **103,** 1012–1020.

Hiscock, N., Fischer, C. P., Pilegaard, H., and Pedersen, B. K. (2003). Vascular endothelial growth factor mRNA expression and arteriovenous balance in response to prolonged, submaximal exercise in humans. *Am. J. Physiol. Heart Circ. Physiol.* **285,** H1759–H1763.

Hudlicka, O., Brown, M., and Egginton, S. (1992). Angiogenesis in skeletal and cardiac muscle. *Physiol. Rev.* **72,** 369–417.

Jain, R. K. (2003). Molecular regulation of vessel maturation. *Nat. Med.* **9,** 685–693.

Jensen, L., Bangsbo, J., and Hellsten, Y. (2004). Effect of high intensity training on capillarization and presence of angiogenic factors in human skeletal muscle. *J. Physiol.* **557,** 571–582.

Joe, A. W., Yi, L., Natarajan, A., Le Grand, F., So, L., Wang, J., Rudnicki, M. A., and Rossi, F. M. (2010). Muscle injury activates resident fibro/adipogenic progenitors that facilitate myogenesis. *Nat. Cell Biol.* **12,** 153–163.

Kardon, G., Harfe, B. D., and Tabin, C. J. (2003). A Tcf4-positive mesodermal population provides a prepattern for vertebrate limb muscle patterning. *Dev. Cell* **5,** 937–944.

Kiel, M. J., and Morrison, S. J. (2008). Uncertainty in the niches that maintain haematopoietic stem cells. *Nat. Rev. Immunol.* **8,** 290–301.

Kim, K. I., Cho, H. J., Hahn, J. Y., Kim, T. Y., Park, K. W., Koo, B. K., Soo, S. C., Kim, C. H., Oh, B. H., Lee, M. M., Park, Y. B., and Kim, H. S. (2006). {beta}-Catenin overexpression augments angiogenesis and skeletal muscle regeneration through dual mechanism of vascular endothelial growth factor-mediated endothelial cell proliferation and progenitor cell mobilization. *Arterioscler. Thromb. Vasc. Biol.* **26,** 91–98.

Kitzmann, M., Bonnieu, A., Duret, C., Vernus, B., Barro, M., Laoudj-Chenivesse, D., Verdi, J. M., and Carnac, G. (2006). Inhibition of notch signaling induces myotube hypertrophy by recruiting a subpopulation of reserve cells. *J. Cell. Physiol.* **208,** 538–548.

Kivela, R., Silvennoinen, M., Lehti, M., Jalava, S., Vihko, V., and Kainulainen, H. (2008). Exercise-induced expression of angiogenic growth factors in skeletal muscle and in capillaries of healthy and diabetic mice. *Cardiovasc. Diabetol.* **7,** 13.

Kuang, S., Kuroda, K., Le Grand, F., and Rudnicki, M. A. (2007). Asymmetric self-renewal and commitment of satellite stem cells in muscle. *Cell* **129,** 999–1010.

Kuang, S., Gillespie, M. A., and Rudnicki, M. A. (2008). Niche regulation of muscle satellite cell self-renewal and differentiation. *Cell Stem Cell* **2,** 22–31.

Kutcher, M. E., and Herman, I. M. (2009). The pericyte: Cellular regulator of microvascular blood flow. *Microvasc. Res.* **77,** 235–246.

Lepper, C., Conway, S. J., and Fan, C. M. (2009). Adult satellite cells and embryonic muscle progenitors have distinct genetic requirements. *Nature* **460,** 627–631.

Li, L., and Clevers, H. (2010). Coexistence of quiescent and active adult stem cells in mammals. *Science* **327,** 542–545.

Li, Q., Ford, M. C., Lavik, E. B., and Madri, J. A. (2006). Modeling the neurovascular niche: VEGF- and BDNF-mediated cross-talk between neural stem cells and endothelial cells: An in vitro study. *J. Neurosci. Res.* **84,** 1656–1668.

Lichtman, M. A. (1981). The ultrastructure of the hemopoietic environment of the marrow: A review. *Exp. Hematol.* **9,** 391–410.

Lloyd, P. G., Prior, B. M., Yang, H. T., and Terjung, R. L. (2003). Angiogenic growth factor expression in rat skeletal muscle in response to exercise training. *Am. J. Physiol. Heart Circ. Physiol.* **284,** H1668–H1678.

Louissaint, A., Jr., Rao, S., Leventhal, C., and Goldman, S. A. (2002). Coordinated interaction of neurogenesis and angiogenesis in the adult songbird brain. *Neuron* **34,** 945–960.

Luque, E., Pena, J., Martin, P., Jimena, I., and Vaamonde, R. (1995). Capillary supply during development of individual regenerating muscle fibers. *Anat. Histol. Embryol.* **24,** 87–89.

Mendez-Ferrer, S., Michurina, T. V., Ferraro, F., Mazloom, A. R., Macarthur, B. D., Lira, S. A., Scadden, D. T., Ma'ayan, A., Enikolopov, G. N., and Frenette, P. S. (2010). Mesenchymal and haematopoietic stem cells form a unique bone marrow niche. *Nature* **466,** 829–834.

Mitchell, K. J., Pannerec, A., Cadot, B., Parlakian, A., Besson, V., Gomes, E. R., Marazzi, G., and Sassoon, D. A. (2010). Identification and characterization of a non-satellite cell muscle resident progenitor during postnatal development. *Nat. Cell Biol.* **12,** 257–266.

Ochoa, O., Sun, D., Reyes-Reyna, S. M., Waite, L. L., Michalek, J. E., McManus, L. M., and Shireman, P. K. (2007). Delayed angiogenesis and VEGF production in CCR2−/− mice during impaired skeletal muscle regeneration. *Am. J. Physiol. Regul. Integr. Comp. Physiol.* **293,** R651–R661.

Olfert, I. M., Breen, E. C., Mathieu-Costello, O., and Wagner, P. D. (2001). Skeletal muscle capillarity and angiogenic mRNA levels after exercise training in normoxia and chronic hypoxia. *J. Appl. Physiol.* **91,** 1176–1184.

Olfert, I. M., Howlett, R. A., Tang, K., Dalton, N. D., Gu, Y., Peterson, K. L., Wagner, P. D., and Breen, E. C. (2009). Muscle-specific VEGF deficiency greatly reduces exercise endurance in mice. *J. Physiol.* **587,** 1755–1767.

Olguin, H. C., and Olwin, B. B. (2004). Pax-7 up-regulation inhibits myogenesis and cell cycle progression in satellite cells: A potential mechanism for self-renewal. *Dev. Biol.* **275,** 375–388.

Oustanina, S., Hause, G., and Braun, T. (2004). Pax7 directs postnatal renewal and propagation of myogenic satellite cells but not their specification. *EMBO J.* **23,** 3430–3439.

Palmer, T. D., Willhoite, A. R., and Gage, F. H. (2000). Vascular niche for adult hippocampal neurogenesis. *J. Comp. Neurol.* **425,** 479–494.

Parmar, K., Mauch, P., Vergilio, J. A., Sackstein, R., and Down, J. D. (2007). Distribution of hematopoietic stem cells in the bone marrow according to regional hypoxia. *Proc. Natl. Acad. Sci. USA* **104,** 5431–5436.

Perez-Ruiz, A., Ono, Y., Gnocchi, V. F., and Zammit, P. S. (2008). {beta}-Catenin promotes self-renewal of skeletal-muscle satellite cells. *J. Cell Sci.* **121,** 1373–1382.

Prior, B. M., Lloyd, P. G., Yang, H. T., and Terjung, R. L. (2003). Exercise-induced vascular remodeling. *Exerc. Sport Sci. Rev.* **31,** 26–33.

Prior, B. M., Yang, H. T., and Terjung, R. L. (2004). What makes vessels grow with exercise training? *J. Appl. Physiol.* **97,** 1119–1128.

Rhoads, R. P., Johnson, R. M., Rathbone, C. R., Liu, X., Temm-Grove, C., Sheehan, S. M., Hoying, J. B., and Allen, R. E. (2009). Satellite cell-mediated angiogenesis in vitro coincides with a functional hypoxia-inducible factor pathway. *Am. J. Physiol. Cell Physiol.* **296,** C1321–C1328.

Richardson, R. S., Noyszewski, E. A., Leigh, J. S., and Wagner, P. D. (1998). Lactate efflux from exercising human skeletal muscle: Role of intracellular PO2. *J. Appl. Physiol.* **85,** 627–634.

Schmalbruch, H., and Hellhammer, U. (1977). The number of nuclei in adult rat muscles with special reference to satellite cells. *Anat. Rec.* **189,** 169–175.

Scholz, D., Thomas, S., Sass, S., and Podzuweit, T. (2003). Angiogenesis and myogenesis as two facets of inflammatory post-ischemic tissue regeneration. *Mol. Cell. Biochem.* **246,** 57–67.

Shea, K. L., Xiang, W., LaPorta, V. S., Licht, J. D., Keller, C., Basson, M. A., and Brack, A. S. (2010). Sprouty1 regulates reversible quiescence of a self-renewing adult muscle stem cell pool during regeneration. *Cell Stem Cell* **6,** 117–129.

Shen, Q., Goderie, S. K., Jin, L., Karanth, N., Sun, Y., Abramova, N., Vincent, P., Pumiglia, K., and Temple, S. (2004). Endothelial cells stimulate self-renewal and expand neurogenesis of neural stem cells. *Science* **304,** 1338–1340.

Shim, W. S., Ho, I. A., and Wong, P. E. (2007). Angiopoietin: A TIE(d) balance in tumor angiogenesis. *Mol. Cancer Res.* **5,** 655–665.

Shinin, V., Gayraud-Morel, B., Gomes, D., and Tajbakhsh, S. (2006). Asymmetric division and cosegregation of template DNA strands in adult muscle satellite cells. *Nat. Cell Biol.* **8,** 677–682.

Takahashi, A., Kureishi, Y., Yang, J., Luo, Z., Guo, K., Mukhopadhyay, D., Ivashchenko, Y., Branellec, D., and Walsh, K. (2002). Myogenic Akt signaling regulates blood vessel recruitment during myofiber growth. *Mol. Cell. Biol.* **22,** 4803–4814.

Tang, K., Breen, E. C., Gerber, H. P., Ferrara, N. M., and Wagner, P. D. (2004). Capillary regression in vascular endothelial growth factor-deficient skeletal muscle. *Physiol. Genomics* **18,** 63–69.

Tatsumi, R., Anderson, J. E., Nevoret, C. J., Halevy, O., and Allen, R. E. (1998). HGF/SF is present in normal adult skeletal muscle and is capable of activating satellite cells. *Dev. Biol.* **194,** 114–128.

Tedesco, F. S., Dellavalle, A., Diaz-Manera, J., Messina, G., and Cossu, G. (2010). Repairing skeletal muscle: Regenerative potential of skeletal muscle stem cells. *J. Clin. Invest.* **120,** 11–19.

Wagatsuma, A. (2007). Endogenous expression of angiogenesis-related factors in response to muscle injury. *Mol. Cell. Biochem.* **298,** 151–159.

Wagner, P. D., Olfert, I. M., Tang, K., and Breen, E. C. (2006). Muscle-targeted deletion of VEGF and exercise capacity in mice. *Respir. Physiol. Neurobiol.* **151,** 159–166.

Walgenbach, K. J., Gratas, C., Shestak, K. C., and Becker, D. (1995). Ischaemia-induced expression of bFGF in normal skeletal muscle: A potential paracrine mechanism for mediating angiogenesis in ischaemic skeletal muscle. *Nat. Med.* **1,** 453–459.

Yamada, M., Tatsumi, R., Yamanouchi, K., Hosoyama, T., Shiratsuchi, S., Sato, A., Mizunoya, W., Ikeuchi, Y., Furuse, M., and Allen, R. E. (2010). High concentrations of HGF inhibit skeletal muscle satellite cell proliferation in vitro by inducing expression of myostatin: A possible mechanism for reestablishing satellite cell quiescence in vivo. *Am. J. Physiol. Cell Physiol.* **298,** C465–C476.

Yun, Z., Lin, Q., and Giaccia, A. J. (2005). Adaptive myogenesis under hypoxia. *Mol. Cell. Biol.* **25,** 3040–3055.

CHAPTER SIX

Nonmyogenic Cells in Skeletal Muscle Regeneration

Ben Paylor, Anuradha Natarajan, Regan-Heng Zhang, *and* Fabio Rossi

Contents

Abstract

Although classical dogma dictates that satellite cells are the primary cell type involved in skeletal muscle regeneration, alternative cell types such as a variety of inflammatory and stromal cells are also actively involved in this process. A model describing myogenic cells as direct contributors to regeneration and

Biomedical Research Center, University of British Colombia, Vancouver, British Colombia, Canada

Current Topics in Developmental Biology, Volume 96

ISSN 0070-2153, DOI: 10.1016/B978-0-12-385940-2.00006-1

nonmyogenic cells from other developmental sources as important accessories has emerged, with similar systems having been described in numerous other tissues in the body. Increasing evidence supports the notion that inflammatory cells function as supportive accessory cells, and are not merely involved in clearing damage following skeletal muscle injury. Additionally, recent studies have highlighted the role of tissue resident mesenchymal cell populations as playing a central role in regulating regeneration. These "accessory" cell populations are proposed to influence myogenesis via direct cell contact and secretion of paracrine trophic factors. The basic foundations of accessory cell understanding should be recognized as a crucial component to all prospects of regenerative medicine, and this chapter intends to provide a comprehensive background on the current literature describing immune and tissue-resident mesenchymal cells' role in skeletal muscle regeneration.

1. Introduction

Skeletal muscle has a great capacity to grow and regenerate during adult life. Its main constituent, myofibers, are long, cylindrical, specialized cells containing multiple nuclei and segments of myofibrils made of both actin and myosin. These basic functional units give skeletal muscle its contractile capacity. Myofiber growth takes place when mononuclear myoblasts fuse together, increasing the number of nuclei and allowing elongation of existing myofibers. In adults, the ancestors to these myoblasts are satellite cells, which reside between the myofibers and the basement membrane of the muscle bundle (Fig. 6.1). Following traumatic myofiber damage or temporal progression of myopathy, regeneration requires the activation of satellite cells, which facilitate this process by either fusing with damaged myofibers or generating new myofibers (Charge and Rudnicki, 2004; Collins *et al.*, 2005; Dhawan and Rando, 2005; Kuang *et al.*, 2007; Mauro, 1961; Morgan and Partridge, 2003; Sacco *et al.*, 2008). Normally quiescent, satellite cells are able to proliferate in response to tissue damage and eventually contribute to myofibers by progressing through a stepwise differentiation process that has been extensively characterized at the cellular and molecular level (Relaix *et al.*, 2005).

The coordination of this process is complex and a number of extracellular signals have been proposed to regulate the steps of myogenic differentiation (Bodine *et al.*, 2001; Brack *et al.*, 2007; Conboy and Rando, 2002; Otto *et al.*, 2008; Serrano *et al.*, 2008). Although classical dogma dictates that satellite cells are the primary cell type involved in skeletal muscle regeneration, increasing evidence implicates alternative cell types, such as a variety of inflammatory and stromal cells being involved in this process (Arnold *et al.*, 2007; Contreras-Shannon *et al.*, 2007a,b; Joe *et al.*, 2010a; Sonnet *et al.*, 2006a,b). Clearly, the presence of molecular signals regulating myogenic

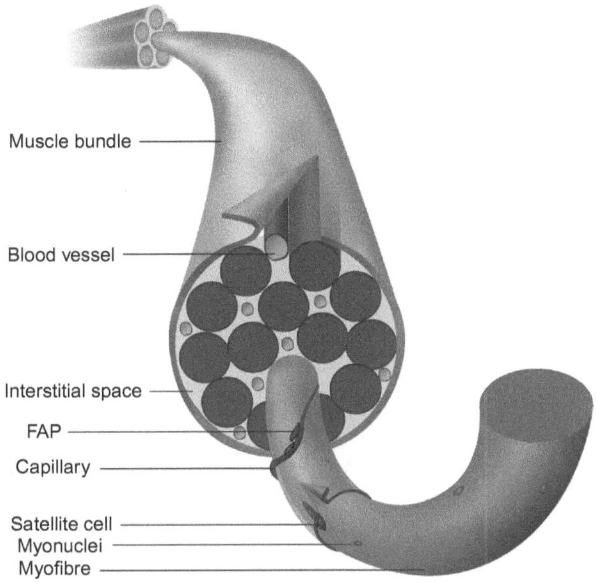

Figure 6.1 Schematic representation of muscle anatomy and the positional relationships between muscle fibers and blood vessels. Blood vessels are dispersed throughout the interstitial space in muscle bundles, with capillaries in close proximity to individual myofibers. Myogenic satellite cells are known to reside under the basal lamina of muscle fibers, which are in close proximity to fibro/adipogenic progenitors (FAP), which are positioned near myofiber-associated capillary vessels at steady state.

cell activity is suggestive of the importance of nonmyogenic cells that produce these signals. A model describing myogenic cells as direct contributors to regeneration and nonmyogenic cells from other developmental sources as important accessories has been proposed (Fig. 6.2; Natarajan *et al.*, 2010; Rodeheffer, 2010). Following this model, it becomes clear that a full understanding of the mechanisms ensuring regeneration proceeds efficiently requires the study of integrated cellular interactions rather than one-sided investigations solely focused on myogenic cells.

2. Accessory Cells in Skeletal Muscle Regeneration

One of the clearest examples of nonmyogenic cells having an accessory role in muscle regeneration arises from widespread observations of immune/inflammatory responses following tissue injury. The presence of immune cells in regenerating tissue was first evidenced (Godman, 1957) by the observation that circulating cells invade skeletal muscle following injury. Further, the relative abundance of distinct leukocyte populations such as

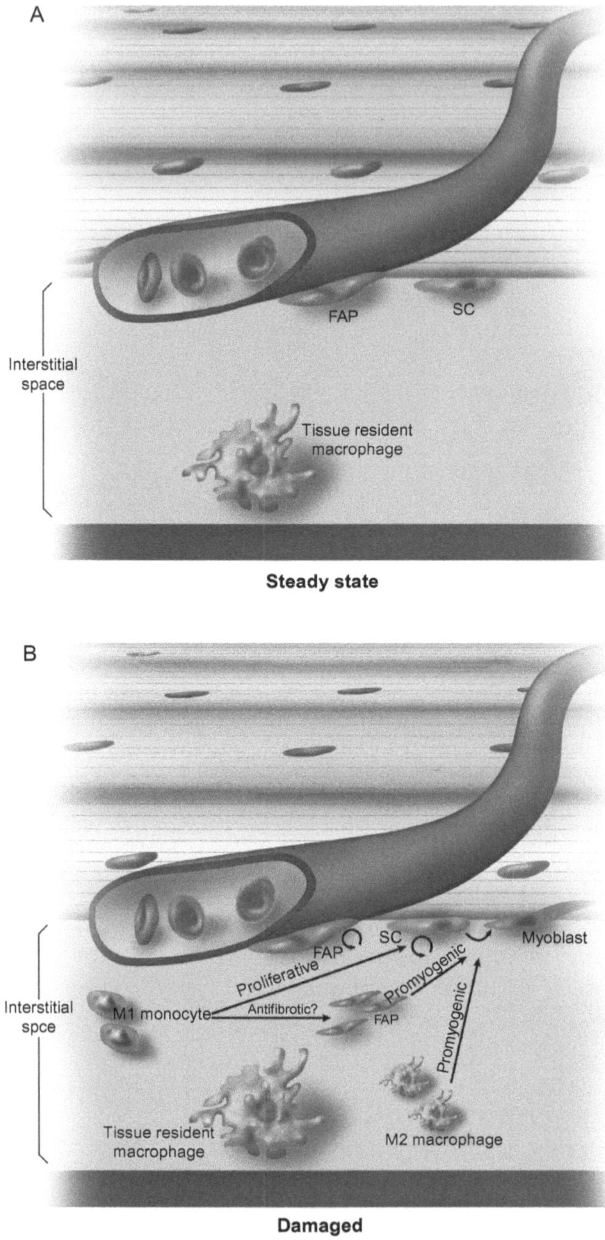

Figure 6.2 (A) Visualization of the interstitial space in skeletal muscle at steady state. Fibro/adipogenic progenitors (FAP) are a population of mesenchymal cells associated with capillary vessels and distinct from satellite cells (SC). (B) In damaged skeletal muscle, significant cross talk occurs between the numerous cell types present. Satellite cells have the greatest myogenic capacity in skeletal muscle, and their capacity to fuse with existing myofibers or generate new myofibers is regulated by both immune- and tissue-resident mesenchymal FAPs. Macrophages, monocytes, and FAPs influence myogenesis by providing both proliferative and promyogenic signals to satellite cells.

neutrophils and various types of macrophages changed during the course of regeneration, suggesting differing temporal roles for each population. Active research on the link between inflammation and regeneration has been limited, but increasing evidence supports the notion that inflammatory cells are not merely involved in clearing damage following skeletal muscle injury, but also have a central role in regulating regeneration.

In addition to inflammatory cells, tissue resident stromal cells are likely to play a significant role in regeneration. For example, mesenchymal stromal cells (MSC) exist in muscle, and are likely to contribute to the restoration of the extracellular matrix (ECM). It seems intuitive that such processes proceed hand-in-hand with myogenesis to achieve functional restoration, suggesting that some form of cross talk exists and coordinate the functions of both myogenic and nonmyogenic cell types involved. In support of this notion, we have recently characterized tissue resident progenitors endowed with both fibroblastic and adipogenic potentials (Joe *et al.*, 2010a; Uezumi *et al.*, 2010). These cells colocalize with satellite cells, proliferate upon muscle injury and are a rich source of trophic factors, suggesting an active role in skeletal muscle regeneration.

3. Accessory Cells in Regeneration: A Mechanism Common to Multiple Tissues

The accessory cell types and their roles described in this chapter have comparable roles in the regenerative processes of other tissues (Cleaver and Melton, 2003). In skeletal muscle, these regulatory effects primarily modulate the proliferation and differentiation of satellite cells—the principal myogenic progenitor present in this tissue. However, similar relationships between "principal" and "accessory" cells can be drawn across multiple organs and tissue types, and it is likely that most progenitors involved in adult regeneration rely on microenvironments comprised of stromal structures, local signals, and accessory cells which are commonly called "niches." It is certain that simply residing in an obstinate physical structure is not enough for a progenitor to operate properly; there must also be a dynamic aspect to the microenvironment that enables or disables progenitors. Following this, the "stem cell niche" was proposed by Scadden and defined as a supportive microenvironment having "both anatomic and functional dimensions" (Scadden, 2006). These dynamic niches are a result of the interplay between a variety of elements. This chapter will focus primarily on supportive cellular constituents that are able to modulate myogenic activity, but it is important to remember that multiple mechanisms underlie communication among different progenitors.

During development, many progenitor populations are regulated by nearby endothelial cells in numerous tissues such as the pancreas, heart, brain, liver, adrenal glands, bone, and blood (Cleaver and Melton, 2003). In the hippocampus, neural precursors colocalize with angioblasts and glial cells to generate neurons (Palmer *et al.*, 2000). In the lung, epithelial repair involves the activation of fibroblasts, endothelial cells, macrophages, and neutrophils, which contribute the release of growth factors and cytokines for successful generation and spread of epithelial cells (Crosby and Waters, 2010). In hematopoiesis, mesenchymal cells contribute to the microenvironment around hematopoietic stem cells (HSCs; Kiel and Morrison, 2008); their increase in number correlates with increased HSC numbers (Calvi *et al.*, 2003; Zhang *et al.*, 2003) and have been shown to regulate the HSC population by secreting factors such as angiopoietin (Arai *et al.*, 2004) and osteopontin (Nilsson *et al.*, 2005; Stier *et al.*, 2005). In some cases, a "third-party" cell type has been involved, such as when osteoblasts use signals to recruit vasculature to the bone marrow, which in turn attracts HSCs to the endosteal niche (Street *et al.*, 2002; Tombran-Tink and Barnstable, 2004). The combination of this angiogenic influence and the induced local recruitment of HSCs stimulate bone regeneration (Street *et al.*, 2002).

4. Primary and Accessory Cell Communication

Soluble molecules are among the most direct mediators, and the most convincing reported examples of regulatory communication between accessory cells and progenitors. Niche regulation of progenitors may also occur through direct contact. For instance, osteoblasts in the endosteum have been proposed to regulate HSC activity through N-cadherin and Notch attachments (Adams *et al.*, 2006; Li and Xie, 2005; Suda *et al.*, 2005), with these molecules potentially providing both a physical attachment site and an extracellular signal.

Despite the numerous examples of niche regulation of progenitors in multiple organ systems during homeostasis, the exact nature of the cells involved in providing signals that coordinate acute responses to damage is still poorly defined. The majority of findings implicating mesenchymal cell types' involvement in regeneration stems from implantation studies, where mixed stromal populations were deliberately transplanted in tissues on the false assumption that they would contain multipotent stem cells that could contribute to regeneration directly. In most cases, it was later found that the proregenerative effects of such procedures were due to paracrine production of trophic factors influencing multiple endogenous cell types. The role of these mesenchymal cells in muscle regeneration has only recently begun to be understood, as studies characterizing the response of endogenous, tissue-resident mesenchymal progenitors to damage have emerged (Joe *et al.*, 2010a; Uezumi *et al.*, 2010). These findings support the notion that production of paracrine trophic factors is an important physiological function of mesenchymal cells *in vivo*. Together with the notion that all tissues contain similar mesenchymal constituents (Crisan *et al.*, 2008), this recent data suggests that MSCs may form a diffuse system carrying out similar functions in vastly different anatomical locations, reminiscent of tissue macrophages or endothelial cells. Further, it is postulated that this system plays an important role in tissue homeostasis and regeneration by providing trophic support for local, tissue specific progenitors. Additionally, the accessory cell types discussed in this chapter are not tissue-specific, and are likely involved in biological phenomena such as inflammation and fibrosis that are common to many regenerative environments. Discoveries in this area will provide insight into the influence of these diffuse accessory systems on regeneration as a general mechanism of tissue sustainability. Thus, the basic foundations of accessory cell understanding should be recognized as a crucial component to all prospects of regenerative medicine.

5. Development Versus Adult Regeneration

Following tissue damage, regeneration in muscle is regulated in part by muscle-specific transcription factors that belong to the basic helix-loop-helix family (bHLH; Le Grand and Rudnicki, 2007; Lluis *et al.*, 2006). Based on the expression of the bHLH transcription factors MyoD, MRF4, Myf5, and myogenin, the process of muscle regeneration can be divided into three stages. The first, the proliferative stage, involves the activation and expansion of satellite cells. MyoD and Myf5 are expressed during this stage, however, they are kept in the inactive state until the cessation of satellite cell proliferation and initiation of the early differentiation stage (Cornelison and Wold, 1997; Cornelison *et al.*, 2000; Fuchtbauer and Westphal, 1992; Grounds *et al.*, 1992; Yablonka-Reuveni and Rivera, 1994). Along with the activation of MyoD and Myf5, the expression of MRF4 and myogenin is initiated during the early differentiation stage and drives the expression of muscle-specific genes that are necessary for progression to the terminal differentiation stage and cell fusion (Cornelison and Wold, 1997; Cornelison *et al.*, 2000; Fuchtbauer and Westphal, 1992; Grounds *et al.*, 1992). The pattern of myogenic regulatory factor induction that ensues after muscle damage closely reiterates embryonic muscle development. However, a major difference between the two settings lies in the contribution of immune cells to the process. While immune cells are scarce in the environment of embryonic myogenesis (Abood and Jones, 1991), they are the predominant cell population in acutely regenerating muscle. Nevertheless, our understanding of the innate immune system response to muscle injury and the complex mechanism(s) through which these cells modulate regeneration is still rudimentary. The problem is made more complex by the fact that the inflammatory responses triggered by acute damage are significantly different from those stimulated by the chronic damage.

6. Regulation of Acute Muscle Regeneration and Repair by Immune Cells

Acute muscle trauma results from a variety of insults including trauma, freezing, overuse, and chemical or biological insults. Regeneration is defined as the functional restoration of the tissue, and it requires degradation of severely injured myofibers, elimination of the degradation products and in some cases generation of new muscle fibers. Repair on the other hand is defined as a process that restores the continuity of the tissue but fails to restore its function (e.g., scar formation). Inflammation, which invariably follows muscle damage, invokes a stereotypic response that involves the sequential invasion of muscle by specific myeloid cell populations, the most abundant of which are neutrophils and macrophages. Similar to other tissues, the primary responders are $Ly6c^{+}/F4/80^{-}$ neutrophils, which appear at the injury site in elevated numbers within 2 h following damage. Their numbers peak between 6- and 24-h postinjury, after which they rapidly decline (Tidball *et al.*, 1999). Following neutrophil influx, phagocytic macrophages become the dominant inflammatory cell type to invade the muscle, reaching elevated numbers within 24 h and reaching a peak around 3 days postinjury, after which they gradually decline (Bondesen *et al.*, 2006; Contreras-Shannon *et al.*, 2007a,b; Ochoa *et al.*, 2007; St. Pierre and Tidball, 1994; Tidball *et al.*, 1999). It is suggestive that the sequential changes in the pattern of myogenic transcription factors expressed during muscle regeneration and repair are concomitant with the changes in the relative abundance of subpopulations of myeloid cells (Yan *et al.*, 2003). However, this observation should not be overinterpreted, as the *in vitro* pattern of gene expression observed myogenic cell cultures devoid of myeloid cells are similar in both sequence and timing to that observed during *in vivo* myogenesis (Cornelison and Wold, 1997; Yablonka-Reuveni and Rivera, 1994). Although both *in vitro* studies and *in vivo* embryonic myogenesis challenge the role of immune cells in regulating myogenic progression, they leave open the possibility that myeloid cells may regulate other important aspects of regeneration such as cell survival or return to quiescence. In addition, myeloid cells could influence myogenic progenitors by acting on "third parties" such as mesenchymal progenitors to attenuate negative effects that these cells may have on regeneration (Fig. 6.2).

6.1. Do neutrophils regulate skeletal muscle regeneration?

Neutrophil activation in response to injury is a highly organized event involving penetration of the basement membrane and subsequent transmigration through venules into the site of damage, which is facilitated by adhesion molecules and integrin signaling (Molteni *et al.*, 2006; Wang *et al.*, 2006). Upon reaching the site of damage, activated neutrophils undergo a respiratory burst and degranulation to rapidly release free radicals and proteases (such as elastase) that target the cellular debris or ECM for degradation (Wiedow and Meyer-Hoffert, 2005). In addition, neutrophils secrete proinflammatory cytokines that further promote recruitment and/or proliferation of macrophages, potentiating their influence on regeneration and repair. Thus, neutrophils infiltrating damaged tissue could set the stage for subsequent steps in regeneration by initiating the clearance of debris and thereby preparing the ECM for further tissue remodeling. However, the role of neutrophils in regeneration has been controversial as the free radicals they release are cytotoxic and could potentially aggravate muscle damage.

Several lines of evidence support the notion that neutrophil-mediated phagocytosis promotes muscle regeneration. Depletion of neutrophils and monocytes prior to damage by snake venom resulted in more tissue debris and slower muscle regeneration, thus allowing one to speculate that the impaired capacity to clear tissue debris could slow the regenerative process (Teixeira *et al.*, 2003). In addition, Summan *et al.* (2006) showed in a freeze injury model that depletion of myeloid cells by intravenous administration of liposomes containing clondronate slows the removal of cellular debris and subsequently muscle regeneration, which further supports the role of neutrophils in regeneration. A caveat of this study is that the clondronate treatment depletes both the neutrophils and macrophages.

Although the role of neutrophils in clearing up debris is critical, the oxidizing environment associated with their infiltration and subsequent respiratory burst may cause collateral damage to healthy muscle during the early inflammatory period. Using a hindlimb suspension model, Nguyen and Tidball (2003a), demonstrated that mice deficient for NADPH oxidase (null mutation of gp91phox) displayed 90% reduction in muscle fiber damage during reloading without concomitant changes in the number of neutrophils and macrophages in the reloaded muscles. This suggests that indeed neutrophils can exacerbate damage through production of free radicals. Furthermore, myofibers underwent normal hypertrophy during the reloading period suggesting that superoxide-mediated signals are not required for myeloid cell recruitment or normal regeneration. Alternatively, blocking of CD11b, a major adhesion receptor for both neutrophils and macrophages, by administering soluble CD11b as a competitive inhibitor or blocking antibodies, reduced both injury and the number of neutrophils entering damaged tissue following needle puncture wounds in rats (Zerria *et al.*,

2006). Similarly, administration of anti-CD11b antibody (M1/70) prior to stretch injury of the tibialis anterior muscle reduced neutrophil infiltration as well as myofiber damage at 24 h postinjury in rabbits (Brickson *et al.*, 2003). Differences in the injury models used could explain the lack of consensus about the role of neutrophils in regenerating muscle and further highlight the need for both careful and standardized investigation of these cells.

6.2. Macrophages and muscle regeneration

Circulating monocytes originating from the bone marrow are the precursors of mature macrophages and can be recruited to peripheral sites and further influenced to differentiate by proinflammatory and immune stimuli. Even though macrophages are well known for their role in innate immunity, these multifaceted cells also play a critical role in sterile injuries, supporting regeneration, tissue remodeling and/or repair depending on their environmental context. Monocyte/macrophages have been implicated as key players in skeletal muscle regeneration (Chazaud *et al.*, 2009). Beyond their role as scavengers, clear evidence supporting a direct influence of macrophages on muscle progenitors stems from *in vitro* studies (Cantini and Carraro, 1995; Massimino *et al.*, 1997). Addition of conditioned media from J774 macrophage cell line to rat and human myoblasts cultures, led to increased expression of MyoD and myogenin, suggesting that macrophages secrete myogenic factors that modulate proliferation and/or differentiation of myoblasts (Cantini *et al.*, 2002). However, another study reported that when cocultured with myoblasts from turkeys and mice, macrophages showed increased proliferation as assessed by BrDU incorporation suggesting the cross talk between these cells types might be bidirectional. Surprisingly, the number of myogenin positive cells was decreased in these cultures (Merly *et al.*, 1999). Although the latter study suggests that secreted factors or cell–cell contact can inhibit muscle differentiation, it is critical to take into consideration that there exist a number of functional phenotypes of macrophages present in these studies.

6.3. Phenotypic plasticity of macrophages during muscle regeneration

Macrophages are well known for their heterogeneity, which is dictated by their function, anatomical location (Gordon and Taylor, 2005) and differing activation states. Muscle damage induces an innate immune response driven by Th1 cytokines such as TNF-α and interferon-γ resulting in a "classical activation program." This results in recruitment and expansion of so-called M1 macrophages, and eventually in the production of both proinflammatory cytokines and reactive oxygen species. Initial studies using rodent hind limb suspension model (St. Pierre and Tidball, 1994) and freeze injury (McLennan, 1996) reported that M1 macrophages expressing CD68 (also known as ED1 or macrosialin) accumulated in skeletal muscle 24 h after damage and were abundant within both the degenerating fibers and perimysium between arterioles and degenerating fibers (McLennan, 1993). In contrast, $CD68^{+}$ cells were rare in undamaged tissue. M1 macrophages entering damaged muscle are commonly defined by expression of a specific combination of surface markers including $CD45^{+}$, $CD11b^{+}$, $F4/80^{low}$, $Ly6c^{+}$, $CCR2^{+}$, and CX_3CR1^{low} (Arnold *et al.*, 2007; Gordon and Taylor, 2005). It has been reported that the binding of oxidized low-density lipoproteins (LDLs) to CD68 activates M1 macrophages phagocytic activity and production of proinflammatory cytokines such as TNFα and numerous other mediators known to have a negative effect on muscle regeneration (Ottnad *et al.*, 1995; Ramprasad *et al.*, 1995). Interestingly, neutrophils invading the damaged muscle prior to macrophages release myeloperoxidases (MPOs; Nguyen and Tidball, 2003b) that oxidize LDLs (Zouaoui Boudjeltia *et al.*, 2004) thereby potentially playing a key role in the activation of M1 macrophages. Indeed, the number of infiltrating neutrophils and macrophages was not altered when mice lacking MPO were subjected to hind limb unloading- and reloading-induced damage, but a significant reduction in fiber injury was observed (Nguyen *et al.*, 2005).

Following the first phase of proinflammatory cytokine production and phagocytosis of apoptotic and necrotic myofibers in which M1 macrophages are the most abundant myelomonocytic cells found in the damaged area, a second phase ensues in which so-called anti-inflammatory M2 macrophages become prevalent. This phenotype usually involves an environment rich in Th2 cytokines and is often associated with chronic inflammation. However, during acute muscle regeneration the switch from proinflammatory to anti-inflammatory state, triggered by environmental cues, has been linked to both the resolution of the inflammation and to efficient tissue regeneration. Specifically, activation of M2 macrophages by IL-4 and IL-10 is thought to promote tissue repair whereas activation by IL-13 results in the attenuation of inflammation (Gordon, 2003). In mice, M2 macrophages are identified by the following expression pattern—$CD45^{+}$ $CD11b^{+}$, $CD206^{+}$, $F4/80^{hi}$, $Ly6c^{-}$, $CCR2^{-}$, $CX3CR1^{hi}$ (Arnold *et al.*, 2007; Gordon and Taylor, 2005). What is the origin of each macrophage subset? M1 macrophages are thought to directly derive from bone marrow- and spleen-resident precursor monocytes that are mobilized by systemic signals to enter the bloodstream and infiltrate peripheral tissues (Serbina and Pamer, 2006; Swirski *et al.*, 2009). The origin of M2 macrophages is more controversial. Arnold *et al.* (2007) demonstrated both *in vitro* and *in vivo* that M2 macrophages can derive from M1 macrophages. However, it is still unclear whether and to which extent the expansion of tissue-resident macrophages can also contribute to the M2 subset. Resident macrophages may indeed play a more significant role than previously appreciated in initiating the cascade of cellular events that follows muscle damage. Recent findings have suggested that macrophages residing in the epimysium/perimysium become activated and increase in number very early in response to damage, secreting chemokines that play a major role in the recruitment of neutrophils and M1 macrophages to the injury site (Brigitte *et al.*, 2010).

As mentioned above, the timing of transition from a proinflammatory to an anti-inflammatory milieu correlates with the commitment of myoblasts to terminal differentiation. The M1 phenotype is observed during the proliferative phase and M2 phenotype during early and late differentiation phases of myogenesis suggesting that factors originating from macrophages may modulate myogenesis, or at least dictate its timing. Consistently, inflammatory macrophages cocultured with myogenic cells stimulate cell proliferation and inhibit differentiation (Arnold *et al.*, 2007), and the medium from proinflammatory macrophages exhibit negative effects on myogenic cell differentiation and fusion by stimulating cell motility (Bondesen *et al.*, 2007). On the contrary, anti-inflammatory macrophages exerted a strong prodifferentiation influence on myogenic cells, which is reminiscent of the *in vivo* pattern (Arnold *et al.*, 2007). These results substantiate the view that macrophage plasticity modulates regeneration by exerting its effect directly on myogenesis. However, the quality of the inflammatory milieu also has a deep influence on other cell types such as stromal progenitors, which can in turn affect myogenic cells.

6.4. Is inflammation necessary for promoting muscle regeneration?

Macrophages transmit survival signals to myoblast in culture in a manner dependent on cell–cell contact (Chazaud *et al.*, 2003; Sonnet *et al.*, 2006a,b). Correlative data suggested that specific subsets of macrophages play critical roles in muscle regeneration. Even though results obtained *in vitro* are notoriously difficult to extrapolate to the *in vivo* situation, a vast majority of *in vivo* studies based on mutants affecting macrophage function support a positive role for these cells in modulating regeneration. Damage in mice lacking chemokine receptor 2 (CCR2) or its ligand, MCP-1, which are necessary for the efficient recruitment of M1 macrophages, resulted in impaired muscle regeneration (Contreras-Shannon *et al.*, 2007a,b; Shireman *et al.*, 2007). Neutrophil recruitment was not affected in either of these mutant mice. Interestingly, during the first 72 h after damage, CCR2−/− mice showed fewer infiltrating macrophages when compared to MCP-1−/− mice and a correspondingly larger defect in regeneration. Recent studies suggest that such differences in the severity of the phenotype observed in the CCR2−/− mice and MCP-1−/− mice correlate with an impaired capillary density observed in the receptor mutant (Martinez *et al.*, 2010a,b), suggesting that M1 macrophages may play additional roles outside of influencing myogenic progression.. Consistent with *in vitro* data demonstrating that different subsets of macrophages play different roles in regulating myogenesis, specific depletion of intramuscular F4/80^{hi} macrophages by DTA delivery in CD11b-DTR mice at the time of muscle injury largely prevented muscle regeneration. This was contrasted by the finding that depletion at a later time point corresponded with a switch from a M1–M2 milieu, and led to a reduction in the diameter of regenerating fibers (Arnold *et al.*, 2007). Altogether, these observations strongly indicate that macrophages play an important role during muscle regeneration. However, as mentioned above, free radicals released by neutrophils or macrophages can also aggravate muscle damage *in vivo*. It could be concluded that inflammatory cell influx following muscle damage is necessary for proper muscle regeneration, but excessive or insufficient inflammation can interfere with the efficient restoration of tissue function.

6.5. Role of mast cells in skeletal muscle regeneration

Like monocytes and macrophages, mast cells arise from hematopoietic precursor cells of bone marrow origins that circulate in the blood and mature after entering peripheral tissue. Although mast cell activation has traditionally been associated with allergic reactions, these cells are recently emerging as key players in other physiological processes such as wound healing, tissue remodeling, angiogenesis, and innate immunity (Abraham and St John, 2010). To date, not much is known about their role in skeletal muscle regeneration, but their position and ubiquitous occurrence in proximity to blood vessels makes them an interesting candidate to regulate early steps in this process. Upon activation, mast cells secrete large amounts of proinflammatory cytokines like TNFα, and the injection of this cytokine into the mouse soleus muscle is sufficient to initiate recruitment of neutrophils and macrophages suggesting a role for these cells during the early stages of damage response (Peterson *et al.*, 2006). Two recent studies shed more light on the role of these cells in injured muscle. Dumont *et al.* (2007) observed a threefold increase in mast cell activation following a hind limb unloading and reloading damage model. When rats were treated with cromolyn, an inhibitor of mast cell activation, the numbers of neutrophils found in damaged muscle was significantly reduced. On the other hand, increased numbers of ED1+ve macrophages were observed (Dumont *et al.*, 2007). Concomitantly, administration of compound 48/80, a stimulator of mast cells, provoked a significant increase in the number of neutrophils in uninjured muscle (Dumont *et al.*, 2007). Similar results were obtained in a mild exercise-induced muscle damage model, which involved muscle lengthening and contractions in rats (Cote *et al.*, 2008). Both studies suggest that mast cells modulate the influx of specific leukocytes into the damage tissue, with opposite effects of this modulation occurring in relation to neutrophils and macrophages. An alternative interpretation could be that the increase in the number of ED1+ve macrophages might represent a compensatory mechanism to efficiently clear up cellular debris. Independent of the interpretation, the role of mast cells during muscle regeneration and repair merits further analysis and research.

6.6. Immune cells during chronic damage

Similar to what is observed in acute damage, immune cells are thought to play a significant role in chronic damage. In addition to macrophages, neutrophils and mast cells, cytotoxic T-lymphocytes and eosinophils have also been involved in the pathogenesis of hereditary, progressive muscle disease. However, only macrophages and eosinophils have been reported to contribute to regeneration. Studies from mdx mice, a genetic mouse model for Duchenne muscular Dystrophy (DMD), showed that inflammatory cell profiles observed at early stages of the disease are similar to those evoked by acute damage. During the first degenerative event, taking place at around 4 weeks of age in the mdx model, Villata *et al.*, observed neutrophils and activated M1 macrophages invading the tissue. Genetic ablation of the iNOS gene in mdx mice significantly reduced muscle membrane lysis *in vivo* indicating that activated macrophages aggravate muscle damage via an NO dependent mechanism (Villalta *et al.*, 2009). Consistent with this notion, depletion of macrophages before the onset of the disease significantly reduced muscle membrane injury (Wehling *et al.*, 2001). Further, treating DMD patients with corticosteroids systemically delayed the disease, indicating that immune cells may play a significant role in modulating the disease pathology (Kinali *et al.*, 2002).

In contrast, M2 macrophages predominate at the later stages of disease in mdx mice (Villalta *et al.*, 2009). Moreover, during the transition from the early muscle necrosis stage to the regenerative stage, M2 macrophages expressed elevated levels of IL-4 and IL-10 that reduce muscle damage by deactivating proinflammatory M1 macrophages and, presumably, support muscle regeneration (Villalta *et al.*, 2009). Paradoxically, it is at these later stages of disease that muscle regeneration is less efficient. Whether this is due to a cell autonomous defect in satellite cells or to an environmental defect it is currently unclear.

7. MSCs and Tissue Regeneration

7.1. Definition of MSCs

Although there is growing body of evidence implicating the importance of immune cells in skeletal muscle regeneration, a second cell type first described in bone marrow has also generated significant research interest. MSCs were first described as a plastic adherent cell population which formed colonies, and were termed colony-forming fibroblasts (CFU-f; Friedenstein *et al.*, 1974). It was then demonstrated that this population could be differentiated *in vitro* into numerous mesenchymal cell types including adipocytes, chondrocytes, and osteoblasts (Friedenstein *et al.*, 1976). This population is highly heterogeneous and whether its apparent multipotentiality is due to the presence of several distinct types of lineage-restricted progenitors or to rare clones with wider developmental potential (hence the often misused term "mesenchymal stem cells") is still controversial. In addition, significant ambiguity is introduced in that it is difficult to know whether MSCs isolated from different groups represent the same cell type or distinct cell types within this heterogeneous population. This is due to variations in nomenclature (e. g., mesenchymal stem cell, multipotent stromal cell, marrow stromal cell, colony-forming unit-fibroblast (CFU-F)), isolation criteria and differentiation protocols across the field (Lindner *et al.*, 2010), making it difficult to compare and contrast study outcomes which ultimately hampers the progress of MSC research as a whole (Ankrum and Karp, 2010). Indeed, similar to MSC studies in other organs, different groups have utilized a variety of different isolation techniques to obtain MSC-like cells from skeletal muscle, resulting in a myriad of differently defined cell populations. This problem is compounded by the fact that cellular expression of various markers changes when cells are placed in culture and characterized using *in vitro* methods. In an effort to provide a set of characteristics that would unambiguously identify this often ill-defined cell population, the International Society for Cellular Therapy (ISCT) described human MSCs using the term MSC and the following criteria: (i) cells are plastic adherent when maintained in culture conditions using tissue culture flasks, (ii) must express CD105, CD73, CD90 and not express CD45, CD34, CD14 or CD11b, CD78α or CD19, and HLA class II, and (iii) must be able to differentiate into osteoblasts, adipocytes, and chondroblasts under standard *in vitro* differentiating conditions (Dominici *et al.*, 2006). It is important to point out that cells need to fulfill these criteria to be officially defined as MSCs, but cells fitting this definition are not necessarily MSCs. Indeed, a set of markers allowing the prospective purification of MSCs, and therefore their direct identification *in vivo* is still lacking. In summary, despite the establishment of criteria to define MSCs, there is a continuing debate over the characterization and definition of the MSC phenotype and substantial variety in techniques used to isolate and induce to differentiate this widely used cell type.

7.2. *In vitro*, *ex vivo* and *in vivo* roles of MSCs

Despite a lack of consensus over their definition, the field of MSC research has expanded dramatically since their initial discovery, mostly due to a great deal of interest in the potential practical applications of this cell type. MSCs are an attractive candidate for tissue engineering and cell therapy due to their inherent plasticity, various potential tissue sources, and lack of ethical or legal constrictions in relation to other stem cell populations of overlapping developmental potential (Abdallah and Kassem, 2008; Jackson *et al.*,

2010). In addition to a well-documented capacity to generate connective tissues, MSCs have also been postulated by numerous studies to generate additional mesenchymal lineages, such as cardiogenic (Hattan *et al.*, 2005; Li *et al.*, 2007; Toma *et al.*, 2002) or myogenic cell types (Jackson *et al.*, 2010). However, there is still a significant debate as to the frequency and importance of MSCs in ensuring tissue homeostasis *in vivo*, as much of the evidence supporting this role comes from *in vitro* data. Independent of their physiological role, the well-documented *in vitro* expansion and differentiation of MSCs does make them an attractive candidate for exploitation in tissue engineering approaches (Khademhosseini *et al.*, 2009), highlighted by the recent generation of an artificial trachea (Macchiarini *et al.*, 2008). In regard with cell therapy, numerous studies have demonstrated the regenerative benefit of MSC transplantation in a variety of different tissues and damage models, which has led to numerous large-scale clinical trials (reviewed in Jackson *et al.*, 2010). It is now believed that the majority of the regenerative benefits arising from MSC transplantation are due to the paracrine secretion of trophic factors by the transplanted cells rather than direct contribution to the regenerating tissue via *de novo* generation of differentiated progeny as was initially thought (Caplan, 2007). Further evidence of their trophic support capacity can be found in a growing body of evidence supporting the concept that mesenchymal cells and the stromal tissue they constitute play an important supportive role to progenitors of diverse developmental origins in a wide variety of systems such as hematopoesis (Méndez-Ferrer *et al.*, 2010) and hair growth (Fernandes *et al.*, 2004). Although MSCs were first described following isolation from bone-marrow, it is now known that they can be obtained from numerous other tissue sources including umbilical cord blood (Erices *et al.*, 2000), adipose tissue (Zuk *et al.*, 2001), salivary glands (Rotter *et al.*, 2008), sinovial fluid (Fan *et al.*, 2009), parathyroid gland (Shih *et al.*, 2009), and dental pulp (Gronthos *et al.*, 2000) supporting the notion that such cells are essentially ubiquitous. Despite the efforts invested in both identifying alternative tissue sources through which MSCs could be obtained and further characterizing of their *in vitro* behavior, there is significant lack of data concerning the normal physiological function of these widespread tissue-resident cells. However, recent work in the context of skeletal muscle has offered an initial insight in the response that MSC mount to tissue damage.

7.3. Skeletal muscle resident MSCs

A number of recent reports have described cell populations in skeletal muscle that demonstrate MSC-like properties, and suggested that these cells may be important to both regular organ function and regeneration following injury. One population, called muscle-derived stem cells (MDSCs) was first harvested from murine skeletal by a selective plating method which enriches slowly adhering MDSCs while eliminating both non adherent and more adherent cell types (Gharaibeh *et al.*, 2008; Qu-Petersen *et al.*, 2002). These cells were demonstrated to be able to differentiate not only along connective tissue lineages, but also to replenish the hematopoietic system of irradiated recipients and to directly contribute to the generation of mature myofibers following transplantation *in vivo* (Cao *et al.*, 2003). A similar population was isolated from human skeletal muscle tissue using FACS to identify cells positive for CD34, CD56, and CD144 (Zheng *et al.*, 2007). This population, termed myoendothelial cells, was shown to reside in close proximity to the vasculature and demonstrated the defined MSC characteristic of tri-potency for adipogenic, osteogenic, and chondrogenic differentiation, in addition to being able to generate myogenic and endothelial progeny. Later work form the same group reported that these pericytic cells are widespread in human tissues, and they retain their myogenic potential even when they originate from other tissues (Crisan *et al.*, 2008). Support for the *in vitro* ability of pericytes to generate the three classical connective tissues (bone, cartilage, and adipose) defining MSC potential stems form a variety of other studies (Collett and Canfield, 2005; Farrington-Rock *et al.*, 2004). More recent evidence strongly suggest that tissue-resident stromal progenitors are vessel-associated cells, and it has been argued that MSCs harvested from various tissue may in fact originate from pericytes (Caplan, 2008; Corselli *et al.*, 2010). It is important to consider that while data in both human and animal models support the notion that a subset of pericytes can generate progeny belonging to the three main connective lineages, all data supporting their contribution to myogenic or other lineages is based on human studies. Without strict lineage tracing experiments, which are only possible in animal models, it is difficult to discern whether or not a cell population's myogenic and mesenchymal potency actually exists *in vivo*. An additional caveat must be added that both MDSCs and pericytes myogenic and mesenchymal properties have been primarily evidenced in cells that had undergone significant *in vitro* expansion, which could lead to reprogramming and selection taking place. This caveat affects numerous studies across a broad array of tissues which claim MSCs capacity for non-lineage restricted transdifferentiation (Phinney and Prockop, 2007). To date, solid *in vivo* evidence supporting the myogenic potential of stromal cell populations as demonstrated with satellite cells and myogenic precursors (MPs) is lacking, making is difficult to assess if these cells actually do have a significant role in tissue homeostasis and response to damage, or instead have an inherent *in vitro* plasticity toward alternative lineages.

7.4. The supportive role of MSCs in skeletal muscle

Side population (SP) cells were first described in the bone marrow, based on their ability to preferentially exclude Hoechst-3333 dye, and it was shown that this cell subset contained most of the HSC activity in this tissue (Goodell *et al.*, 1996). Using this approach to identify immature cells, a nonsatellite stem cell population resident in skeletal muscle was first described by Gussoni *et al.* (1999) who further demonstrated that these cells were capable of differentiation into several different lineages following intravenous injection. More recently, Kallestad and McLoon (2010) investigated this means of identifying progenitor cell populations, highlighting the problems in using Hoechst 33342 dye exclusion and stated that sorting cells *ex vivo* using accepted precursor cell markers will yield superior results in terms of cell homogeneity and potential for transplantation therapy. To investigate the heterogeneity of skeletal muscle-derived SP cells, Uezumi *et al.* (2006) subdivided the SP cell fraction using CD31 and CD45 markers to exclude endothelial and hematopoietic cells. They found that the relatively rare CD31(−)CD45(−)SP fraction expressed mesenchymal lineage markers and demonstrated connective potential *in vitro*, as well as being able to differentiate into myofibers after intramuscular transplantation *in vivo*, albeit at a level which was limited when compared to satellite cells. Further work by the same group (Motohashi *et al.*, 2008) demonstrated that the CD31(−)CD45(−)SP was still a heterogeneous population containing both myogenic cells as well as another type of progenitor responsible for the connective potential, and postulated that these latter cells may play critical roles during muscle regeneration other than directly contributing to muscle tissue. They clearly demonstrated that myogenic engraftment is improved by cotransplantation of CD31(−)CD45(−)SP cells with myoblasts, and that CD31(−)CD45(−)SP cells promote muscle regeneration by stimulating the proliferation and migration of myoblasts, rather than generating myofibers. Numerous other groups contributed to the search for specific markers to identify skeletal muscle-derived nonmyogenic cells (Bosnakovski *et al.*, 2008; Hidestrand *et al.*, 2008; Kafadar *et al.*, 2009). Bosnakovski *et al.* (2008) used transgenic Pax7-ZsGreen reporter mice, and identified a ZsGreen-negative fraction that could be specifically identified using Sca1 and PDGFRα. This population (CD45−, CD34−, Sca1$^+$, PDGFRα^+) was shown to be relatively homogenous and did not exhibit myogenic activity in culture. Another study (Hidestrand *et al.*, 2008) suggested that these Sca1$^+$ nonmyogenic cells contribute to fibrosis implicating them in the fibroadipocytic depositions which are observed in aged skeletal muscle. These results are supported by recent data from two studies (Joe *et al.*, 2010b; Uezumi *et al.*, 2010), which investigated the *in vivo* behavior of this cell population (CD31−, CD45−, Sca1$^+$, PDGFRα^+, CD34$^+$) in both normal tissue homeostasis and a variety of damage models. These studies clearly demonstrated through use of *in vivo* lineage tracing and transplantation experiments that a mesenchymal cell population exists in skeletal muscle that can give rise to both adipocytes and fibroblasts *in vivo*, leading Joe *et al.* (2010b) to label these cells as fibro/adipogenic progenitors (FAP). However, the question of whether these cells also possess osteogenic potential, therefore aligning them with the classical definition of MSC, is still unresolved. Uezumi *et al.* (2010) reported *in vitro* expression of alkaline phosphatase, an early bone marker, following treatment with BMP2 but Joe *et al.* (2010a) failed to observe nodule formation in osteogenic conditions often used for MSC differentiation. Most importantly, these cells respond to damage by undergoing a dramatic expansion and play an important role even when tissue function is fully restored and no adipocytes or fibrocytes are generated, as evidenced by their promyogenic effect on cultured myoblasts. In healthy, efficiently regenerating tissue, a rapid contraction of FAP number follows the expansion phase, suggesting that these cells are activated to provide a transient source of trophic signals for proliferating myogenic progenitors following muscle damage. This work provided one of the clearest demonstrations to date that the paracrine trophic support role reported for these cells and for MSCs in general in transplantation assays may well reflect one of their main physiological functions. Which signals or combination of signals these cells provide to myogenic progenitors is currently under active investigation, and known promyogenic factors such as IL-6 (Serrano *et al.*, 2008; Wang *et al.*, 2008) or Wnt (Seale *et al.*, 2003) family members have been proposed as candidates (Rodeheffer, 2010). Thus strong evidence exists, both *in vitro* and *in vivo*, for a Janus-like role of this cell population with the same stromal progenitors providing paracrine support to tissue-specific stem/progenitor cells during normal muscle regeneration, but rapidly differentiating and leading to fibrosis and adipocytic infiltration to ensure the structural and metabolic continuity of the tissue when regeneration fails, as often happens in aging and chronic disease (Natarajan *et al.*, 2010). With similar tissue-resident cell populations having been found in fat depots (Joe *et al.*, 2009), and under investigation in multiple other tissues including heart and lungs this duplicity of functions in tissue regeneration and repair could be hypothesized to be a process common to several organ systems and disease states. Considering that FAPs are vessel associated in resting tissue (Figs. 6.1 and 6.2), and that they are the source of the vast majority of CFU-Fs obtained form skeletal muscle, it is tempting to propose that such cells may acquire a more pronounced osteogenic potential during extended *in vitro* culture, and thus represent the ancestry of classically defined MSCs.

8. Conclusion

In summary, significant evidence exists indicating that at least two nonmyogenic cell types, inflammatory cells and stromal cells, play important accessory roles during myogenic regeneration. At least some of these roles are carried out by directly influencing the progression of myogenic progenitors through the activation/proliferation/differentiation steps that lead to the generation of mature progeny. Importantly, both inflammatory cells and stromal cells are also involved in regulating the balance between regeneration and repair, and at least in this context are likely to influence each other via the production of extracellular signals. Thus, a global view of regenerating muscle tissue is likely to include complex three way communications among myogenic, inflammatory, and stromal cells, subtly modulating each other's activity to coordinate their functions and adapt their response to specific local as well as systemic requirements (Fig. 6.2B). It is clear that, if we hope to understand regeneration in all its complexity, the cellular participants in this process as well as their interactions need to be studied as a whole.

REFERENCES

Abdallah, B. M., and Kassem, M. (2008). Human mesenchymal stem cells: From basic biology to clinical applications. *Gene Ther.* **15,** 109–116.

Abood, E. A., and Jones, M. M. (1991). Macrophages in developing mammalian skeletal muscle: Evidence for muscle fibre death as a normal developmental event. *Acta. Anat.* **140,** 201–212.

Abraham, S. N., and St John, A. L. (2010). Mast cell-orchestrated immunity to pathogens. *Nat. Rev. Immunol.* **10,** 440–452.

Adams, G. B., Chabner, K. T., Alley, I. R., Olson, D. P., Szczepiorkowski, Z. M., Poznansky, M. C., Kos, C. H., Pollak, M. R., Brown, E. M., and Scadden, D. T. (2006). Stem cell engraftment at the endosteal niche is specified by the calcium-sensing receptor. *Nature* **439,** 599–603.

Ankrum, J., and Karp, J. M. (2010). Mesenchymal stem cell therapy: Two steps forward, one step back. *Trends Mol. Med.* **16,** 203–209.

Arai, F., Hirao, A., Ohmura, M., Sato, H., Matsuoka, S., Takubo, K., Ito, K., Koh, G. Y., and Suda, T. (2004). Tie2/angiopoietin-1 signaling regulates hematopoietic stem cell quiescence in the bone marrow niche. *Cell* **118,** 149–161.

Arnold, L., Henry, A., Poron, F., Baba-Amer, Y., van Rooijen, N., Plonquet, A., Gherardi, R. K., and Chazaud, B. (2007). Inflammatory monocytes recruited after skeletal muscle injury switch into antiinflammatory macrophages to support myogenesis. *J. Exp. Med.* **204,** 1057–1069.

Bodine, S. C., *et al.* (2001). Akt/mTOR pathway is a crucial regulator of skeletal muscle hypertrophy and can prevent muscle atrophy in vivo. *Nat. Cell Biol.* **3,** 1014–1019.

Bondesen, B. A., Mills, S. T., and Pavlath, G. K. (2006). The COX-2 pathway regulates growth of atrophied muscle via multiple mechanisms. *Am. J. Physiol. Cell Physiol.* **290,** C1651–C1659.

Bondesen, B. A., Jones, K. A., Glasgow, W. C., and Pavlath, G. K. (2007). Inhibition of myoblast migration by prostacyclin is associated with enhanced cell fusion. *FASEB J.* **21,** 3338–3345.

Bosnakovski, D., Xu, Z., Li, W., Thet, S., Cleaver, O., Perlingeiro, R. C. R., and Kyba, M. (2008). Prospective isolation of skeletal muscle stem cells with a Pax7 reporter. *Stem Cells* **26,** 3194–3204.

Brack, A. S., Conboy, M. J., Roy, S., Lee, M., Kuo, C. J., Keller, C., and Rando, T. A. (2007). Increased Wnt signaling during aging alters muscle stem cell fate and increases fibrosis. *Science* **317,** 807–810.

Brickson, S., Ji, L. L., Schell, K., Olabisi, R., St. Pierre Schneider, B., and Best, T. M. (2003). M1/70 attenuates blood-borne neutrophil oxidants, activation, and myofiber damage following stretch injury. *J. Appl. Physiol.* **95,** 969–976.
Brigitte, M., Schilte, C., Plonquet, A., Baba-Amer, Y., Henri, A., Charlier, C., Tajbakhsh, S., Albert, M., Gherardi, R. K., and Chretien, F. (2010). Muscle resident macrophages control the immune cell reaction in a mouse model of notexin-induced myoinjury. *Arthritis Rheum.* **62,** 268–279.
Calvi, L. M., *et al.* (2003). Osteoblastic cells regualte the haematopoietic stem cell niche. *Nature* **425,** 841–846.
Cantini, M., and Carraro, U. (1995). Macrophage-released factor stimulates selectively myogenic cells in primary muscle culture. *J. Neuropathol. Exp. Neurol.* **54,** 121–128.
Cantini, M., Giurisato, E., Radu, C., Tiozzo, S., Pampinella, F., Senigaglia, D., Zaniolo, G., Mazzoleni, F., and Vitiello, L. (2002). Macrophage-secreted myogenic factors: A promising tool for greatly enhancing the proliferative capacity of myoblasts in vitro and in vivo. *Neurol. Sci.* **23,** 189–194.
Cao, B., Zheng, B., Jankowski, R. J., Kimura, S., Ikezawa, M., Deasy, B., Cummins, J., Epperly, M., Qu-Petersen, Z., and Huard, J. (2003). Muscle stem cells differentiate into haematopoietic lineages but retain myogenic potential. *Nat. Cell Biol.* **5,** 640–646.
Caplan, A. I. (2007). Adult mesenchymal stem cells for tissue engineering versus regenerative medicine. *J. Cell. Physiol.* **213,** 341–347.
Caplan, A. I. (2008). All MSCs are pericytes? *Cell Stem Cell* **3,** 229–230.
Charge, S. B., and Rudnicki, M. A. (2004). Cellular and molecular regulation of muscle regeneration. *Physiol. Rev.* **84,** 209–238.
Chazaud, B., Sonnet, C., Lafuste, P., Bassez, G., Rimaniol, A. C., Poron, F., Authier, F. J., Dreyfus, P. A., and Gherardi, R. K. (2003). Satellite cells attract monocytes and use macrophages as a support to escape apoptosis and enhance muscle growth. *J. Cell Biol.* **163,** 1133–1143.
Chazaud, B., Brigitte, M., Yacoub-Youssef, H., Arnold, L., Gherardi, R., Sonnet, C., Lafuste, P., and Chretien, F. (2009). Dual and beneficial roles of macrophages during skeletal muscle regeneration. *Exerc. Sport Sci. Rev.* **37,** 18–22.
Cleaver, O., and Melton, D. A. (2003). Endothelial signaling during development. *Nat. Med.* **9,** 661–668.
Collett, G. D. M., and Canfield, A. E. (2005). Angiogenesis and pericytes in the initiation of ectopic calcification. *Circ. Res.* **96,** 930–938.
Collins, C. A., *et al.* (2005). Stem cell function, self-renewal, and behavioral heterogeneity of cells from the adult muscle satellite cell niche. *Cell* **122,** 289–301.
Conboy, I. M., and Rando, T. A. (2002). The regulation of Notch signaling controls satellite cell activation and cell fate determination in postnatal myogenesis. *Dev. Cell* **3,** 397–409.
Contreras-Shannon, V., *et al.* (2007a). Fat accumulation with altered inflammation and regeneration in skeletal muscle of CCR2−/− mice following ischemic injury. *Am. J. Physiol. Cell Physiol.* **292,** C953–C967.
Contreras-Shannon, V., Ochoa, O., Reyes-Reyna, S. M., Sun, D., Michalek, J. E., Kuziel, W. A., McManus, L. M., and Shireman, P. K. (2007b). Fat accumulation with altered inflammation and regeneration in skeletal muscle of CCR2−/− mice following ischemic injury. *Am. J. Physiol. Cell Physiol.* **292,** C953–C967.
Cornelison, D. D., and Wold, B. J. (1997). Single-cell analysis of regulatory gene expression in quiescent and activated mouse skeletal muscle satellite cells. *Dev. Biol.* **191,** 270–283.
Cornelison, D. D., Olwin, B. B., Rudnicki, M. A., and Wold, B. J. (2000). MyoD(−/−) satellite cells in single-fiber culture are differentiation defective and MRF4 deficient. *Dev. Biol.* **224,** 122–137.
Corselli, M., Chen, C.-W., Crisan, M., Lazzari, L., and Péault, B. (2010). Perivascular ancestors of adult multipotent stem cells. *Arterioscler. Thromb. Vasc. Biol.* **30,** 1104–1109.

Cote, C. H., Tremblay, M. H., Duchesne, E., and Lapoite, B. M. (2008). Inflammation-induced leukocyte accumulation in injured skeletal muscle: Role of mast cells. *Muscle Nerve* **37,** 754–763.

Crisan, M., Yap, S., Casteilla, L., Chen, C.-W., Corselli, M., Park, T. S., Andriolo, G., Sun, B., Zheng, B., Zhang, L., Norotte, C., Teng, P.-N., *et al.* (2008). A perivascular origin for mesenchymal stem cells in multiple human organs. *Cell Stem Cell* **3,** 301–313.

Crosby, L. M., and Waters, C. M. (2010). Epithelial repair mechanisms in the lung. *Am. J. Physiol. Lung Cell. Mol. Physiol.* **298,** L715–L731.

Dhawan, J., and Rando, T. A. (2005). Stem cells in postnatal myogenesis: Molecular mechanisms of stellite cell quiescence, activation and replenishment. *Trends Cell Biol.* **15,** 666–673.

Dominici, M., Le Blanc, K., Mueller, I., Slaper-Cortenbach, I., Marini, F., Krause, D., Deans, R., Keating, A., Prockop, D., and Horwitz, E. (2006). Minimal criteria for defining multipotent mesenchymal stromal cells. The International Society for Cellular Therapy position statement. *Cytotherapy* **8,** 315–317.

Dumont, N., Lepage, K., Cote, C. H., and Frenette, J. (2007). Mast cells can modulate leukocyte accumulation and skeletal muscle function following hindlimb unloading. *J. Appl. Physiol.* **103,** 97–104.

Erices, A., Conget, P., and Minguell, J. J. (2000). Mesenchymal progenitor cells in human umbilical cord blood. *Br. J. Haematol.* **109,** 235–242.

Fan, J., Varshney, R. R., Ren, L., Cai, D., and Wang, D. A. (2009). Synovium-derived mesenchymal stem cells: A new cell source for musculoskeletal regeneration. *Tissue Eng. Part B Rev.* **15,** 75–86.

Farrington-Rock, C., Crofts, N. J., Doherty, M. J., Ashton, B. A., Griffin-Jones, C., and Canfield, A. E. (2004). Chondrogenic and adipogenic potential of microvascular pericytes. *Circulation* **110,** 2226–2232.

Fernandes, K. J., McKenzie, I. A., Mill, P., Smith, K. M., Akhavan, M., Barnabe-Heider, F., Biernaskie, J., Junek, A., Kobayashi, N. R., Toma, J. G., Kaplan, D. R., Labosky, P. A., *et al.* (2004). A dermal niche for multipotent adult skin-derived precursor cells. *Nat. Cell Biol.* **6,** 1082–1093.

Friedenstein, A. J., Deriglasova, U. F., Kulagina, N. N., Panasuk, A. F., Rudakowa, S. F., Luria, E. A., and Ruadkow, I. A. (1974). Precursors for fibroblasts in different populations of hematopoietic cells as detected by the in vitro colony assay method. *Exp. Hematol.* **2,** 83–92.

Friedenstein, A. J., Gorskaja, J. F., and Kulagina, N. N. (1976). Fibroblast precursors in normal and irradiated mouse hematopoietic organs. *Exp. Hematol.* **4,** 267–274.

Fuchtbauer, E. M., and Westphal, H. (1992). MyoD and myogenin are coexpressed in regenerating skeletal muscle of the mouse. *Dev. Dyn.* **193,** 34–39.

Gharaibeh, B., Lu, A., Tebbets, J., Zheng, B., Feduska, J., Crisan, M., Peault, B., Cummins, J., and Huard, J. (2008). Isolation of a slowly adhering cell fraction containing stem cells from murine skeletal muscle by the preplate technique. *Nat. Protoc.* **3,** 1501–1509.

Godman, G. C. (1957). On the regeneration and redifferentiation of mammalian striated muscle. *J. Morphol.* **100,** 27–81.

Goodell, M. A., Brose, K., Paradis, G., Conner, A. S., and Mulligan, R. C. (1996). Isolation and functional properties of murine hematopoietic stem cells that are replicating in vivo. *J. Exp. Med.* **183,** 1797–1806.

Gordon, S. (2003). Alternative activation of macrophages. *Nat. Rev. Immunol.* **3,** 23–35.

Gordon, S., and Taylor, P. R. (2005). Monocyte and macrophage heterogeneity. *Nat. Rev. Immunol.* **5,** 953–964.

Gronthos, S., Mankani, M., Brahim, J., Robey, P. G., and Shi, S. (2000). Postnatal human dental pulp stem cells (DPSCs) in vitro and in vivo. *Proc. Natl. Acad. Sci. USA* **97,** 13625–13630.

Grounds, M. D., Garrett, K. L., Lai, M. C., Wright, W. E., and Beilharz, M. W. (1992). Identification of skeletal muscle precursor cells in vivo by use of MyoD1 and myogenin probes. *Cell Tissue Res.* **267,** 99–104.

Gussoni, E., Soneoka, Y., Strickland, C. D., Buzney, E. A., Khan, M. K., Flint, A. F., Kunkel, L. M., and Mulligan, R. C. (1999). Dystrophin expression in the mdx mouse restored by stem cell transplantation. *Nature* **401,** 390–394.

Hattan, N., Kawaguchi, H., Ando, K., Kuwabara, E., Fujita, J., Murata, M., Suematsu, M., Mori, H., and Fukuda, K. (2005). Purified cardiomyocytes from bone marrow mesenchymal stem cells produce stable intracardiac grafts in mice. *Cardiovasc. Res.* **65,** 334–344.

Hidestrand, M., Richards-Malcolm, S., Gurley, C. M., Nolen, G., Grimes, B., Waterstrat, A., Zant, G. V., and Peterson, C. A. (2008). Sca-1-expressing nonmyogenic cells contribute to fibrosis in aged skeletal muscle. *J. Gerontol. A Biol. Sci. Med. Sci.* **63,** 566–579.

Jackson, W. M., Nesti, L. J., and Tuan, R. S. (2010). Potential therapeutic applications of muscle-derived mesenchymal stem and progenitor cells. *Expert Opin. Biol. Ther.* **10,** 505–517.

Joe, A. W. B., Yi, L., Even, Y., Vogl, A. W., and Rossi, F. M. V. (2009). Depot-specific differences in adipogenic progenitor abundance and proliferative response to high-fat diet. *Stem Cells* **27,** 2563–2570.

Joe, A. W., Yi, L., Natarajan, A., Le Grand, F., So, L., Wang, J., Rudnicki, M. A., and Rossi, F. M. (2010). Muscle injury activates resident fibro/adipogenic progenitors that facilitate myogenesis. *Nat. Cell Biol.* **12,** 153–163.

Kafadar, K. A., Yi, L., Ahmad, Y., So, L., Rossi, F., and Pavlath, G. K. (2009). Sca-1 expression is required for efficient remodeling of the extracellular matrix during skeletal muscle regeneration. *Dev. Biol.* **326,** 47–59.

Kallestad, K. M., and Mcloon, L. K. (2010). Defining the heterogeneity of skeletal muscle-derived side and main population cells isolated immediately ex vivo. *J. Cell. Physiol.* **222,** 676–684.

Khademhosseini, A., Vacanti, J. P., and Langer, R. (2009). Progress in tissue engineering. *Sci. Am.* **300,** 64–71.

Kiel, M. J., and Morrison, S. J. (2008). Uncertainty in the niches that maintain haematopoietic stem cells. *Nat. Rev. Immunol.* **8,** 290–301.

Kinali, M., Mercuri, E., Main, M., Muntoni, F., and Dubowitz, V. (2002). An effective, low-dosage, intermittent schedule of prednisolone in the long-term treatment of early cases of Duchenne dystrophy. *Neuromuscul. Disord.* **12**(Suppl. 1), S169–S174.

Kuang, S., Kuroda, K., Le Grand, F., and Rudnicki, M. A. (2007). Asymmetric self-renewal and commitment of satellite stem cells in muscle. *Cell* **129,** 999–1010.

Le Grand, F., and Rudnicki, M. A. (2007). Skeletal muscle satellite cells and adult myogenesis. *Curr. Opin. Cell Biol.* **19,** 628–633.

Li, L., and Xie, T. (2005). Stem cell niche: Structure and function. *Annu. Rev. Cell Dev. Biol.* **21,** 605–631.

Li, X., Yu, X., Lin, Q., Deng, C., Shan, Z., Yang, M., and Lin, S. (2007). Bone marrow mesenchymal stem cells differentiate into functional cardiac phenotypes by cardiac microenvironment. *J. Mol. Cell. Cardiol.* **42,** 295–303.

Lindner, U., Kramer, J., Rohwedel, J., and Schlenke, P. (2010). Mesenchymal stem or stromal cells: Toward a better understanding of their biology? *Transfus. Med. Hemother. offizielles Organ Dtsch Ges. Transfusionsmed. Immunhamatologie* **37,** 75–83.

Lluis, F., Perdiguero, E., Nebreda, A. R., and Munoz-Canoves, P. (2006). Regulation of skeletal muscle gene expression by p38 MAP kinases. *Trends Cell Biol.* **16,** 36–44.

Macchiarini, P., Jungebluth, P., Go, T., Asnaghi, M. A., Rees, L. E., Cogan, T. A., Dodson, A., Martorell, J., Bellini, S., Parnigotto, P. P., Dickinson, S. C., Hollander, A. P., *et al.* (2008). Clinical transplantation of a tissue-engineered airway. *Lancet* **372,** 2023–2030.

Martinez, C. O., McHale, M. J., Wells, J. T., Ochoa, O., Michalek, J. E., McManus, L. M., and Shireman, P. K. (2010). Regulation of skeletal muscle regeneration by CCR2-activating chemokines is directly related to macrophage recruitment. *Am. J. Physiol. Regul. Integr. Comp. Physiol.* **299**(3), R832–42.

Massimino, M. L., Rapizzi, E., Cantini, M., Libera, L. D., Mazzoleni, F., Arslan, P., and Carraro, U. (1997). $ED2^+$ macrophages increase selectively myoblast proliferation in muscle cultures. *Biochem. Biophys. Res. Commun.* **235,** 754–759.

Mauro, A. (1961). Satellite cell of skeletal muscle fibers. *J. Biophys. Biochem. Cytol.* **9,** 493–495.

McLennan, I. S. (1993). Resident macrophages (ED2- and ED3-positive) do not phagocytose degenerating rat skeletal muscle fibres. *Cell Tissue Res.* **272,** 193–196.

McLennan, I. S. (1996). Degenerating and regenerating skeletal muscles contain several subpopulations of macrophages with distinct spatial and temporal distributions. *J. Anat.* **188**(Pt 1), 17–28.

Méndez-Ferrer, S., Michurina, T. V., Ferraro, F., Mazloom, A. R., Macarthur, B. D., Lira, S. A., Scadden, D. T., Ma'ayan, A., Enikolopov, G. N., and Frenette, P. S. (2010). Mesenchymal and haematopoietic stem cells form a unique bone marrow niche. *Nature* **466,** 829–834.

Merly, F., Lescaudron, L., Rouaud, T., Crossin, F., and Gardahaut, M. F. (1999). Macrophages enhance muscle satellite cell proliferation and delay their differentiation. *Muscle Nerve* **22,** 724–732.

Molteni, R., Fabbri, M., Bender, J. R., and Pardi, R. (2006). Pathophysiology of leukocyte-tissue interactions. *Curr. Opin. Cell Biol.* **18,** 491–498.

Morgan, J. E., and Partridge, T. A. (2003). Muscle satellite cells. *Int. J. Biochem. Cell Biol.* **35,** 1151–1156.

Motohashi, N., Uezumi, A., Yada, E., Fukada, S.-I., Fukushima, K., Imaizumi, K., Miyagoe-Suzuki, Y., and Takeda, S.i (2008). Muscle CD31(−) CD45(−) side population cells promote muscle regeneration by stimulating proliferation and migration of myoblasts. *Am. J. Pathol.* **173,** 781–791.

Natarajan, A., Lemos, D. R., and Rossi, F. M. (2010). Fibro/adipogenic progenitors: A double-edged sword in skeletal muscle regeneration. *Cell Cycle* **9,** 11.

Nguyen, H. X., and Tidball, J. G. (2003a). Expression of a muscle-specific, nitric oxide synthase transgene prevents muscle membrane injury and reduces muscle inflammation during modified muscle use in mice. *J. Physiol.* **550,** 347–356.

Nguyen, H. X., and Tidball, J. G. (2003b). Null mutation of gp91phox reduces muscle membrane lysis during muscle inflammation in mice. *J. Physiol.* **553,** 833–841.

Nguyen, H. X., Lusis, A. J., and Tidball, J. G. (2005). Null mutation of myeloperoxidase in mice prevents mechanical activation of neutrophil lysis of muscle cell membranes in vitro and in vivo. *J. Physiol.* **565,** 403–413.

Nilsson, S. K., Johnston, H. M., Whitty, G. A., Williams, B., Webb, R. J., Denhardt, D. T., Bertoncello, I., Bendall, L. J., Simmons, P. J., and Haylock, D. N. (2005). Osteopontin, a key component of the hematopoietic stem cell niche and regulator of primitive hematopoietic progenitor cells. *Blood* **106,** 1232–1239.

Ochoa, O., Sun, D., Reyes-Reyna, S. M., Waite, L. L., Michalek, J. E., McManus, L. M., and Shireman, P. K. (2007). Delayed angiogenesis and VEGF production in CCR2−/− mice during impaired skeletal muscle regeneration. *Am. J. Physiol. Regul. Integr. Comp. Physiol.* **293,** R651–R661.

Ottnad, E., Parthasarathy, S., Sambrano, G. R., Ramprasad, M. P., Quehenberger, O., Kondratenko, N., Green, S., and Steinberg, D. (1995). A macrophage receptor for oxidized low density lipoprotein distinct from the receptor for acetyl low density lipoprotein: Partial purification and role in recognition of oxidatively damaged cells. *Proc. Natl. Acad. Sci. USA* **92,** 1391–1395.

Otto, A., Schmidt, C., Luke, G., Allen, S., Valasek, P., Muntoni, F., Lawrence-Watt, D., and Patel, K. (2008). Canonical Wnt signalling induces satellite-cell proliferation during adult skeletal muscle regeneration. *J. Cell Sci.* **121,** 2939–2950.

Palmer, T. D., Willhoite, A. R., and Gage, F. H. (2000). Vascular niche for adult hippocampal neurogenesis. *J. Comp. Neurol.* **425,** 479–494.

Peterson, J. M., Feeback, K. D., Baas, J. H., and Pizza, F. X. (2006). Tumor necrosis factor-alpha promotes the accumulation of neutrophils and macrophages in skeletal muscle. *J. Appl. Physiol.* **101,** 1394–1399.

Phinney, D. G., and Prockop, D. J. (2007). Concise review: Mesenchymal stem/multipotent stromal cells: The state of transdifferentiation and modes of tissue repair–current views. *Stem Cells* **25,** 2896–2902.

Qu-Petersen, Z., Deasy, B., Jankowski, R., Ikezawa, M., Cummins, J., Pruchnic, R., Mytinger, J., Cao, B., Gates, C., Wernig, A., and Huard, J. (2002). Identification of a novel population of muscle stem cells in mice: Potential for muscle regeneration. *J. Cell Biol.* **157,** 851–864.

Ramprasad, M. P., Fischer, W., Witztum, J. L., Sambrano, G. R., Quehenberger, O., and Steinberg, D. (1995). The 94- to 97-kDa mouse macrophage membrane protein that recognizes oxidized low density lipoprotein and phosphatidylserine-rich liposomes is identical to macrosialin, the mouse homologue of human CD68. *Proc. Natl. Acad. Sci. USA* **92,** 9580–9584.

Relaix, F., Rocancourt, D., Mansouri, A., and Buckingham, M. (2005). A Pax3/Pax7-dependent population of skeletal muscle progenitor cells. *Nature* **435,** 948–953.

Rodeheffer, M. S. (2010). Tipping the scale: Muscle versus fat. *Nat. Cell Biol.* **12,** 102–104.

Rotter, N., Oder, J., Schlenke, P., Lindner, U., Bohrnsen, F., Kramer, J., Rohwedel, J., Huss, R., Brandau, S., Wollenberg, B., and Lang, S. (2008). Isolation and characterization of adult stem cells from human salivary glands. *Stem Cells Dev.* **17,** 509–518.

Sacco, A., Doyonnas, R., Kraft, P., Vitorovic, S., and Blau, H. M. (2008). Self-renewal and expansion of single transplanted muscle stem cells. *Nature* **456,** 502–506.

Scadden, D. T. (2006). The stem-cell niche as an entity of action. *Nature* **441,** 1075–1079.

Seale, P., Polesskaya, A., and Rudnicki, M. A. (2003). Adult stem cell specification by Wnt signaling in muscle regeneration. *Cell Cycle* **2,** 418–419.

Serbina, N. V., and Pamer, E. G. (2006). Monocyte emigration from bone marrow during bacterial infection requires signals mediated by chemokine receptor CCR2. *Nat. Immunol.* **7,** 311–317.

Serrano, A. L., Baeza-Raja, B., Perdiguero, E., Jardi, M., and Munoz-Canoves, P. (2008). Interleukin-6 is an essential regulator of satellite cell-mediated skeletal muscle hypertrophy. *Cell Metab.* **7,** 33–44.

Shih, Y. R., Kuo, T. K., Yang, A. H., Lee, O. K., and Lee, C. H. (2009). Isolation and characterization of stem cells from the human parathyroid gland. *Cell Prolif.* **42,** 461–470.

Shireman, P. K., Contreras-Shannon, V., Ochoa, O., Karia, B. P., Michalek, J. E., and McManus, L. M. (2007). MCP-1 deficiency causes altered inflammation with impaired skeletal muscle regeneration. *J. Leukoc. Biol.* **81,** 775–785.

Sonnet, C., *et al.* (2006a). Human macrophages rescue myoblasts and myotubes from apoptosis through a set of adhesion molecular systems. *J. Cell Sci.* **119,** 2497–2507.

Sonnet, C., Lafuste, P., Arnold, L., Brigitte, M., Poron, F., Authier, F. J., Chretien, F., Gherardi, R. K., and Chazaud, B. (2006b). Human macrophages rescue myoblasts and myotubes from apoptosis through a set of adhesion molecular systems. *J. Cell Sci.* **119,** 2497–2507.

St. Pierre, B. A., and Tidball, J. G. (1994). Differential response of macrophage subpopulations to soleus muscle reloading after rat hindlimb suspension. *J. Appl. Physiol.* **77,** 290–297.

Stier, S., Ko, Y., Forkert, R., Lutz, C., Neuhaus, T., Grunewald, E., Cheng, T., Dombkowski, D., Calvi, L. M., Rittling, S. R., and Scadden, D. T. (2005). Osteopontin is a hematopoietic stem cell niche component that negatively regulates stem cell pool size. *J. Exp. Med.* **201,** 1781–1791.

Street, J., Bao, M., deGuzman, L., Bunting, S., Peale, F. V., Jr., Ferrara, N., Steinmetz, H., Hoeffel, J., Cleland, J. L., Daugherty, A., van Bruggen, N., Redmond, H. P., *et al.* (2002). Vascular endothelial growth factor stimulates bone repair by promoting angiogenesis and bone turnover. *Proc. Natl. Acad. Sci. USA* **99,** 9656–9661.

Suda, T., Arai, F., and Hirao, A. (2005). Hematopoietic stem cells and their niche. *Trends Immunol.* **26,** 426–433.

Summan, M., Warren, G. L., Mercer, R. R., Chapman, R., Hulderman, T., Van Rooijen, N., and Simeonova, P. P. (2006). Macrophages and skeletal muscle regeneration: A clodronate-containing liposome depletion study. *Am. J. Physiol. Regul. Integr. Comp. Physiol.* **290,** R1488–R1495.

Swirski, F. K., Nahrendorf, M., Etzrodt, M., Wildgruber, M., Cortez-Retamozo, V., Panizzi, P., Figueiredo, J. L., Kohler, R. H., Chudnovskiy, A., Waterman, P., Aikawa, E., Mempel, T. R., *et al.* (2009). Identification of splenic reservoir monocytes and their deployment to inflammatory sites. *Science* **325,** 612–616.

Teixeira, C. F., Zamuner, S. R., Zuliani, J. P., Fernandes, C. M., Cruz-Hofling, M. A., Fernandes, I., Chaves, F., and Gutierrez, J. M. (2003). Neutrophils do not contribute to local tissue damage, but play a key role in skeletal muscle regeneration, in mice injected with Bothrops asper snake venom. *Muscle Nerve* **28,** 449–459.

Tidball, J. G., Berchenko, E., and Frenette, J. (1999). Macrophage invasion does not contribute to muscle membrane injury during inflammation. *J. Leukoc. Biol.* **65,** 492–498.

Toma, C., Pittenger, M. F., Cahill, K. S., Byrne, B. J., and Kessler, P. D. (2002). Human mesenchymal stem cells differentiate to a cardiomyocyte phenotype in the adult murine heart. *Circulation* **105,** 93–98.

Tombran-Tink, J., and Barnstable, C. J. (2004). Osteoblasts and osteoclasts express PEDF, VEGF-A isoforms, and VEGF receptors: Possible mediators of angiogenesis and matrix remodeling in the bone. *Biochem. Biophys. Res. Commun.* **316,** 573–579.

Uezumi, A., Ojima, K., Fukada, S.-I., Ikemoto, M., Masuda, S., Miyagoe-Suzuki, Y., and Takeda, S.i. (2006). Functional heterogeneity of side population cells in skeletal muscle. *Biochem. Biophys. Res. Commun.* **341,** 864–873.

Uezumi, A., Fukada, S.-I., Yamamoto, N., Takeda, S.i., and Tsuchida, K. (2010). Mesenchymal progenitors distinct from satellite cells contribute to ectopic fat cell formation in skeletal muscle. *Nat. Cell Biol.* **12,** 143–152.

Villalta, S. A., Nguyen, H. X., Deng, B., Gotoh, T., and Tidball, J. G. (2009). Shifts in macrophage phenotypes and macrophage competition for arginine metabolism affect the severity of muscle pathology in muscular dystrophy. *Hum. Mol. Genet.* **18,** 482–496.

Wang, S., Voisin, M. B., Larbi, K. Y., Dangerfield, J., Scheiermann, C., Tran, M., Maxwell, P. H., Sorokin, L., and Nourshargh, S. (2006). Venular basement membranes contain specific matrix protein low expression regions that act as exit points for emigrating neutrophils. *J. Exp. Med.* **203,** 1519–1532.

Wang, X., Wu, H., Zhang, Z., Liu, S., Yang, J., Chen, X., and Fan, M. (2008). Effects of interleukin-6, leukemia inhibitory factor, and ciliary neurotrophic factor on the proliferation and differentiation of adult human myoblasts. *Cell. Mol. Neurobiol.* **28,** 113–124.

Wehling, M., Spencer, M. J., and Tidball, J. G. (2001). A nitric oxide synthase transgene ameliorates muscular dystrophy in mdx mice. *J. Cell Biol.* **155,** 123–131.

Wiedow, O., and Meyer-Hoffert, U. (2005). Neutrophil serine proteases: Potential key regulators of cell signalling during inflammation. *J. Intern. Med.* **257,** 319–328.

Yablonka-Reuveni, Z., and Rivera, A. J. (1994). Temporal expression of regulatory and structural muscle proteins during myogenesis of satellite cells on isolated adult rat fibers. *Dev. Biol.* **164,** 588–603.

Yan, Z., Choi, S., Liu, X., Zhang, M., Schageman, J. J., Lee, S. Y., Hart, R., Lin, L., Thurmond, F. A., and Williams, R. S. (2003). Highly coordinated gene regulation in mouse skeletal muscle regeneration. *J. Biol. Chem.* **278,** 8826–8836.

Zerria, K., Jerbi, E., Hammami, S., Maaroufi, A., Boubaker, S., Xiong, J. P., Arnaout, M. A., and Fathallah, D. M. (2006). Recombinant integrin CD11b A-domain blocks polymorphonuclear cells recruitment and protects against skeletal muscle inflammatory injury in the rat. *Immunology* **119,** 431–440.

Zhang, J., Niu, C., Ye, L., Huang, H., He, X., Tong, W. G., Ross, J., Haug, J., Johnson, T., Feng, J. Q., Harris, S., Wiedemann, L. M., *et al.* (2003). Identification of the haematopoietic stem cell niche and control of the niche size. *Nature* **425,** 836–841.

Zheng, B., Cao, B., Crisan, M., Sun, B., Li, G., Logar, A., Yap, S., Pollett, J. B., Drowley, L., Cassino, T., Gharaibeh, B., Deasy, B. M., *et al.* (2007). Prospective identification of myogenic endothelial cells in human skeletal muscle. *Nat. Biotechnol.* **25,** 1025–1034.

Zouaoui Boudjeltia, K., Moguilevsky, N., Legssyer, I., Babar, S., Guillaume, M., Delree, P., Vanhaeverbeek, M., Brohee, D., Ducobu, J., and Remacle, C. (2004). Oxidation of low density lipoproteins by myeloperoxidase at the surface of endothelial cells: An additional mechanism to subendothelium oxidation. *Biochem. Biophys. Res. Commun.* **325,** 434–438.

Zuk, P. A., Zhu, M., Mizuno, H., Huang, J., Futrell, J. W., Katz, A. J., Benhaim, P., Lorenz, H. P., and Hedrick, M. H. (2001). Multilineage cells from human adipose tissue: Implications for cell-based therapies. *Tissue Eng.* **7,** 211–228.

CHAPTER SEVEN

Cellular and Molecular Mechanisms Regulating Fibrosis in Skeletal Muscle Repair and Disease

Antonio L. Serrano,* Christopher J. Mann,* Berta Vidal,* Esther Ardite,* Eusebio Perdiguero,* *and* Pura Muñoz-Cánoves*,†

Contents

Abstract

The repair of an injured tissue is a complex biological process involving the coordinated activities of tissue-resident and infiltrating cells in response to local and systemic signals. Following acute tissue injury, inflammatory cell infiltration and activation/proliferation of resident stem cells is the first line of defense to restore tissue homeostasis. However, in the setting of chronic tissue damage, such as in Duchenne Muscular Dystrophy, inflammatory infiltrates persist, the ability of stem cells (satellite cells) is blocked and fibrogenic cells are continuously activated, eventually leading to the conversion of muscle into nonfunctional fibrotic tissue. This review explores our current understanding of

* Department of Experimental and Health Sciences, Cell Biology Unit, CIBERNED, Pompeu Fabra University, Barcelona, Spain
† ICREA, Barcelona, Spain

Current Topics in Developmental Biology, Volume 96
ISSN 0070-2153, DOI: 10.1016/B978-0-12-385940-2.00007-3

the cellular and molecular mechanisms underlying efficient muscle repair that are dysregulated in muscular dystrophy-associated fibrosis and in aging-related muscle dysfunction.

1. Introduction

Fibrosis is the aberrant deposition of extracellular matrix (ECM) components during wound healing leading to loss of organ/tissue architecture and function, and is thus an important cause of morbidity and mortality worldwide (Wynn, 2008). Fibrotic diseases are often associated with chronic forms of disease pathology and occur in a large variety of vital organs and tissues, including lung, liver, kidney, intestine, heart, skeletal muscle, skin, eyes, and even in fluid tissues such as bone marrow (BM). Associated diseases include pulmonary fibroses, systemic sclerosis, liver cirrhosis, cardiovascular disease, muscular dystrophies, progressive kidney disease, macular degeneration, and BM fibrosis, respectively, among others. Despite the diverse range of tissues susceptible to fibrosis, all fibrotic reactions share common underlying cellular and molecular mechanisms, namely cell/tissue degeneration, leukocyte infiltration, persistence of inflammation in the tissue, and proliferation of cells with a fibroblast-like phenotype. The interplay and imbalance of different cell types sustain production of numerous factors including growth factors, proteolytic enzymes, angiogenic factors, and fibrogenic cytokines, which together stimulate the deposition of connective tissue elements that progressively remodel, destroy, and replace normal tissue architecture. However, despite these common elements, there are still many unknown initiators and contributors to fibrotic pathways as well as important differences between distinct tissue systems. Thus, improving our understanding of the mechanisms, cell types, and factors involved is critical for developing treatment strategies for these diseases.

In skeletal muscle, fibrosis is most often associated with the muscular dystrophies, a clinically and molecularly heterogenous group of diseases. These diseases are primarily characterized by skeletal muscle wasting that compromises patient mobility leading to wheelchair dependency. In the most severe cases, such as Duchenne muscular dystrophy (DMD), which is caused by lack of the dystrophin gene, muscle loss and fibrosis also cause premature death through respiratory and cardiac failure (Emery, 2002). In many dystrophies, including DMD, the mutation affects proteins that form a link between the cytoskeleton and the basal lamina, generally resulting in disassembly of the whole complex and consequent fragility of the sarcolemma, especially during intense contractile activity. In turn, this results in increased calcium entry and focal or diffuse damage to the fiber, though the molecular mechanisms are not yet elucidated in detail (Blake *et al.*, 2002).

In normal muscle repair after acute injury, such as experimental animal models, sports injuries, and in chronic human diseases such as DMD, damaged, or dead fibers are first removed by inflammatory cells and then repaired or replaced by tissue-resident muscle stem cells known as satellite cells (Fig. 7.1; Mauro, 1961). Until a few years ago, satellite cells were the only regenerative cells with myogenic potential known to be present in postnatal life. However, in DMD and many other dystrophies, newly generated fibers are also prone to further degeneration since they retain the underlying molecular defect and undergo constant cycles of fiber

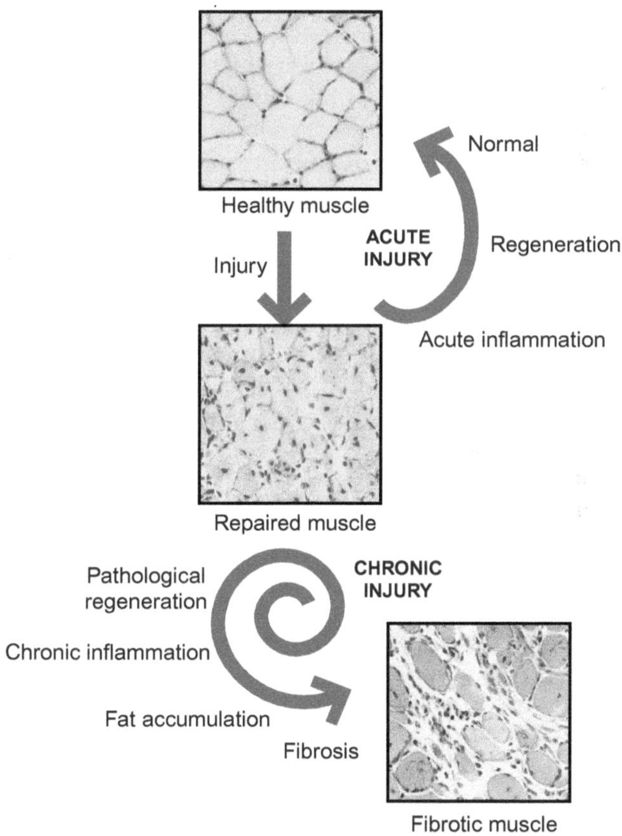

Figure 7.1 Processes involved in normal and fibrotic muscle regeneration. Acute injury to healthy muscle (top) involves rapid and controlled inflammatory events that remove dead and damaged myofibers and promote activation of resident satellite cells that mediate replacement of injured muscle (centre). However, in conditions of chronic injury, such as Duchenne muscular dystrophy, chronic inflammatory events result in excess accumulation of extracellular matrix components that in turn prevent efficient myogenic repair and lead to replacement of muscle with fibrotic/scar tissue in addition to some fat accumulation (bottom).

degeneration in association with chronic inflammation (Porter *et al.*, 2002). Thus, with time, the satellite cell population either becomes exhausted or unable to mediate repair and the muscle tissue is progressively replaced by adipose and especially fibrotic tissue. Fibrosis and loss of muscle tissue in dystrophies not only reduces motile and contractile functions, but also reduces the amount of target tissue available for therapeutic intervention. For example, cell and gene therapy applications to replace the missing dystrophin gene often require efficient transduction of semifunctional muscle to achieve their therapeutic effects (Muir and Chamberlain, 2009). In addition, increased fibrotic tissue can also act as an additional barrier to transduction efficiency. Not surprisingly, and despite ongoing advancements in the lab and the clinic, there is currently no effective therapy for DMD. The only relatively effective pharmacotherapy for DMD is corticosteroids, which prolong muscle strength and walking capacity in early years, but eventually lead to undesired secondary effects (Angelini, 2007). Further, there is also no effective clinical treatment to combat or attenuate muscle fibrosis in DMD patients. However, recent studies using the *mdx* mouse model of DMD have focused more attention on elucidating cellular and molecular mechanisms underlying skeletal muscle fibrosis associated with dystrophin deficiency. Importantly, these studies have tested several pharmacological agents that target muscle fibrosis and strongly suggest that combating fibrosis development could ameliorate DMD progression and increase the success of new cell- and gene-based therapies.

The goal of this review is therefore to cover the impact of fibrosis after normal and aberrant muscle repair, aging, and disease, particularly in DMD. Indeed, DMD-associated fibrosis bears all hallmarks of chronic organ/tissue damage and similarities and differences with other organ systems are bidirectionally informative although beyond the scope of this review. Instead, we will focus on the underlying mechanisms leading to fibrosis development during dystrophy progression, such as the its etiology, the identity and role of different cell types, the key cytokines and growth factors, animal models developed to study muscle fibrosis, and finally the mechanism and potential of antifibrotic therapies.

2. Muscle Tissue Regeneration: To Repair or Pathologically Scar

Tissue repair normally occurs very quickly after mechanical trauma, exposure to toxins or infections which cause tissue injury. However, rapid resolution of tissue injury requires a sequential and well-orchestrated series of events. Perturbation of any of these stages can result in unsuccessful muscle regeneration, typically characterized by persistent myofiber degeneration, inflammation, and pathological scarring or fibrosis, which is essentially an excessive accumulation of ECM components (reviewed in Grounds *et al.*, 2005; Kaariainen *et al.*, 2000; Wynn, 2008). The key events leading to normal and defective/fibrotic muscle repair are detailed in Fig. 7.2 and as follows.

Immediately after skeletal muscle injury, cytokines and growth factors are released from both the injured blood vessels and infiltrating inflammatory cells (reviewed by Chazaud *et al.*, 2009; Tidball and Villalta, 2010). These factors stimulate migration of inflammatory cells to and at the site of injury, as well as mediating proliferation and cell survival, whereas invading inflammatory cells are also responsible for phagocytosing cell debris. The specific role of many damage signals, growth factors, and inflammatory molecules on satellite cell behavior is still being investigated (Green *et al.*, 2009; Tidball and Villalta, 2010), but the next critical stage of repair is the formation of new muscle fibers via these cells. This begins with their activation since satellite cells normally lie beneath the muscle basal lamina of muscle fibers in a quiescent state. Extensive proliferation follows activation, with some cells undergoing self-renewal to replenish the satellite cell pool whereas most undergo commitment and subsequent differentiation, whereby myoblasts fuse either to themselves, or to the damaged myofibers to replace the lost muscle.

In addition to inflammatory and satellite cells, efficient muscle repair also requires the migration and proliferation of fibroblasts to produce new and temporary ECM components, such as collagen type I, collagen type III, fibronectin, elastin, proteoglycans, and laminin, in order to stabilize the tissue and serve as a scaffold for new fibers. In addition to new ECM components, satellite cells also utilize the basement membranes of preexisting necrotic fibers to maintain a similar myofiber position. Basement membranes and temporary ECM components are also critical for guiding the formation of neuromuscular junctions (NMJs; Lluri *et al.*, 2006). Formation and degradation of the ECM is performed by several proteases and their specific inhibitors which are expressed during tissue repair. ECM degradation also leads to the generation of protein fragments that provide important biological activities needed to facilitate normal tissue repair at an injury site (Chen and Li, 2009). Aberrant or dysregulated ECM accumulation during repair is thus the classical definition of fibrosis. Finally, in addition to ECM remodeling, angiogenesis facilitates development of a new vascular network at the site of injury while newly formed muscle fibers undergo growth and maturation.

Although this summary underscores the process in skeletal muscle, similar processes are known to occur in other organs and tissues. Overall it emphasizes that fibrosis-associated pathology is a complicated sequence of mechanisms whose key is the sustained interaction between activated immune and structural cells with both soluble and cell-associated factors

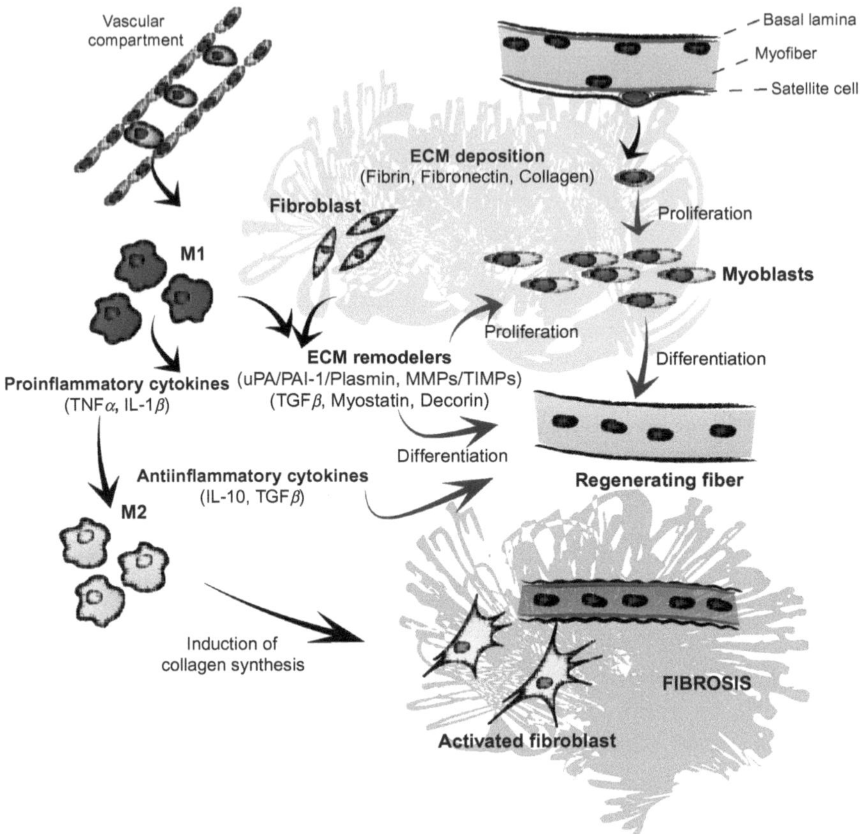

Figure 7.2 Cellular and soluble effectors of fibrosis in skeletal muscle repair. Replacement of damaged muscle fibers is mediated by tissue-resident stem cells, the satellite cells, which are normally quiescent and located under the basal lamina of muscle fibers (top right). Acute and chronic injury leads to activation, proliferation, and differentiation of these cells, however, their final ability to mediate repair is modified by the extent and type of injury and consequently their interaction with various cellular and soluble mediators. Tissue-resident and extravasating peripheral macrophages play an important early role, with classically activated M1 macrophages acting first to clear the damage and alternatively activated M2 macrophages (left) acting at later stages in tissue repair and fibrosis. M1 and M2 macrophages release a range of pro- and anti-inflammatory cytokines, respectively, that act on both myogenic cells and fibroblasts (center). Activated fibroblasts (termed myofibroblasts) respond to damage and growth factor/cytokine signals by releasing a host of agents that modulate extracellular matrix (ECM) formation, including TGFβ, metalloproteinases (MMPs) and their inhibitors, the TIMPs, different types of collagen and other molecules. Aberrations in the intensity, the kinetics, or the interaction of these various factors, such as in aging and the muscular dystrophies, leads to excessive ECM accumulation and replacement of muscle with fibrotic tissue (bottom right). (See Color Insert.)

that are able to give rise to pronounced impairment of tissue or organ function and even death. As in other fibrotic conditions, sustained or aberrant activation of the wound-healing process is the most common and relevant mechanism for inducing fibrosis. In skeletal muscle and particularly in DMD, this may be further defined by the following key features: (i) persistence of muscle tissue damage in conjunction with different degrees of necrosis and apoptosis; (ii) the recruitment and persistence of inflammatory cells which release profibrotic growth factors and cytokines; (iii) recruitment and activation of ECM-producing cells; and (iv) qualitative and quantitative changes in the ECM which limit the capacity for repair in the presence of persistent degeneration. In this review, we will overview these key points and discuss how the persistence of muscle damage-driven inflammation is a dominant promoter of fibrosis in skeletal muscle. Indeed, damage to tissues, and to muscle tissue in particular, can result from a variety of acute or chronic stimuli, but pathogenic fibrosis typically results from chronic inflammatory reactions.

3. Fibrosis Development in Dystrophic Muscle by the TGFβ Family of Growth Factors

The tissue environment of skeletal muscle after injury is a complex and dynamic mixture of both invading and tissue-resident inflammatory, myogenic, and fibroblastic cells (see Fig. 7.2 and also below). Not surprisingly, many studies have reported evidence for an intense and elaborate cross-talk between these cells, both involving and inducing the release of cytokines and other proinflammatory mediators that are likely to contribute to the overall environment observed in chronic fibrogenesis. Indeed, most profibrogenic polypeptides are produced by infiltrating immune, inflammatory, mesenchymal, and tissue-specialized cells, thus facilitating paracrine profibrogenic effects which will perpetuate inflammation-driven fibrotic processes (reviewed in Wynn, 2008).

One of the most potent profibrogenic cytokines amongst these effectors, and a major regulator of tissue wound healing and fibrosis development *in vivo*, is transforming growth factor-beta (TGFβ). There are three TGFβ isoforms, TGFβ1, TGFβ2, and TGFβ3, all of which are initially generated as latent precursors (Zhou *et al.*, 2006). Active TGFβ is liberated from the precursor by proteolysis, enabling it to bind to a heterodimeric receptor complex which consists of one TGFβ type I receptor molecule, termed activin linked kinase 5 (ALK5), and one TGFβ type II receptor. In the canonical TGFβ pathway, ligand binding leads ALK5 to phosphorylate Smad2 and 3, which in turn bind Smad4 and together they translocate into the nucleus and activate transcription (Bonniaud *et al.*, 2005).

This ALK5/Smad pathway is generally considered the primary pathway responsible for TGFβ signaling in normal fibroblasts. Similarly, recombinant TGFβ, which is fibrogenic in distinct tissues and organs in most *in vitro* and *in vivo* models acts predominantly through this canonical pathway. However, TGFβ has also been shown to signal via several additional pathways, including Ras/MEK/ERK, which requires the heparan sulfate-containing proteoglycan (HSPG) syndecan 4; p38, which requires the HSPG betaglycan; c-abl signaling; and JNK which requires focal adhesion kinase (FAK) and TGFβ activated kinase 1 (TAK1; Shi and Massague, 2003). Signaling via these pathways appears to modify gene expression in a promoter-selective fashion. For example, FAK, JNK, and TAK1 are required for myofibroblast differentiation and alpha smooth muscle actin (α-SMA) expression whereas ERK is required for collagen type I expression. However, p38 appears not to be involved with the fibrogenic activity of TGFβ. Thus, it is likely that additional signaling pathways are abnormally activated in muscle fibroblasts in a fashion independent of the canonical TGFβ pathway.

TGFβ is expressed in normal muscle after injury as well as in the dystrophic muscle of DMD patients and *mdx* mice (Zhou *et al.*, 2006), where it has been shown to have the potential to induce fibrosis around myofibers, probably via stimulation of fibroblasts to produce ECM proteins such as collagen and fibronectin. Of equal importance, TGFβ is also able to reduce production of enzymes like collagenase which degrade the ECM, whilst simultaneously augmenting production of proteins like tissue inhibitors of metalloproteinases (TIMPs) and plasminogen activator inhibitor type-1 (PAI-1) which act by inhibiting other enzymes which degrade the ECM (see below). Direct injection of TGFβ into skeletal muscle *in vivo* was shown to stimulate myogenic cells to express TGFβ in an autocrine fashion and to induce connective tissue formation within the injected area (reviewed in Brandan *et al.*, 2008; Zhu *et al.*, 2007). Moreover, myoblasts transfected with vectors that induce TGFβ expression can differentiate into myofibroblastic cells after intramuscular transplantation, a process inhibited by decorin, a small leucine-rich proteoglycan that can bind TGFβ and inhibit its activity (Li *et al.*, 2004, 2005). Taken together, these studies highlight the critical role of TGFβ in the initiation of fibrotic processes and in the induction of myogenic cells to differentiate into myofibroblastic cells in injured muscle.

Antagonism of TGFβ signaling by several therapeutic agents has been shown to inhibit fibrosis and improve muscle regeneration in several experimental and animal models although no agents have been shown to convincingly reduce fibrosis once formed. For example, direct immunomodulation of TGFβ inhibited connective tissue accumulation and fibrosis progression in *mdx* diaphragm, with the caveat that inflammation was shown to increase (Andreetta *et al.*, 2006). The aforementioned molecule decorin is another example, as is losartan, an angiotensin II type 1 receptor antagonist that is widely used as an antihypertensive medication. Angiotensin II is able to both directly stimulate TGFβ production while also enhancing TGFβ signaling by increasing Smad2 levels and the nuclear translocation of phosphorylated Smad3. Thus, by preventing this stimulating action of natural angiotensin II, losartan can inhibit TGFβ signaling. Importantly, reductions in tissue fibrosis after injury were detected *in vivo* in both decorin- and angiotensin receptor blocker-treated mice (Bedair *et al.*, 2008). As suggested above, decorin treatment was also able to improve muscle regeneration and prevent TGFβ-induced differentiation of myogenic cells into fibrotic cells in injured muscle, whereas extended treatment of *mdx* mice with losartan significantly reduced *mdx* diaphragm fibrosis with no adverse effects. Similar results were reported after treatment of animals with suramin, which acts by competing with TGFβ for receptor binding (Chan *et al.*, 2003). Halofuginone inhibits phosphorylation and activation of Smad3 and is thus a potent antifibrotic agent. Administration of halofuginone to aged *mdx* mice improved cardiac and respiratory function by drastically decreasing expression of collagen and significantly reducing levels of phosphorylated Smad3 (Roffe *et al.*, 2010). An important and interesting observation following halofuginone treatment was its ability to enhance myogenesis, a feature which might be relevant to improve muscle regeneration and function in muscular dystrophies (Huebner *et al.*, 2008).

When connective tissue is damaged, fibroblasts migrate into the wound and begin to produce and remodel the ECM in response to profibrotic cytokines such as TGFβ. Although necessary and fundamental to tissue homeostasis and normal wound repair, the fibroblast is clearly a critical intermediate in chronic fibrotic diseases and chronic inflammation is widely accepted to lead to their apparently unregulated activity. One additional and important aspect of fibroblast biology is that, once activated, they may be identified by the increased expression of α-SMA. α-SMA is a contractile protein organized into stress fibers which connect to the ECM through specialized structures called "mature" and "super-mature" focal adhesions, in addition to intercellular gap and adherens junctions. As a result, these α-SMA stress fibers can contract in order to exert mechanical tension on the ECM which in turn provides a mechanically resistant support. This contractile phenotype lends activated cells the name "*myofibroblast*," and they are associated with tissue repair and fibrosis in many tissues and organs ranging from muscle, skin, lung, bone, and cartilage (Hinz *et al.*, 2007). Persistence of myofibroblasts is believed to be responsible for a diverse array of fibrotic diseases including pulmonary fibrosis and scleroderma. However, despite their relevance in these diseases, it is still contested whether myofibroblasts really do exist in fibrotic skeletal muscle, or whether they are mature fibroblasts actively producing ECM components. Nevertheless, for consistency with the literature, we will use the term "myofibroblast" in the muscle context from hereon. There are potentially multiple origins of myofibroblasts. For example, they may arise by differentiation of local, resident fibroblasts in response to specific effectors. Alternatively, they may arrive by migration from the circulation or from neighboring sites and potentially undergo further reprogramming in the damaged area. Therefore, understanding the origin of myofibroblasts is of great importance for developing new methods to combat the fibrotic process observed in diverse fibrogenic diseases.

Although fibroblasts are the major collagen-producing cells, C2C12 myoblasts and myofiber-associated satellite cells have also been shown to express significant levels of interstitial collagens I and III, which decrease during the process of differentiation (Alexakis *et al.*, 2007). Although collagen can markedly suppress differentiation of C2C12 cells, collagen III expression is retained in aged *mdx* myogenic cells, whereas collagen I becomes restricted to myofiber-associated nonmyogenic cells. This suggests that reprogramming of myoblasts to myofibroblasts with progressing age may act via positive feedback (Alexakis *et al.*, 2007). Several recent studies have also investigated the induction of fibroblastic phenotypes in myogenic cells in other models. One group demonstrated Smad-dependent upregulation of sphingosine kinase-1 (SK1) in C2C12 myoblasts by TGFβ, while pharmacological or siRNA-mediated inhibition of SK1 prevented induction of fibrotic markers by TGFβ. Rho/Rho kinase signaling was also found to be implicated in the TGFβ-mediated transition of myoblasts into myofibroblasts downstream of SK1 activation (Cencetti *et al.*, 2010). Downregulation of Notch2 expression has also been linked to nonmuscle fibrotic tissue and TGFβ-dependent induction of myofibroblast markers in C2C12 myoblasts. Overexpression of active Notch2 in C2C12 cells inhibited the ability of TGFβ to induce expression of α-SMA and collagen I, whereas more surprisingly, transient knockdown of Notch2 by siRNA in cultured myoblasts resulted in differentiation of C2C12 myoblasts into myofibroblastic cells expressing fibrotic proteins such as α-SMA and collagen I, even in the absence of TGFβ. Finally, Notch3 was revealed to be counterregulated by Notch2, suggesting that the later is able to inhibit differentiation of myoblasts into myofibroblasts by limiting Notch3 expression (Ono *et al.*, 2007). The Notch pathway has also been strongly implicated in aging-associated fibrosis (see below).

Connective tissue growth factor (CTGF) is another molecule that has been shown to reproduce many of the profibrotic effects of TGFβ on C2C12 cells. Elevated levels of CTGF have been reported in skeletal muscle from DMD patients, dystrophic dogs, and *mdx* mice (Sun *et al.*, 2008; Vial *et al.*, 2008). Myoblasts and myotubes both produce and respond to CTGF *in vitro* by downregulating muscle-specific proteins like MyoD and desmin whilst increasing the accumulation of ECM components such as fibronectin and collagen I (Vial *et al.*, 2008). These results suggest a role for TGFβ and CTGF in skeletal muscle remodeling by inducing fibrosis, by inhibiting myogenesis, and by promoting dedifferentiation of myoblasts into myofibroblast-like cells (Vial *et al.*, 2008).

Another family of growth factors able to induce fibroblastic changes in muscle is the PDGF family which includes PDGF-AA, PDGF-AB, PDGFBB, PDGF-CC, and PDGF-DD. These molecules bind two different PDGF receptors, α and β, and *in vivo* are able to attract neutrophils, macrophages, fibroblasts, and smooth muscle cells to proliferate and migrate into the wound site. *In vitro*, PDGF production can be induced by TGFβ while also stimulating fibroblasts to contract collagen matrices and differentiate into myofibroblasts (Bonner, 2004). As described above, signaling via c-abl is an alternative to the noncanonical Smad pathway used by TGFβ to mediate its fibrogenic effects, and PDGF is also able to induce c-abl kinase activity. Not surprisingly, blocking c-abl activity has antifibrotic effects. For example, use of the antineoplastic agent imatinib which selectively and competitively blocks the ATP binding site of c-abl and several other tyrosine kinases, including PDGF receptors and c-kit, has been shown to reduce tissue fibrosis in many experimental mouse models of fibrotic disorders. In addition to TGFβ, PDGF and its receptors are upregulated in inflammatory cells and regenerating fibers of DMD patients and *mdx* mice. Consequently, use of imatinib in *mdx* mice resulted in reduced fibrosis in the diaphragm. Additional effects of imatinib in the *mdx* model included attenuation of skeletal muscle necrosis and inflammation with concomitant improvements to muscle function (Huang *et al.*, 2009). Complementary studies demonstrated that imatinib ameliorated dystrophy *mdx* mice after exercise; however, treated mice also exhibited an important weight loss (Bizario *et al.*, 2009). Taken together, these studies suggest that TGFβ, CTGF, and PDGF are likely to cooperate in maintaining tissue repair and fibrogenic responses in fibroblasts within regenerating skeletal muscle.

Myostatin, which is also known as growth differentiation factor 8 (GDF8), is a TGFβ family member structurally related to activins. It is specifically expressed in the skeletal muscle lineage and is a major negative regulator of muscle growth such that myostatin deficiency increases skeletal muscle mass, myofiber diameter, and strength (Glass, 2010; Kollias and McDermott, 2008). Normal myostatin signaling can be counterregulated by the extracellular protein follistatin (Brandan *et al.*, 2008).

However, myostatin not only regulates the growth of muscle cells but also directly regulates muscle fibroblast activation and hence fibrosis progression (Li *et al.*, 2008). Several groups have now reported improvements to regeneration and decreased fibrosis in the absence of myostatin. One reason for this could be that by binding to the activin receptor IIB, which is expressed on muscle fibroblasts, myostatin is able to directly stimulate proliferation of muscle fibroblasts and induce their differentiation into myofibroblasts, as shown both *in vitro* and *in vivo*. In fibroblasts, myostatin stimulates activation of Smad, p38 MAPK, and AKT pathways to stimulate ECM protein synthesis. In one study, it was shown that myostatin-deficient mice developed significantly less fibrosis and exhibited better muscle regeneration after injury than wild-type mice (Zhu *et al.*, 2007). In agreement with this, *mdx* mice lacking myostatin had less fibrosis in the diaphragm and were stronger and more muscular than their *mdx* counterparts, suggesting improved muscle regeneration (Wagner *et al.*, 2002). As a potential therapeutic approach, immunological neutralization of myostatin in *mdx* mice was also shown to attenuate dystrophy by improving muscle regeneration and reducing ECM accumulation (McCroskery *et al.*, 2005). The converse of these experiments was also true. Injection of recombinant myostatin protein *in vivo* stimulates myofibers to express TGFβ, whereas it is also able to stimulate TGFβ secretion from C2C12 myoblasts *in vitro*, which can have autocrine profibrotic effects as described above. Furthermore, TGFβ also induces myostatin production, revealing a collaborative action of both profibrotic cytokines on muscle cells. Finally, one important consideration is that the physiological role of myostatin on cardiac muscle appears significantly different than in skeletal muscle. Myostatin does not induce cardiac hypertrophy and does not modulate cardiac fibrosis in *mdx* mice (Cohn *et al.*, 2007).

The actions of several other growth factors may also work via modulating myostatin function. For example, the antifibrotic, proregenerative effects of decorin, may derive not only from its ability to neutralize the effects of TGFβ but also from its capacity to antagonize the action of myostatin on both fibroblasts and myoblasts. Decorin is also able to upregulate expression of the myostatin inhibitor follistatin (Zhu *et al.*, 2007). Overexpression of insulin-like growth factor 1 (IGF-1) and simultaneous loss of myostatin *in vivo* have been observed to have interesting synergistic effects on myofiber growth and fibrosis reduction. Local expression of a muscle-restricted IGF-1 transgene accelerates the regenerative process of injured skeletal muscle, modulating the inflammatory response, and limiting fibrosis (Pelosi *et al.*, 2007). However, the mechanisms by which these two genes combine to regulate fibrosis and muscle regeneration and any potential physiological interconnection of these processes in injured muscle are still unknown and under investigation. Overall, myostatin has distinct fibrogenic properties which, taken together with other signaling systems, suggests the existence of a coregulatory relationship between TGFβ, CTGF, myostatin, and decorin.

4. From Muscle Injury to the Chronic Inflammatory Response and Pathological Muscle Fibrosis

Sterile tissue damage is usually associated with the release of intracellular components that can act as specific signals for the initiation of a repair process, first by signaling to tissue-resident macrophages and subsequently by directly and/or indirectly recruiting peripheral cells. That is, the earliest phases of tissue repair are generally characterized by local activation of the innate immune system although the original immunogenic stimuli are not always known (see also below; Green *et al.*, 2009). Macrophages have a prominent role in the innate immune response to infection and/or tissue injury through their ability to phagocytose particles such as bacteria or cellular debris and to secrete proinflammatory cytokines (Chazaud *et al.*, 2009). In addition to tissue-resident macrophages, invasion of the damaged site involves both polymorphonuclear leukocytes (e.g., neutrophils) and blood monocytes which also differentiate into macrophages (Tidball, 2005). In other tissues, other inflammatory cell types such as mast cells and T cells have also been shown to play a key role in repair and fibrogenesis, but to date only a limited number of studies have investigated the role of these cells in muscle repair and DMD (Nahirney *et al.*, 1997). However, before discussing in detail the roles played by some of these cell types in muscular dystrophies and muscle repair, it is important to emphasize two key ideas when considering cell types involved in tissue repair and fibrosis. Firstly, some form of inflammatory response is necessary to effectively repair damaged tissues and fibrosis is often generally considered as an imbalance between pro- and antifibrotic cell types and their associated effectors. Secondly, despite considerable research and debate around the issue of the etiology and pathophysiology of fibrosis, it is becoming more widely accepted that the *chronic* nature of the inflammatory response to tissue damage is a key driver of unrestrained wound healing and the fibrotic response in diverse organs and tissues.

Macrophages are the major inflammatory cell type present in injured muscle (Chazaud *et al.*, 2009). In regenerating and dystrophic muscle they function by clearing myofiber debris and modulating the regeneration process in part by secreting cytokines. There is now a wealth of evidence that the nature, duration, and intensity of the inflammatory response in muscle damage and regeneration can critically influence the outcome of the muscle repair process or, alternatively, of fibrosis, particularly during the progression of some dystrophies (reviewed in Tidball, 2005; Tidball and Villalta, 2010; Wynn, 2008). For example, interfering with the transient inflammatory response after acute injury may negatively affect the phagocytosis of dead and damaged fibers and thus impede formation of new tissue, whereas modulating the chronically high level of inflammation in dystrophic muscle can be beneficial in reducing both muscle degeneration and fibrosis, while simultaneously promoting regeneration (Segawa *et al.*, 2008).

An important development in our understanding of muscle repair and fibrosis has been the demonstration that macrophages constitute a heterogenous population in regenerating muscle after injury, with their activities appear opposing in nature, being either proinflammatory or anti-inflammatory and following different kinetics of activity (Arnold *et al.*, 2007).

A nomenclature system for polarized macrophages has also been proposed (Mantovani *et al.*, 2004; Wynn and Barron, 2010). Macrophages are now referred to as classically and alternatively activated macrophages, or M1 and M2 macrophages, respectively (Fig. 7.3).

Classically activated (M1) or *proinflammatory macrophages* arise from exposure to the Th1 cytokines interferon-gamma (IFN-γ) and tumor necrosis factor-alpha (TNFα) in addition to lipopolysaccharide (LPS)/endotoxin

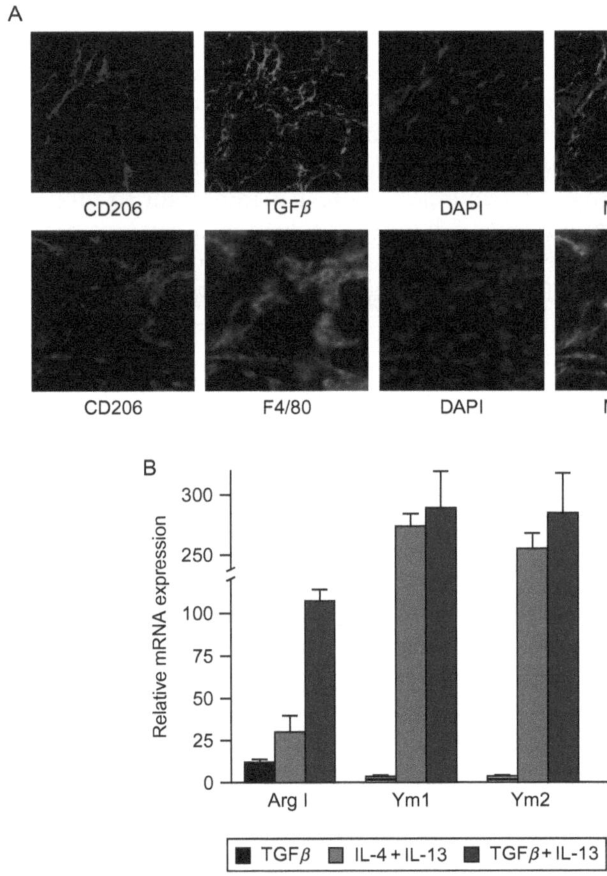

Figure 7.3 Alternatively activated (M2) macrophages are associated with development of fibrosis in dystrophic skeletal muscle. (A) Alternatively activated (M2) macrophages coexpress CD206 (red), TGFβ (green; top line), and F4/80 (green; bottom line) in the fibrotic diaphragm of a 12-month-old *mdx* mouse. Nuclei are stained with DAPI in blue. (B) Macrophages from *mdx* mice were isolated and treated with recombinant TGFβ, IL-4 + IL-13, or TGFβ+IL-13 for 24 h and expression of key alternatively activated macrophage markers Arginase-1, Ym1, and Ym2 were analyzed by qRT–PCR. (See Color Insert.)

(Mantovani *et al.*, 2004; Wynn, 2004; Wynn and Barron, 2010). M1 macrophages play a key role in acute inflammatory process and are thus considered the prototypical type of macrophage. They are found during the initial stages after muscle damage in association with recruited monocytes, where their extensive phagocytic potential helps remove necrotic material in addition to their ability to process and present antigens and activate T cells. M1 macrophages are also characterized by their ability to produce high levels of proinflammatory cytokines such as TNFα and IL-1β. In addition to proinflammatory cytokines, they also produce large amounts of IL-12 and express inducible nitric oxide synthase (iNOS; also known as NOS2), which is necessary to efficiently metabolize L-arginine in order to generate the large amounts of NO that are involved in the killing of intracellular pathogens.

M2 macrophages show an additional layer of complexity to M1 macrophages in that they are currently divided into three distinct subtypes which reflect different functional specializations. M2a macrophages, or *alternatively activated macrophages*, are activated by the Th2 cytokines IL-4 and IL-13 and are most commonly associated with tissue repair, wound healing and fibrosis. M2b macrophages are *anti-inflammatory macrophages* due to their release of Th2 cytokines. As they are activated by immune complexes and/or toll-like receptors (TLRs), they appear to play a less important role in normal muscle repair. Finally, M2c macrophages are also considered to be *anti-inflammatory* as they play a key role in deactivating the M1 phenotype in addition to having proliferative effects on nonmyeloid cells. M2c macrophages also release anti-inflammatory cytokines and are activated by IL-10. Thus, classically activated M1 macrophages are usually found at early stages after muscle injury, closely followed by M2c anti-inflammatory macrophages and finally M2a alternatively activated macrophages are abundant at advanced stages of the regeneration process in association with healing and tissue repair (Arnold *et al.*, 2007). Aberrations in the intensity or duration of macrophage kinetics can have profound effects on muscle regeneration and fibrosis.

In terms of tissue fibrosis, M2a macrophages are generally considered the most important. They express specific cell surface markers such as the mannose receptor CD206 and the type II IL-1 decoy receptor in addition to releasing a range of regulatory cytokines such as IL-10 and the soluble IL-1 receptor antagonist (IL-1Ra) as well as many profibrotic molecules like TGFβ, fibronectin, proline, several TIMPs and chemokine (C–C motif) ligand 17 (CCL17). CCL17 in particular has been shown to enhance fibrosis in several mouse models of pulmonary disease by binding to the CC chemokine receptor 4 (CCR4; Yogo *et al.*, 2009). Finally, one additional reason for the ability of M2 macrophages to neutralize the M1 proinflammatory response is via their expression of high levels of arginase 1 (ARG1) which directly competes with M1-associated iNOS for L-arginine (Pesce *et al.*, 2009).

In healthy individuals, M2 macrophages are predominantly found in the placenta and lung, where they promote tolerance to self-antigens and curb inflammation. However, more recently these cells have been shown to have an increasingly important role in other pathologic conditions as diverse as tumorigenesis, asthma, allergy, and certain fibrotic pathologies (Wynn, 2008). Perhaps not surprisingly, the presence of alternatively activated M2a macrophages increases progressively with age in fibrotic *mdx* diaphragms (Vidal *et al.*, 2008). Similarly, fibrinogen depletion in *mdx* mice results in reduced fibrosis, concomitant with a significant decrease in the number of M2a macrophages in the diaphragm. In DMD patients, fibrosis has also been associated with increased numbers of alternatively activated macrophages (Desguerre *et al.*, 2009).

Several groups have used different *in vitro* and *in vivo* animal models to begin to unravel the role of macrophages on myogenesis, muscle repair, fibrosis and the development and treatment DMD. *In vitro*, proinflammatory macrophages have a positive influence on myoblast proliferation while repressing differentiation. On the other hand, anti-inflammatory macrophages were shown to stimulate myoblast differentiation and fusion. Importantly, *in vivo* depletion of either of these macrophage cell types had negative effects on the muscle repair process after injury (Arnold *et al.*, 2007). Other similar studies of macrophage depletion *in vivo* or in models of impaired macrophage recruitment, such as in urokinase plasminogen activator (uPA)-deficient mice (Segawa *et al.*, 2008; Suelves *et al.*, 2002, 2007; Wehling *et al.*, 2001), have revealed critical functions of macrophages in the regulation of fibrogenesis in dystrophic muscle. These studies have been supported by data from other groups in *mdx* mice using a variety of anti-inflammatory agents acting on cytokines such as TNFα and their cellular receptors, or on other major proinflammatory pathways such as NF-κB (Pelosi *et al.*, 2007; Peterson and Guttridge, 2008; Radley *et al.*, 2008). The general outcome of these studies has been that appropriate modulation of macrophage activity can ameliorate progression of dystrophy.

The roles of macrophage polarization on tissue repair and fibrosis have also investigated more directly in several recent studies. One group reported that deletion of two CREB-binding sites from the C/EBPβ promoter specifically impaired M2, but not M1, gene expression and interfered with the later stages of injury-induced muscle regeneration. Mutant mice were shown to be able to remove necrotic tissue from injured muscle, but exhibited severe defects in myofiber regeneration (Ruffell *et al.*, 2009). The C/EBPβ promoter mutation also resulted in a reduction of ARG1 expression in macrophages. This additional alteration was hypothesized to reroute arginine metabolism away from arginase-mediated polyamine synthesis, which promotes tissue regeneration, toward iNOS-mediated NO production, which was shown previously by another group to promote degradation of the key myogenic transcription factor MyoD (Di Marco *et al.*, 2005).

Further support for this idea has come from a study which showed that similar shifts in macrophage polarization and macrophage competition for arginine metabolism can influence the severity of muscle pathology in *mdx* dystrophic mice (Villalta *et al.*, 2009). One recent paper has shown a potentially important clinical link between arginine metabolism by M2a macrophages and fibrosis (Tidball and Villalta, 2010). In this study, Th2 cytokines were shown to increase expression and activity of arginase by M2 macrophages in different *mdx* mice, with intriguing differences in the effects of arginase-2 deletion in different muscles. More importantly from the clinical perspective, this study showed that long term dietary supplementation with arginine increased fibrosis of skeletal and cardiac muscle, in contrast to other reports showing benefits from short-term supplementation. As many DMD patients have been given dietary arginine based on these earlier studies (Evangeliou *et al.*, 2009), it is now possible that negative effects on fibrosis could have been promoted. Taken together, these studies indicate the critical roles of macrophage polarization both in muscle repair and fibrogenesis, particularly in dystrophic muscle. Although experimental evidence is still lacking, it could be speculated that M2a macrophages could exert opposing roles in regenerating muscle after injury and in dystrophic muscle, by promoting myofiber growth and/or by increasing collagen production, respectively.

Macrophages are not the only cells of the immune system shown to play a role in muscle repair and dystrophy. Tissue repair and fibrosis is also tightly regulated by the T helper (Th) cell response. Similarly to macrophages, Th lymphocytes can differentiate into different functional subsets, namely Th1 and Th2 cells which are able to orchestrate the host-response via distinct profiles of cytokine production (Wynn, 2004). CD4+ Th1 cells produce cytokines that promote cell-mediated immunity, including IFN-γ, TNF, IL-12, and IL-2, all of which have been demonstrated to be antifibrotic cytokines. In contrast, CD4+ Th2 cells promote humoral immunity and produce IL-4, IL-5, IL-6, and IL-13 which have been demonstrated as profibrotic cytokines. Th1 cytokines inhibit the development of Th2 cells and, vice versa, Th2 cytokines inhibit the development of Th1 cells. Clearly, alterations or imbalances in these pathways have the potential to skew repair via anti- or profibrotic pathways. One well-studied example of this is demonstrated by the importance of Th2 cytokines in the development of liver fibrogenesis (Wynn, 2008).

T cell-produced cytokines also have a regulatory function on macrophage polarization. For example, loss of uPA proteolytic activity in transgenic knockout mice resulted in reduced macrophage and T lymphocyte infiltration of injured muscle, correlating with a greater persistence of myofiber degeneration (Lluis *et al.*, 2001). Moreover, *scid/mdx* mice which are deficient in functional T and B lymphocytes present with markedly reduced diaphragm fibrosis at 1 year of age, concomitant with decreased skeletal muscle levels of activated TGFβ protein compared with normal *mdx* mice (Farini *et al.*, 2007).

Lack of functional T cells alone in *nu/nu/mdx* mice was also shown to lead to reduced diaphragm fibrosis at 3 months, supporting a pathogenic role for T cells in *mdx* muscle and revealing this lymphocyte subclass as an important source of TGFβ1 (Morrison *et al.*, 2000).

Another group more recently identified a specific subpopulation of T cells expressing the Vβ8.1/8.2 T cell receptor (TCR) which were both enriched in *mdx* muscle and produced high levels of osteopontin, a cytokine that promotes immune cell migration and survival (Vetrone *et al.*, 2009). Osteopontin levels were shown to be elevated in DMD patients and *mdx* mice after disease onset. Importantly, loss of osteopontin in *mdx* double mutant mice resulted in reduced infiltration of NKT-like cells, cells which express both T and natural killer (NK) cell markers, and neutrophils, in addition to reduced levels of TGFβ. These results correlated well with improved muscle strength and reduced diaphragm and cardiac fibrosis (Vetrone *et al.*, 2009). Other immune-associated proteins in addition to osteopontin have been shown to modify muscle repair and dystrophy progression. Eosinophil-derived major basic protein (MBP) was demonstrated to regulate muscular dystrophy (Wehling-Henricks *et al.*, 2008). The same group previously showed a role of perforin in mediating muscle damage and repair in association with eosinophilia, which interestingly was also linked to Th2 responses (Cai *et al.*, 2000).

However, not all studies have shown such definitive results and the implication of lymphocytes and their subtypes in muscle repair and fibrosis clearly requires further investigation. For example, thymectomy at 1 month of age to force near-complete postnatal depletion of circulating T cells in *mdx* mice, followed by anti-CD4 and/or -CD8 antibody treatment, failed to improve diaphragm fibrosis at 6 months of age (Farini *et al.*, 2007; Morrison *et al.*, 2005; Spencer *et al.*, 2001). In another study, M2 macrophages were conversely shown to exert an influence on CD4+ Th cells. Here, ARG1-expressing macrophages suppressed Th2 cytokine-driven inflammation and fibrosis in the liver induced by *Schistosoma mansoni* infection (Wynn, 2004), adding more complexity to the mechanisms regulating inflammation and fibrosis development. Therefore, future studies are still needed to elucidate whether distinct type of Th responses and macrophage subtypes operate in dystrophic muscle and how they mediate their interaction.

5. Alteration of the ECM Proteolytic Environment Leads to Fibrosis Development in Dystrophic Muscle

As previously mentioned, fibrosis is often defined as a wound healing response that has gone out of control, or an excessive deposition of ECM that prevents normal regeneration following tissue injury. However, appropriate ECM deposition is not only critical for efficient repair, but the ECM surrounding skeletal muscle has an important role in maintaining the structure in addition to reinforcing the contractile function of muscle. Furthermore, the ECM sequesters and presents heparin-binding growth factors such as hepatocyte growth factor (HGF) and fibroblast growth factor (FGF) to the fibers, as well as participates in signaling to the differentiated fibers through dystroglycan and sarcoglycan complexes (Cornelison, 2008; Durbeej and Campbell, 2002).

Analysis of individual components of the ECM in transgenic knockout models after muscle injury or after pharmacological intervention has provided important insight into the function of the ECM in normal conditions and during muscle regeneration. In normal muscle, the external lamina is composed of collagen IV, laminin, and HSPGs, while the interstitial matrix that surrounds them contains collagens I, III, and V, fibronectin and perlecan (Cornelison, 2008; Grounds *et al.*, 2005; Kaariainen *et al.*, 2000). The systematic and timely process of breakdown and replacement of these layers is thus critical for ensuring full and rapid repair as well as avoiding fibrosis.

The first stage after muscle injury is the formation of a hematoma between damaged fibers. This hematoma is rapidly invaded by inflammatory cells. As described above, macrophages, in particular, play a critical early role, phagocytosing and clearing myofiber debris and blood clot components. Fibrin and fibronectin extravasate from the circulation and cross-link at the hematoma site to form a primary matrix. This is necessary in order to provide a scaffold and anchorage site that strengthens newly formed connective tissue against contractile forces and for both infiltrating cells and activated resident cells to start rebuilding the muscle (Kaariainen *et al.*, 2000; Vidal *et al.*, 2008). It is at this time that activated fibroblasts begin secreting a wide variety of growth factors and ECM components, such as fibronectin, collagen I and III, and proteoglycans, which mediate expansion of the connective tissue and promote cell proliferation and migration. The final stage of normal repair is muscle growth which requires appropriate degradation and/or regulation of this temporary ECM to allow expansion of the fibers. Although some matrix deposition is necessary, excessive and/or permanent collagenous tissue deposition around the myofiber is potentially fibrotic and pathogenic. In DMD and some other muscle pathologies, it ultimately leads to substitution of skeletal and cardiac muscle by fibrous tissue.

One key component of the provisional primary ECM in injured skeletal muscle is fibrinogen and its end-product fibrin, which are collectively referred to as fibrin/ogen. Excessive and persistent fibrin deposition is deleterious for myofiber repair. Transgenic mice with genetic loss of the fibrinolytic proteases uPA and plasmin (see also below) have been shown to exhibit impaired muscle regeneration due to defective fibrinolysis and fibrin accumulation after muscle injury (Lluis *et al.*, 2001; Suelves *et al.*, 2002). Fibrin also accumulates in the muscles of human DMD patients and in the diaphragm of *mdx* mice (Suelves *et al.*, 2007; Vidal *et al.*, 2008). Importantly, fibrin accumulation also correlates with increasing age and fibrosis progression. However, fibrin/ogen depletion in these animals can rescue muscle regeneration. Due to fibrin/ogen's capacity to bind the integrin receptor Mac-1 on classically activated M1 macrophages (Vidal *et al.*, 2008), intramuscular fibrin/ogen deposits might serve as important temporo-spatial cues for mediating the inflammatory response. For example, it has been shown that following engagement of fibrin/ogen with Mac-1, macrophages in *mdx* muscle induce the expression of the profibrotic cytokine TGFβ. Increases in the expression of proinflammatory chemokines and cytokines such as IL-1β, TNFα, IL-6, and MIP-2, which are known to promote muscle degeneration (Tidball, 2005) also occur via similar mechanisms. Together, these reactions may in turn potentiate alternative activation of macrophages. As described above, M2a macrophages have been shown to promote fibrosis in several different pathogenic conditions (Wynn, 2008). By binding the αVβ3 integrin receptor on fibroblasts, fibrin/ogen can also directly stimulate the expression of collagen (Vidal *et al.*, 2008). This data highlights the integrative concept that fibrin/ogen is not merely a structural component of the transient matrix formed after injury, but is able to further promote tissue inflammation and fibrosis in persistent chronic injuries such as muscular dystrophies. Further, these results indicate that the persistence of components of the ECM produced immediately after injury are deleterious for muscle repair, at least in part, by promoting fibrosis development.

Decorin is the main proteoglycan present in the ECM of adult muscle, whereas biglycan expression is lower, although both proteins are increased in *mdx* muscle (Casar *et al.*, 2004; Zanotti *et al.*, 2005). As described above, both are able to modulate TGFβ functions and fibrosis (Brandan *et al.*, 2008). In particular, loss of biglycan in transgenic mice does not greatly affect regenerative capacity, despite inducing delays in new myofiber growth (Casar *et al.*, 2004), whereas decorin delivery to regenerating muscle with adeno-associated viral (AAV) vectors prevents TGFβ activation and fibrosis (Li *et al.*, 2007). Finally, latent TGFβ-binding protein 4 protein (LTBP4), which regulates the release and bioavailability of TGFβ from the ECM, has also been identified as a modulator of fibrosis in the context of muscular dystrophy in mice (Heydemann *et al.*, 2009). Although detailed studies of the role of ECM-associated TGFβ-regulatory proteins in human muscle disease and fibrosis are still lacking, the available evidence from mice reinforces the importance of ECM-mediated control of TGFβ signaling in the regulation of the fibrosis.

In addition to direct synthesis of ECM components, efficient muscle repair also requires factors which positively and negatively regulate controlled proteolytic degradation of the ECM. These muscle-associated proteolytic molecules include a broad family of matrix metalloproteinases

(MMPs), calcium-dependent enzymes which specifically degrade collagens and noncollagenous substrates, and their inhibitors, the TIMPs. Modulators of proteolysis also include molecules of the plasminogen activation (PA) system (reviewed in Chen and Li, 2009; Nagamine *et al.*, 2005). MMPs, alone or in conjunction with the plasminogen/plasmin system, have the ability to degrade ECM components which is essential for cell migration and tissue remodeling.

The large family of MMPs includes various collagenases (MMP-1, MMP-8, and MMP-13), gelatinases (MMP-2 and MMP-9), stromelysins (MMP-3, MMP-7, MMP-10, and MMP-11) in addition to membrane-type enzymes (MMP-14, MMP-15, MMP-16, MMP-17, MMP-24, and MMP-25) and the metalloelastase MMP-12 (Chen and Li, 2009). MMPs are released from both damaged muscle and infiltrating cells in order to disrupt the fiber basement membrane and thus facilitate recruitment of myogenic, inflammatory, vascular, and fibroblastic cells to damaged tissue. However, MMP function is not only controlled by expression and release by damaged muscle and inflammatory cells. Net MMP activity also reflects the relative amounts of activated enzyme. Activation requires proteolytic cleavage of the inactive precursor by either membrane-type matrix metalloproteinase 1 (MT1-MMP; Ohtake *et al.*, 2006) or plasmin, and their corresponding inhibitors, especially TIMPs (Visse and Nagase, 2003). For example, increased levels of TIMP-1 and TIMP-2 mediate a net decrease in protease activity and thus matrix accumulation. The serine protease plasmin can directly activate a group of MMP prodomains, such as proMMP-1, proMMP-3, proMMP-9, proMMP-10, and proMMP-13, at least in *in vitro* studies. One difference is the activation of proMMP-2 which also involves hydrolysis by MT1-MMP during plasmin stimulation. In some cases, active MMPs also further activate the proMMPs of other MMPs thereby forming a positive-feedback mechanism.

MMPs play an obvious role in fibrosis development, since collagen accumulation will occur when the rate of synthesis by activated fibroblasts is greater than the rate of breakdown by MMPs/collagenases, that is, when the balance between TIMP and MMP activity favors TIMPs. In addition to fibrosis, MMPs and serine proteases have been proposed to have many additional roles in skeletal muscle repair and satellite cell function by mediating ECM remodeling after injury. MT1-MMP, for example, has a presumptive role in maintaining myofiber integrity since transgenic mice deficient in MT1-MMP exhibited smaller and centrally nucleated myofibers compared to controls (Ohtake *et al.*, 2006), whereas MMP-13 and MMP-1 have also been shown to participate in ECM remodeling during muscle repair (Chen and Li, 2009; Kaar *et al.*, 2008; Wu *et al.*, 2003). The activity of the gelatinase MMP-2 has been shown to affect new fiber formation by promoting degradation of collagen IV in the basement membrane during myoblast proliferation, migration, and fusion. Another gelatinase, MMP-9, has been proposed to play a key role in satellite cell activation during the initial phase of muscle regeneration in addition to its role in inflammation. Perhaps more importantly, skeletal muscle of adult *mdx* mice express high levels of latent and active forms of MMP-2 and MMP-9. Inhibition of MMP-9 activity was able to significantly improve regeneration and contractile functions by reducing muscle structural deterioration, necrosis inflammation, and fibrosis (Chen and Li, 2009; Fukushima *et al.*, 2007; Kherif *et al.*, 1999; Li *et al.*, 2009). Similar results were achieved in aged *mdx* mice after transplantation of modified tendon fibroblasts expressing MMP-9 and PlGF (placenta growth factor). These cells were able to restore a vascular network and reduce collagen deposition and thus raise the exciting possibility of using similar cell-based or gene therapies for improving muscle function in currently untreatable advanced-stage dystrophic patients (Gargioli *et al.*, 2008). Finally, stem cell antigen-1 (Sca-1) has also been shown to modify muscle regeneration via modulating ECM remodeling. Sca-1-deficient mice have defects in muscle regeneration since Sca-1 is a negative regulator of myoblast proliferation and differentiation *in vitro*, but these mice also demonstrate enhanced fibrosis following injury due to reductions in MMP activity. It is still not clear if Sca-1 is acting directly or indirectly to upregulate the activity of MMPs, which could in turn lead to increased matrix breakdown and efficient muscle regeneration while halting fibrosis (Kafadar *et al.*, 2009). Sca-1 also plays a role in the maintenance of progenitor cells.

The action of MMPs are likely to complement the roles of the PA system in ECM remodeling during skeletal muscle regeneration due to their interactions and their capacity to degrade multiple ECM proteins. The PA system itself comprises an inactive, extracellular serine protease, plasminogen, which is converted into the active enzyme, plasmin, by two plasminogen activators (PAs): tissue-type plasminogen activator (tPA) and urokinase-type plasminogen activator (uPA). Inhibition of the PA system occurs at the level of the PAs by plasminogen activator inhibitor 1 (PAI-1), or at the level of plasmin, by α2-antiplasmin (Nagamine *et al.*, 2005). The major function of the PA system is the degradation of fibrin (Nagamine *et al.*, 2005), but it also has an important role in muscle ECM remodeling by cleaving ECM-associated molecules, liberating and/or activating latent forms of bioactive molecules such as growth factors, angiogenic factors and cytokines, in particular bFGF, TGFβ, VEGF, and HGF/SF, as well as certain MMPs, as described above. Many of these molecules may also play a role in promoting the activation of quiescent satellite cells and regulating subsequent myogenic programs.

Not surprisingly then, components of the PA system play important, yet distinct roles in muscle regeneration after injury, based upon the different muscle alterations observed in knockout mice. While it has been shown that both uPA and plasmin activities are necessary for skeletal muscle regeneration, tPA activity is dispensable, indicating that no redundancy exists in muscle (Lluis *et al.*, 2001; Suelves *et al.*, 2002). In contrast, a negative role for PAI-1 in muscle regeneration was suggested by various improvements to muscle repair observed after injury in PAI-1 deficient animals (Suelves *et al.*, 2005). The PA system has also been shown to have an increasingly important role in muscular dystrophies. For example, greater expression of uPA has been found in *mdx* muscle, while conversely, genetic loss of uPA exacerbated dystrophy and reduced muscle function in *mdx* mice (Suelves *et al.*, 2007). Recent data also suggests a role of PAI-1 in dystrophic muscle of humans and mice and a potential role for the uPA/PAI-1 balance in muscle fibrosis (Ardite, Vidal, and Munoz-Cánoves, unpublished results). Satellite cells derived from human DMD patients produce more uPA receptor and PAI-1 and less uPA than normal satellite cells (Fibbi *et al.*, 2001). BM-derived uPA also appears to play a key role in *mdx* muscle repair. Transplantation experiments suggest that BM-derived uPA might act in a number of different ways. Firstly, it prevents excessive deposition of fibrin through its normal role in fibrin-degradation. Secondly, it promotes myoblast migration, presumably by increasing the activation or availability of growth factors. Finally, BM-derived uPA increases infiltration of BM-derived inflammatory cells, especially macrophages (see also above) into the damaged tissue. However, an interesting observation underpinning these results was that genetic loss of the uPA receptor in *mdx* mice failed to exacerbate muscular dystrophy, suggesting that uPA exerts its proteolytic effects independently of its receptor (Suelves *et al.*, 2007).

From all of this emerging data, it is tempting to speculate that the pronounced fibrosis in human DMD muscles relate to altered net proteolytic activity in the dystrophic muscles due to imbalances in expression and activity of PA/MMP system components (Fibbi *et al.*, 2001; Zanotti *et al.*, 2007). This in turn could result in aberrant activation of latent TGFβ that can have serious fibrotic consequences as detailed already above. Although proteolytic processing by plasmin and MMP are important factors, the full gamut of mechanisms leading to TGFβ activation are complex and not well understood, though also include exposure to reactive oxygen species (ROS). Similarly, conformational changes to the latent TGFβ complex induced by certain molecular interactions also appear to be a prominent mechanism for TGFβ activation *in vivo*. For example, integrins such as thrombospondin have been demonstrated to be major activators of TGFβ *in vivo* in this way.

6. Age-Associated Changes in Muscle Function and Fibrosis

Due to better access to public health and nutrition, the mean age of the world's population is increasing, as is the overall lifespan of individuals, particularly in the developed world. One of the consequences of this increased longevity is specific health problems associated with aging. Arguably one of the principal factors that causes and/or exacerbates these age-related health problems is the loss of physical performance and quality of life due to a decline in muscle function. In particular, aging is associated with sarcopenia, which is defined as a significant reduction in skeletal muscle mass and loss of strength in the elderly. Fibrosis too has been shown to increase with aging and at times the two processes share related mechanisms. Here, we briefly summarize key aspects of these processes and the key pathways that lead to age-related changes in muscle with the major emphasis on fibrosis and muscular dystrophies. Age-related research of skeletal muscle is of growing importance not only to our understanding of muscle physiology as a natural extension of developmental biology but also for its relation to general health concerns and future studies will likely make important contributions to both fields.

Among the casual factors involved in the process of age-related skeletal muscle atrophy and weakness are modifications of the central and peripheral nervous system, altered hormonal status, inflammatory factors, and altered caloric and protein intake. The net consequence of these alterations is firstly a progressive atrophy and loss of individual muscle fibers associated with a concomitant loss of motor units (Snijders *et al.*, 2009). Secondly, a parallel reduction in muscle "quality" is observed due to the infiltration of fat and other noncontractile material (Ryall *et al.*, 2008).

As already described above, skeletal muscle connective tissue is primarily comprised of collagen and its primary function is to maintaining muscle structure and transfer force during contraction to the bone. Aberrant muscle repair and/or chronic disease have been shown to increase the predominance of the ECM, to the detriment of muscle function, but a wealth of data has now implicated aging as another important contributing factor. Data from different animal studies and man have shown that the collagen concentration of intact skeletal muscle increases with age. Collagen modifications such as nonenzymatically regulated cross-links, known as advanced glycation end (AGE) products, also increase muscle connective tissue protein stiffness and thus contribute to impaired muscle function in the elderly (Haus *et al.*, 2007). Factors that have to date been associated with increased collagen deposition in aged muscle bear many parallels to causative factors in chronic disease and aberrant muscle repair, although clearly the etiology of the defect is likely to be different. In addition to collagens, components of the transient ECM formed after acute injury such as fibrin/ogen have also been found to be chronically deposited in aged dystrophic mice, whereas their removal reduces fibrosis formation (Vidal *et al.*, 2008). Mediators of age-related fibrosis include altered cell signaling, defects in specific cell populations, and changes in regulation of growth and soluble factors, though, in many examples, overlap between more than one of these factors has also been described.

One of the principal studies in the field has come from Brack and colleagues who elegantly proved that systemic influences that change with age are important in mediating the fibrotic responses of aged muscles (Brack *et al.*, 2007). These authors showed that the increase of collagen deposition with aging in regenerating muscles was accompanied by a greater percentage of fibrogenic cells that arose by the conversion of myogenic into nonmyogenic cells, as demonstrated by genetic lineage tracing experiments. This age-related fibrogenic conversion was associated with increased activity of the canonical Wnt signaling pathway and could be abrogated experimentally by treatment of mice with Wnt inhibitors. In contrast, Wnt3A stimulation negatively modulated cellular proliferation in young regenerating muscles and increased fibrosis. Thus, aging was associated with alterations in the systemic environment that increased Wnt signaling in myogenic progenitors which facilitated their conversion to a fibrogenic fate. Importantly, since these effects were reversible this work therefore provides the strategic basis for interventions aiming at improving tissue repair and reducing fibrosis in pathological conditions such as aging and muscular dystrophies.

Glycogen synthase kinase-3beta (GSK3β) was recently shown by the same group to be a pivotal molecule in the determination of cell fate in muscle stem cell progeny by integrating cross-talk between Wnt and Notch signaling (Brack *et al.*, 2008). Moreover, BCL9/9-2, the mammalian ortholog of the *Drosophila* transcriptional coactivator *Legless*, is required for both the activation of the canonical Wnt pathway in adult myogenic progenitors and for their Wnt-mediated commitment to differentiation and effective muscle regeneration (Brack *et al.*, 2009). However, a role for GSK3β and/or BCL9 on Wnt-mediated cell fate changes from a myogenic to a fibrogenic lineage in resting satellite cells (Brack *et al.*, 2007) remains to be investigated.

Many different myogenic and nonmyogenic cell types have been shown to be key mediators of aging-associated fibrosis and muscle changes. In several studies, muscle stem cells and even myofibers themselves have been implicated to be an active element in the pathogenesis of fibrotic pathologies. One study suggested that satellite cells are a source of collagen and that in pathological states such as muscular dystrophy, this production is exacerbated. Moreover, type I, but not type III collagen production was further dysregulated in older *mdx* myogenic cells when compared to aged wild-type cells (Alexakis *et al.*, 2007). These experiments suggest that fibrosis progression in dystrophic muscle involves alterations in the mechanisms controlling matrix production. This in turn generates a positive feedback that may contribute to the reprogramming of myoblasts to a profibrotic function, leading to the premature and exaggerated formation of nondegraded connective tissue. However, details of the mechanisms driving the reprogramming are still under investigation.

Another study has suggested a major role for myofibers themselves in controlling the homeostasis of the ECM during age-related sarcopenia (Lahoute *et al.*, 2008). In this study, a mouse model with postnatal, muscle-specific deletion of serum response factor (SRF) reproduced at a premature age some of the muscular features associated with aging in humans, including type IIB myofiber atrophy, sarcomere disorganization, and endomysial fibrosis. Impaired functional and morphological regeneration, in addition to persistent and increased fibrosis, was also observed in this mouse model after cardiotoxin-induced injury. As SRF in this mouse model was selectively ablated in postmitotic myofibers, the impaired regeneration and increased fibrosis were attributed to secondary modifications of the stem cell niche and a modified environment that favored the appearance and maintenance of fibrosis. Reinforcing this data was an observed reduction in SRF protein expression with aging in mouse and human muscle samples, suggesting a physiological role for this factor in the pathogenesis of sarcopenia.

A profibrotic role for a Sca-1^{+}, nonmyogenic (MyoD^{-}), and nonimmunohematopoietic (CD45^{-}) cell population has been postulated in aged mice (Hidestrand *et al.*, 2008). These cells can be seen in aged regenerating muscle or can be clonally derived from myoblast cultures from aged animals and late passage C2C12 cells, but are infrequent in young animals, where they appear to retain a greater myogenic potential and express MyoD. Importantly, these cells overexpress fibrosis-associated genes in a manner that may be regulated by Wnt2. Another recent study identified higher numbers of a muscle-resident stromal cell (mrSC) population in the muscles of *mdx* mice compared to age-matched controls (Trensz *et al.*, 2010). Wnt3a was shown to promote both proliferation of and collagen expression by cultured mrSCs, while showing exactly the opposite effects on cultured myoblasts. Injection of Wnt3a in the *tibialis anterior* muscles of adult wild-type mice significantly enhanced the mrSC population and collagen deposition. Conversely, an injection of the Wnt antagonist DKK1 into the skeletal muscles of *mdx* mice significantly reduced fibrosis. Thus, in agreement with other studies described above, the canonical Wnt pathway expands the population of mrSCs and stimulates their production of collagen. These results further emphasize the Wnt pathway, and potentially mrSCs, as a target for therapies to counteract fibrosis formation during aging and various myopathies.

Finally, a novel type of muscle-resident cell responsive to damage has been recently described in skeletal muscle. These cells termed fibro/adipogenic progenitors (FAPs) have been proposed to be a source of prodifferentiation signals for myoblasts during the process of muscle regeneration and, more importantly, show a strong tendency to generate myofibroblasts and adipose cells (Joe *et al.*, 2010; Uezumi *et al.*, 2010). Thus, while FAPs show limited myogenic capacity, both *in vitro* or *in vivo*, if the regenerative process fails for any reason, the FAP population will persist in the tissue and potentially differentiate into adipocytes. This supports the view that local environmental signals play a role in the control FAP fate and that they may play a significant role in the successful regeneration of healthy muscle. Whether FAP conversion also occurs during aging, and whether systemic factors, including Wnt ligands also contribute to this process await confirmation.

The role of TGFβ in fibrosis has already been detailed above, and it also likely that it plays an important role in aging-associated fibrosis and muscle impairment. For example, TGFβ levels have been reported to be increased during aging and could have been proposed to work via the activation of SK1 signaling to trigger fibrotic conversion in aged muscle (Cencetti *et al.*, 2010). Microarray analyses have also revealed that myogenic cells from aged animals exhibit major alterations in the expression of many genes dependent on activation of the TGFβ signaling pathway (Beggs *et al.*, 2004). In particular, PAI-1, fibronectin, and CTGF, which are known to be directly upregulated by TGFβ, were found to be upregulated in aged myogenic cells, together with increased basal level of phosphorylated Smad2/3. Collectively, this data suggests that TGFβ signaling is constitutively active in aged myogenic progenitors and may explain the increased fibrosis present in aged muscles.

Analysis in mice of local alterations of the aged microniche of muscle stem cells has also shown that the high levels of TGFβ and its effector pSmad3, in both differentiated muscle fibers and satellite cells, are reciprocal to the levels of active Notch, which is more abundant in the young microniche (Carlson *et al.*, 2008). Excessive pSmad3 levels in aged muscles attenuates the regenerative capacity of the tissue since it binds to the promoters and stimulating the expression of several cyclin-dependent kinase (CDK) inhibitors, such as p15, p16, p21, and p27, which are negative regulators of cell-cycle progression. Importantly, this imbalance of TGFβ/pSmad3-Notch could be restored by forced activation of Notch. Similar imbalances between reduced Notch activation and increased TGFβ/pSmad3 signaling have been recently reported in aged human muscles, further reinforcing that Notch activation depends positively on MAPK/ERK activity (Carlson *et al.*, 2009).

However, despite these reported increases in systemic TGFβ levels with aging both in mice and humans, there seems to be distinct roles of this cytokine on the inhibition of the myogenic potential of muscle when compared with local changes in the tissue niches (Carlson *et al.*, 2009). Besides its effects on inhibiting myogenesis, the relative contribution of increased local TGFβ production in aged muscles and the elevated systemic levels of TGFβ to the aging-related fibrosis remain to be investigated.

Finally, several other factors have also been shown to modify fibrosis in aging and muscular dystrophy. For example, disintegrin and the metalloproteinase

ADAM12 were thought to play beneficial roles in the context of muscular dystrophy through their effects on upregulation of utrophin and plasma membrane stabilization (Jorgensen *et al.*, 2007). However, chronically high levels of ADAM12 resulted in inhibition of satellite cell responses and delayed myoblast differentiation that subsequently led to loss of skeletal muscle and both accelerated fibrosis and adipogenesis in aged *mdx* mice. High levels of the T-cell derived cytokine osteopontin in serum and muscle have also been correlated with increased fibrosis at later stages of muscular dystrophy. Genetically modified *mdx* mice lacking osteopontin showed reduced inflammation and fibrosis and diminished TGFβ levels, supporting an immunomodulatory and profibrotic role for osteopontin in dystrophic muscle (Vetrone *et al.*, 2009). In conclusion, it is clear that multiple changes both systemically and locally in muscle contribute to loss of muscle quality and function in the elderly. Understanding these factors will have benefits not only for improving life quality of healthy patients, but also for treating muscle disorders since ample evidence now suggests that the aging process exacerbates the fibrotic phenotype in situations in which normal regeneration is chronically inhibited.

7. Concluding Remarks

Abnormal muscle repair with persistent fibrosis, instead of efficient regeneration, plays a prominent role in the clinical decline and life expectancy in severe muscular dystrophies, especially in DMD. Hence, if fibrotic repair processes in dystrophic muscle could be transformed, to some extent, into regeneration, thereby preserving muscle integrity, this would considerably improve health of affected individuals. Persistent fibrosis also represents a major obstacle for successful gene- and cell-based therapies for DMD which aim to restore or replace the dystrophin gene. Thus, modifications of the muscle environment aimed to halt and/or reduce fibrosis in DMD appear crucial for attenuating the progression of the disease as well as for improving gene delivery and stem cell engraftment in otherwise untreatable patients. Although many advances have been made in deciphering the variety of pathways involved in normal muscle regeneration and muscular-dystrophy-associated fibrosis, they have so far not led to any substantial or efficient advances in antifibrotic therapies for DMD patients. It is likely that single-agent therapies, such as administration of a growth factor or antagonizing one single signaling pathway, will have a very low impact on fibrosis at advanced disease stages in a clinical setting. This is because of either the redundancy of growth factors, cellular participants, ECM components, and signaling pathways of the muscle regeneration/fibrosis process, or because of the rapid neutralization or elimination of individual agents.

Therefore, using a combination of strategies, including systemically delivered antifibrotic agents and gene-corrected cells, which could additionally integrate host environmental inputs and convert them into biological transmitters, might be critical in the design of future avenues to combat fibrosis and muscular dystrophy progression.

ACKNOWLEDGMENTS

We would like to thank all the members of our laboratory for helpful discussions. We gratefully acknowledge funding from MDA, MICINN (PLE2009-0124, SAF2009-09782, FIS-PS09/01267), AFM, Marató-TV3, and EU/FP7 (Myoage, Optistem, and Endostem).

REFERENCES

Alexakis, C., Partridge, T., and Bou-Gharios, G. (2007). Implication of the satellite cell in dystrophic muscle fibrosis: A self-perpetuating mechanism of collagen overproduction. *Am. J. Physiol. Cell Physiol.* **293,** C661–C669.

Andreetta, F., Bernasconi, P., Baggi, F., Ferro, P., Oliva, L., Arnoldi, E., Cornelio, F., Mantegazza, R., and Confalonieri, P. (2006). Immunomodulation of TGF-beta 1 in mdx mouse inhibits connective tissue proliferation in diaphragm but increases inflammatory response: Implications for antifibrotic therapy. *J. Neuroimmunol.* **175,** 77–86.

Angelini, C. (2007). The role of corticosteroids in muscular dystrophy: A critical appraisal. *Muscle Nerve* **36,** 424–435.

Arnold, L., Henry, A., Poron, F., Baba-Amer, Y., van Rooijen, N., Plonquet, A., Gherardi, R. K., and Chazaud, B. (2007). Inflammatory monocytes recruited after skeletal muscle injury switch into antiinflammatory macrophages to support myogenesis. *J. Exp. Med.* **204,** 1057–1069.

Bedair, H. S., Karthikeyan, T., Quintero, A., Li, Y., and Huard, J. (2008). Angiotensin II receptor blockade administered after injury improves muscle regeneration and decreases fibrosis in normal skeletal muscle. *Am. J. Sports Med.* **36,** 1548–1554.

Beggs, M. L., Nagarajan, R., Taylor-Jones, J. M., Nolen, G., Macnicol, M., and Peterson, C. A. (2004). Alterations in the TGFbeta signaling pathway in myogenic progenitors with age. *Aging Cell* **3,** 353–361.

Bizario, J. C., Cerri, D. G., Rodrigues, L. C., Oliveira, G. L., Nomizo, A., de Araujo, D. D., Fukuhara, P. S., Ribeiro, J. C., de Castro, F. A., and Costa, M. C. (2009). Imatinib mesylate ameliorates the dystrophic phenotype in exercised mdx mice. *J. Neuroimmunol.* **212,** 93–101.

Blake, D. J., Weir, A., Newey, S. E., and Davies, K. E. (2002). Function and genetics of dystrophin and dystrophin-related proteins in muscle. *Physiol. Rev.* **82,** 291–329.

Bonner, J. C. (2004). Regulation of PDGF and its receptors in fibrotic diseases. *Cytokine Growth Factor Rev.* **15,** 255–273.

Bonniaud, P., Margetts, P. J., Ask, K., Flanders, K., Gauldie, J., and Kolb, M. (2005). TGF-beta and Smad3 signaling link inflammation to chronic fibrogenesis. *J. Immunol.* **175,** 5390–5395.

Brack, A. S., Conboy, M. J., Roy, S., Lee, M., Kuo, C. J., Keller, C., and Rando, T. A. (2007). Increased Wnt signaling during aging alters muscle stem cell fate and increases fibrosis. *Science* **317,** 807–810.

Brack, A. S., Conboy, I. M., Conboy, M. J., Shen, J., and Rando, T. A. (2008). A temporal switch from notch to Wnt signaling in muscle stem cells is necessary for normal adult myogenesis. *Cell Stem Cell* **2,** 50–59.

Brack, A. S., Murphy-Seiler, F., Hanifi, J., Deka, J., Eyckerman, S., Keller, C., Aguet, M., and Rando, T. A. (2009). BCL9 is an essential component of canonical Wnt signaling that mediates the differentiation of myogenic progenitors during muscle regeneration. *Dev. Biol.* **335,** 93–105.

Brandan, E., Cabello-Verrugio, C., and Vial, C. (2008). Novel regulatory mechanisms for the proteoglycans decorin and biglycan during muscle formation and muscular dystrophy. *Matrix Biol.* **27,** 700–708.

Cai, B., Spencer, M. J., Nakamura, G., Tseng-Ong, L., and Tidball, J. G. (2000). Eosinophilia of dystrophin-deficient muscle is promoted by perforin-mediated cytotoxicity by T cell effectors. *Am. J. Pathol.* **156,** 1789–1796.

Carlson, M. E., Hsu, M., and Conboy, I. M. (2008). Imbalance between pSmad3 and Notch induces CDK inhibitors in old muscle stem cells. *Nature* **454,** 528–532.

Carlson, M. E., Conboy, M. J., Hsu, M., Barchas, L., Jeong, J., Agrawal, A., Mikels, A. J., Agrawal, S., Schaffer, D. V., and Conboy, I. M. (2009). Relative roles of TGF-beta1 and Wnt in the systemic regulation and aging of satellite cell responses. *Aging Cell* **8,** 676–689.

Casar, J. C., McKechnie, B. A., Fallon, J. R., Young, M. F., and Brandan, E. (2004). Transient up-regulation of biglycan during skeletal muscle regeneration: Delayed fiber growth along with decorin increase in biglycan-deficient mice. *Dev. Biol.* **268,** 358–371.

Cencetti, F., Bernacchioni, C., Nincheri, P., Donati, C., and Bruni, P. (2010). Transforming growth factor-beta1 induces transdifferentiation of myoblasts into myofibroblasts via up-regulation of sphingosine kinase-1/S1P3 axis. *Mol. Biol. Cell* **21,** 1111–1124.

Chan, Y. S., Li, Y., Foster, W., Horaguchi, T., Somogyi, G., Fu, F. H., and Huard, J. (2003). Antifibrotic effects of suramin in injured skeletal muscle after laceration. *J. Appl. Physiol.* **95,** 771–780.

Chazaud, B., Brigitte, M., Yacoub-Youssef, H., Arnold, L., Gherardi, R., Sonnet, C., Lafuste, P., and Chretien, F. (2009). Dual and beneficial roles of macrophages during skeletal muscle regeneration. *Exerc. Sport Sci. Rev.* **37,** 18–22.

Chen, X., and Li, Y. (2009). Role of matrix metalloproteinases in skeletal muscle: Migration, differentiation, regeneration and fibrosis. *Cell. Adh. Migr.* **3,** 337–341.

Cohn, R. D., Liang, H. Y., Shetty, R., Abraham, T., and Wagner, K. R. (2007). Myostatin does not regulate cardiac hypertrophy or fibrosis. *Neuromuscul. Disord.* **17,** 290–296.

Cornelison, D. D. (2008). Context matters: In vivo and in vitro influences on muscle satellite cell activity. *J. Cell. Biochem.* **105,** 663–669.

Desguerre, I., Mayer, M., Leturcq, F., Barbet, J. P., Gherardi, R. K., and Christov, C. (2009). Endomysial fibrosis in Duchenne muscular dystrophy: A marker of poor outcome associated with macrophage alternative activation. *J. Neuropathol. Exp. Neurol.* **68,** 762–773.

Di Marco, S., Mazroui, R., Dallaire, P., Chittur, S., Tenenbaum, S. A., Radzioch, D., Marette, A., and Gallouzi, I. E. (2005). NF-kappa B-mediated MyoD decay during muscle wasting requires nitric oxide synthase mRNA stabilization, HuR protein, and nitric oxide release. *Mol. Cell. Biol.* **25,** 6533–6545.

Durbeej, M., and Campbell, K. P. (2002). Muscular dystrophies involving the dystrophin–glycoprotein complex: An overview of current mouse models. *Curr. Opin. Genet. Dev.* **12,** 349–361.

Emery, A. E. (2002). The muscular dystrophies. *Lancet* **359,** 687–695.

Evangeliou, A., Vasilaki, K., Karagianni, P., and Nikolaidis, N. (2009). Clinical applications of creatine supplementation on paediatrics. *Curr. Pharm. Biotechnol.* **10,** 683–690.

Farini, A., Meregalli, M., Belicchi, M., Battistelli, M., Parolini, D., D'Antona, G., Gavina, M., Ottoboni, L., Constantin, G., Bottinelli, R., and Torrente, Y. (2007). T and B lymphocyte depletion has a marked effect on the fibrosis of dystrophic skeletal muscles in the scid/mdx mouse. *J. Pathol.* **213,** 229–238.

Fibbi, G., Barletta, E., Dini, G., Del Rosso, A., Pucci, M., Cerletti, M., and Del Rosso, M. (2001). Cell invasion is affected by differential expression of the urokinase plasminogen activator/urokinase plasminogen activator receptor system in muscle satellite cells from normal and dystrophic patients. *Lab. Invest.* **81,** 27–39.

Fukushima, K., Nakamura, A., Ueda, H., Yuasa, K., Yoshida, K., Takeda, S., and Ikeda, S. (2007). Activation and localization of matrix metalloproteinase-2 and -9 in the skeletal muscle of the muscular dystrophy dog (CXMDJ). *BMC Musculoskelet. Disord.* **8,** 54.

Gargioli, C., Coletta, M., De Grandis, F., Cannata, S. M., and Cossu, G. (2008). PlGF-MMP-9-expressing cells restore microcirculation and efficacy of cell therapy in aged dystrophic muscle. *Nat. Med.* **14,** 973–978.

Glass, D. J. (2010). Signaling pathways perturbing muscle mass. *Curr. Opin. Clin. Nutr. Metab. Care* **13,** 225–229.

Green, D. R., Ferguson, T., Zitvogel, L., and Kroemer, G. (2009). Immunogenic and tolerogenic cell death. *Nat. Rev. Immunol.* **9,** 353–363.

Grounds, M. D., Sorokin, L., and White, J. (2005). Strength at the extracellular matrix–muscle interface. *Scand. J. Med. Sci. Sports* **15,** 381–391.

Haus, J. M., Carrithers, J. A., Trappe, S. W., and Trappe, T. A. (2007). Collagen, cross-linking, and advanced glycation end products in aging human skeletal muscle. *J. Appl. Physiol.* **103,** 2068–2076.

Heydemann, A., Ceco, E., Lim, J. E., Hadhazy, M., Ryder, P., Moran, J. L., Beier, D. R., Palmer, A. A., and McNally, E. M. (2009). Latent TGF-beta-binding protein 4 modifies muscular dystrophy in mice. *J. Clin. Invest.* **119,** 3703–3712.

Hidestrand, M., Richards-Malcolm, S., Gurley, C. M., Nolen, G., Grimes, B., Waterstrat, A., Zant, G. V., and Peterson, C. A. (2008). Sca-1-expressing nonmyogenic cells contribute to fibrosis in aged skeletal muscle. *J. Gerontol. A Biol. Sci. Med. Sci.* **63,** 566–579.

Hinz, B., Phan, S. H., Thannickal, V. J., Galli, A., Bochaton-Piallat, M. L., and Gabbiani, G. (2007). The myofibroblast: One function, multiple origins. *Am. J. Pathol.* **170,** 1807–1816.

Huang, P., Zhao, X. S., Fields, M., Ransohoff, R. M., and Zhou, L. (2009). Imatinib attenuates skeletal muscle dystrophy in mdx mice. *FASEB J.* **23,** 2539–2548.

Huebner, K. D., Jassal, D. S., Halevy, O., Pines, M., and Anderson, J. E. (2008). Functional resolution of fibrosis in mdx mouse dystrophic heart and skeletal muscle by halofuginone. *Am. J. Physiol. Heart Circ. Physiol.* **294,** H1550–H1561.

Joe, A. W., Yi, L., Natarajan, A., Le Grand, F., So, L., Wang, J., Rudnicki, M. A., and Rossi, F. M. (2010). Muscle injury activates resident fibro/adipogenic progenitors that facilitate myogenesis. *Nat. Cell Biol.* **12,** 153–163.

Jorgensen, L. H., Jensen, C. H., Wewer, U. M., and Schroder, H. D. (2007). Transgenic overexpression of ADAM12 suppresses muscle regeneration and aggravates dystrophy in aged mdx mice. *Am. J. Pathol.* **171,** 1599–1607.

Kaar, J. L., Li, Y., Blair, H. C., Asche, G., Koepsel, R. R., Huard, J., and Russell, A. J. (2008). Matrix metalloproteinase-1 treatment of muscle fibrosis. *Acta Biomater.* **4,** 1411–1420.

Kaariainen, M., Jarvinen, T., Jarvinen, M., Rantanen, J., and Kalimo, H. (2000). Relation between myofibers and connective tissue during muscle injury repair. *Scand. J. Med. Sci. Sports* **10,** 332–337.

Kafadar, K. A., Yi, L., Ahmad, Y., So, L., Rossi, F., and Pavlath, G. K. (2009). Sca-1 expression is required for efficient remodeling of the extracellular matrix during skeletal muscle regeneration. *Dev. Biol.* **326,** 47–59.

Kherif, S., Lafuma, C., Dehaupas, M., Lachkar, S., Fournier, J.-G., Verdiere-Sahuque, M., Fardeau, M., and Alameddine, H. S. (1999). Expression of matrix metalloproteinases 2 and 9 in regenerating skeletal muscle: A study in experimentally injured and mdx muscles. *Dev. Biol.* **205,** 158–170.

Kollias, H. D., and McDermott, J. C. (2008). Transforming growth factor-beta and myostatin signaling in skeletal muscle. *J. Appl. Physiol.* **104,** 579–587.

Lahoute, C., Sotiropoulos, A., Favier, M., Guillet-Deniau, I., Charvet, C., Ferry, A., Butler-Browne, G., Metzger, D., Tuil, D., and Daegelen, D. (2008). Premature aging in skeletal muscle lacking serum response factor. *PLoS ONE* **3,** e3910.

Li, Y., Foster, W., Deasy, B. M., Chan, Y., Prisk, V., Tang, Y., Cummins, J., and Huard, J. (2004). Transforming growth factor-beta1 induces the differentiation of myogenic cells into fibrotic cells in injured skeletal muscle: A key event in muscle fibrogenesis. *Am. J. Pathol.* **164,** 1007–1019.

Li, Y., Fu, F. H., and Huard, J. (2005). Cutting-edge muscle recovery: Using antifibrosis agents to improve healing. *Phys. Sportsmed.* **33,** 44–50.

Li, Y., Li, J., Zhu, J., Sun, B., Branca, M., Tang, Y., Foster, W., Xiao, X., and Huard, J. (2007). Decorin gene transfer promotes muscle cell differentiation and muscle regeneration. *Mol. Ther.* **15,** 1616–1622.

Li, Z. B., Kollias, H. D., and Wagner, K. R. (2008). Myostatin directly regulates skeletal muscle fibrosis. *J. Biol. Chem.* **283,** 19371–19378.

Li, H., Mittal, A., Makonchuk, D. Y., Bhatnagar, S., and Kumar, A. (2009). Matrix metalloproteinase-9 inhibition ameliorates pathogenesis and improves skeletal muscle regeneration in muscular dystrophy. *Hum. Mol. Genet.* **18,** 2584–2598.

Lluis, F., Roma, J., Suelves, M., Parra, M., Aniorte, G., Gallardo, E., Illa, I., Rodriguez, L., Hughes, S. M., Carmeliet, P., Roig, M., and Munoz-Canoves, P. (2001). Urokinase-dependent plasminogen activation is required for efficient skeletal muscle regeneration in vivo. *Blood* **97,** 1703–1711.

Lluri, G., Langlois, G. D., McClellan, B., Soloway, P. D., and Jaworski, D. M. (2006). Tissue inhibitor of metalloproteinase-2 (TIMP-2) regulates neuromuscular junction development via a beta1 integrin-mediated mechanism. *J. Neurobiol.* **66,** 1365–1377.

Mantovani, A., Sica, A., Sozzani, S., Allavena, P., Vecchi, A., and Locati, M. (2004). The chemokine system in diverse forms of macrophage activation and polarization. *Trends Immunol.* **25,** 677–686.

Mauro, A. (1961). Satellite cells of skeletal fibers. *J. Biophys. Biochem. Cytol.* **9,** 493–495.

McCroskery, S., Thomas, M., Platt, L., Hennebry, A., Nishimura, T., McLeay, L., Sharma, M., and Kambadur, R. (2005). Improved muscle healing through enhanced regeneration and reduced fibrosis in myostatin-null mice. *J. Cell Sci.* **118,** 3531–3541.

Morrison, J., Lu, Q. L., Pastoret, C., Partridge, T., and Bou-Gharios, G. (2000). T-cell-dependent fibrosis in the mdx dystrophic mouse. *Lab. Invest.* **80,** 881–891.

Morrison, J., Palmer, D. B., Cobbold, S., Partridge, T., and Bou-Gharios, G. (2005). Effects of T-lymphocyte depletion on muscle fibrosis in the mdx mouse. *Am. J. Pathol.* **166,** 1701–1710.

Muir, L. A., and Chamberlain, J. S. (2009). Emerging strategies for cell and gene therapy of the muscular dystrophies. *Expert Rev. Mol. Med.* **11,** e18.

Nagamine, Y., Medcalf, R. L., and Munoz-Canoves, P. (2005). Transcriptional and posttranscriptional regulation of the plasminogen activator system. *Thromb. Haemost.* **93,** 661–675.

Nahirney, P. C., Dow, P. R., and Ovalle, W. K. (1997). Quantitative morphology of mast cells in skeletal muscle of normal and genetically dystrophic mice. *Anat. Rec.* **247,** 341–349.

Ohtake, Y., Tojo, H., and Seiki, M. (2006). Multifunctional roles of MT1-MMP in myofiber formation and morphostatic maintenance of skeletal muscle. *J. Cell Sci.* **119,** 3822–3832.
Ono, Y., Sensui, H., Okutsu, S., and Nagatomi, R. (2007). Notch2 negatively regulates myofibroblastic differentiation of myoblasts. *J. Cell. Physiol.* **210,** 358–369.
Pelosi, L., Giacinti, C., Nardis, C., Borsellino, G., Rizzuto, E., Nicoletti, C., Wannenes, F., Battistini, L., Rosenthal, N., Molinaro, M., and Musaro, A. (2007). Local expression of IGF-1 accelerates muscle regeneration by rapidly modulating inflammatory cytokines and chemokines. *FASEB J.* **21,** 1393–1402.
Pesce, J. T., Ramalingam, T. R., Mentink-Kane, M. M., Wilson, M. S., El Kasmi, K. C., Smith, A. M., Thompson, R. W., Cheever, A. W., Murray, P. J., and Wynn, T. A. (2009). Arginase-1-expressing macrophages suppress Th2 cytokine-driven inflammation and fibrosis. *PLoS Pathog.* **5,** e1000371.
Peterson, J. M., and Guttridge, D. C. (2008). Skeletal muscle diseases, inflammation, and NF-kappaB signaling: Insights and opportunities for therapeutic intervention. *Int. Rev. Immunol.* **27,** 375–387.
Porter, J. D., Khanna, S., Kaminski, H. J., Rao, J. S., Merriam, A. P., Richmonds, C. R., Leahy, P., Li, J., Guo, W., and Andrade, F. H. (2002). A chronic inflammatory response dominates the skeletal muscle molecular signature in dystrophin-deficient mdx mice. *Hum. Mol. Genet.* **11,** 263–272.
Radley, H. G., Davies, M. J., and Grounds, M. D. (2008). Reduced muscle necrosis and long-term benefits in dystrophic mdx mice after cV1q (blockade of TNF) treatment. *Neuromuscul. Disord.* **18,** 227–238.
Roffe, S., Hagai, Y., Pines, M., and Halevy, O. (2010). Halofuginone inhibits Smad3 phosphorylation via the PI3K/Akt and MAPK/ERK pathways in muscle cells: Effect on myotube fusion. *Exp. Cell Res.* **316,** 1061–1069.
Ruffell, D., Mourkioti, F., Gambardella, A., Kirstetter, P., Lopez, R. G., Rosenthal, N., and Nerlov, C. (2009). A CREB-C/EBPbeta cascade induces M2 macrophage-specific gene expression and promotes muscle injury repair. *Proc. Natl. Acad. Sci. USA* **106,** 17475–17480.
Ryall, J. G., Schertzer, J. D., and Lynch, G. S. (2008). Cellular and molecular mechanisms underlying age-related skeletal muscle wasting and weakness. *Biogerontology* **9,** 213–228.
Segawa, M., Fukada, S., Yamamoto, Y., Yahagi, H., Kanematsu, M., Sato, M., Ito, T., Uezumi, A., Hayashi, S., Miyagoe-Suzuki, Y., Takeda, S., Tsujikawa, K., *et al.* (2008). Suppression of macrophage functions impairs skeletal muscle regeneration with severe fibrosis. *Exp. Cell Res.* **314,** 3232–3244.
Shi, Y., and Massague, J. (2003). Mechanisms of TGF-beta signaling from cell membrane to the nucleus. *Cell* **113,** 685–700.
Snijders, T., Verdijk, L. B., and van Loon, L. J. (2009). The impact of sarcopenia and exercise training on skeletal muscle satellite cells. *Ageing Res. Rev.* **8,** 328–338.
Spencer, M. J., Montecino-Rodriguez, E., Dorshkind, K., and Tidball, J. G. (2001). Helper (CD4(+)) and cytotoxic (CD8(+)) T cells promote the pathology of dystrophin-deficient muscle. *Clin. Immunol.* **98,** 235–243.
Suelves, M., Lopez-Alemany, R., Lluis, F., Aniorte, G., Serrano, E., Parra, M., Carmeliet, P., and Munoz-Canoves, P. (2002). Plasmin activity is required for myogenesis in vitro and skeletal muscle regeneration in vivo. *Blood* **99,** 2835–2844.
Suelves, M., Vidal, B., Ruiz, V., Baeza-Raja, B., Diaz-Ramos, A., Cuartas, I., Lluis, F., Parra, M., Jardi, M., Lopez-Alemany, R., Serrano, A. L., and Munoz-Canoves, P. (2005). The plasminogen activation system in skeletal muscle regeneration: Antagonistic roles of urokinase-type plasminogen activator (uPA) and its inhibitor (PAI-1). *Front. Biosci.* **10,** 2978–2985.

Suelves, M., Vidal, B., Serrano, A. L., Tjwa, M., Roma, J., Lopez-Alemany, R., Luttun, A., de Lagran, M. M., Diaz-Ramos, A., Jardi, M., Roig, M., Dierssen, M., *et al.* (2007). uPA deficiency exacerbates muscular dystrophy in MDX mice. *J. Cell Biol.* **178,** 1039–1051.
Sun, G., Haginoya, K., Wu, Y., Chiba, Y., Nakanishi, T., Onuma, A., Sato, Y., Takigawa, M., Iinuma, K., and Tsuchiya, S. (2008). Connective tissue growth factor is overexpressed in muscles of human muscular dystrophy. *J. Neurol. Sci.* **267,** 48–56.
Tidball, J. G. (2005). Inflammatory processes in muscle injury and repair. *Am. J. Physiol. Regul. Integr. Comp. Physiol.* **288,** R345–R353.
Tidball, J. G., and Villalta, S. A. (2010). Regulatory interactions between muscle and the immune system during muscle regeneration. *Am. J. Physiol. Regul. Integr. Comp. Physiol.* **298,** R1173–R1187.
Trensz, F., Haroun, S., Cloutier, A., Richter, M. V., and Grenier, G. (2010). A muscle resident cell population promotes fibrosis in hindlimb skeletal muscles of mdx mice through the wnt canonical pathway. *Am. J. Physiol. Cell Physiol.* **299**(5), C939–C947.
Uezumi, A., Fukada, S., Yamamoto, N., Takeda, S., and Tsuchida, K. (2010). Mesenchymal progenitors distinct from satellite cells contribute to ectopic fat cell formation in skeletal muscle. *Nat. Cell Biol.* **12,** 143–152.
Vetrone, S. A., Montecino-Rodriguez, E., Kudryashova, E., Kramerova, I., Hoffman, E. P., Liu, S. D., Miceli, M. C., and Spencer, M. J. (2009). Osteopontin promotes fibrosis in dystrophic mouse muscle by modulating immune cell subsets and intramuscular TGF-beta. *J. Clin. Invest.* **119,** 1583–1594.
Vial, C., Zuniga, L. M., Cabello-Verrugio, C., Canon, P., Fadic, R., and Brandan, E. (2008). Skeletal muscle cells express the profibrotic cytokine connective tissue growth factor (CTGF/CCN2), which induces their dedifferentiation. *J. Cell. Physiol.* **215,** 410–421.
Vidal, B., Serrano, A. L., Tjwa, M., Suelves, M., Ardite, E., De Mori, R., Baeza-Raja, B., Martinez de Lagran, M., Lafuste, P., Ruiz-Bonilla, V., Jardi, M., Gherardi, R., *et al.* (2008). Fibrinogen drives dystrophic muscle fibrosis via a TGFbeta/alternative macrophage activation pathway. *Genes Dev.* **22,** 1747–1752.
Villalta, S. A., Nguyen, H. X., Deng, B., Gotoh, T., and Tidball, J. G. (2009). Shifts in macrophage phenotypes and macrophage competition for arginine metabolism affect the severity of muscle pathology in muscular dystrophy. *Hum. Mol. Genet.* **18,** 482–496.
Visse, R., and Nagase, H. (2003). Matrix metalloproteinases and tissue inhibitors of metalloproteinases: Structure, function, and biochemistry. *Circ. Res.* **92,** 827–839.
Wagner, K. R., McPherron, A. C., Winik, N., and Lee, S. J. (2002). Loss of myostatin attenuates severity of muscular dystrophy in mdx mice. *Ann. Neurol.* **52,** 832–836.
Wehling, M., Spencer, M. J., and Tidball, J. G. (2001). A nitric oxide synthase transgene ameliorates muscular dystrophy in mdx mice. *J. Cell Biol.* **155,** 123–131.
Wehling-Henricks, M., Sokolow, S., Lee, J. J., Myung, K. H., Villalta, S. A., and Tidball, J. G. (2008). Major basic protein-1 promotes fibrosis of dystrophic muscle and attenuates the cellular immune response in muscular dystrophy. *Hum. Mol. Genet.* **17,** 2280–2292.
Wu, N., Jansen, E. D., and Davidson, J. M. (2003). Comparison of mouse matrix metalloproteinase 13 expression in free-electron laser and scalpel incisions during wound healing. *J. Invest. Dermatol.* **121,** 926–932.
Wynn, T. A. (2004). Fibrotic disease and the T(H)1/T(H)2 paradigm. *Nat. Rev. Immunol.* **4,** 583–594.
Wynn, T. A. (2008). Cellular and molecular mechanisms of fibrosis. *J. Pathol.* **214,** 199–210.
Wynn, T. A., and Barron, L. (2010). Macrophages: Master regulators of inflammation and fibrosis. *Semin. Liver Dis.* **30,** 245–257.

Yogo, Y., Fujishima, S., Inoue, T., Saito, F., Shiomi, T., Yamaguchi, K., and Ishizaka, A. (2009). Macrophage derived chemokine (CCL22), thymus and activation-regulated chemokine (CCL17), and CCR4 in idiopathic pulmonary fibrosis. *Respir. Res.* **10,** 80.

Zanotti, S., Negri, T., Cappelletti, C., Bernasconi, P., Canioni, E., Di Blasi, C., Pegoraro, E., Angelini, C., Ciscato, P., Prelle, A., Mantegazza, R., Morandi, L., *et al.* (2005). Decorin and biglycan expression is differentially altered in several muscular dystrophies. *Brain* **128,** 2546–2555.

Zanotti, S., Saredi, S., Ruggieri, A., Fabbri, M., Blasevich, F., Romaggi, S., Morandi, L., and Mora, M. (2007). Altered extracellular matrix transcript expression and protein modulation in primary Duchenne muscular dystrophy myotubes. *Matrix Biol.* **26,** 615–624.

Zhou, L., Porter, J. D., Cheng, G., Gong, B., Hatala, D. A., Merriam, A. P., Zhou, X., Rafael, J. A., and Kaminski, H. J. (2006). Temporal and spatial mRNA expression patterns of TGF-beta1, 2, 3 and TbetaRI, II, III in skeletal muscles of mdx mice. *Neuromuscul. Disord.* **16,** 32–38.

Zhu, J., Li, Y., Shen, W., Qiao, C., Ambrosio, F., Lavasani, M., Nozaki, M., Branca, M. F., and Huard, J. (2007). Relationships between transforming growth factor-beta1, myostatin, and decorin: Implications for skeletal muscle fibrosis. *J. Biol. Chem.* **282,** 25852–25863.

CHAPTER EIGHT

Ferlin Proteins in Myoblast Fusion and Muscle Growth

Avery D. Posey Jr.,* Alexis Demonbreun,† *and* Elizabeth M. McNally*,†,‡

Contents

Abstract

Myoblast fusion contributes to muscle growth in development and during regeneration of mature muscle. Myoblasts fuse to each other as well as to multinucleate myotubes to enlarge the myofiber. The molecular mechanisms of myoblast fusion are incompletely understood. Adhesion, apposition, and membrane fusion are accompanied by cytoskeletal rearrangements. The ferlin family of proteins is implicated in human muscle disease and has been implicated in fusion events in muscle, including myoblast fusion, vesicle trafficking and membrane repair. Dysferlin was the first mammalian ferlin identified and it is now known that there are six different ferlins. Loss-of-function mutations in the dysferlin gene lead to limb girdle muscular dystrophy and the milder disorder Miyoshi Myopathy. Dysferlin is a membrane-associated protein that has been implicated in resealing disruptions in the muscle plasma membrane. Newer

* Genomics and Systems Biology, Committee on Genetics, The University of Chicago, Chicago, Illinois, USA
† Department of Medicine, The University of Chicago, Chicago, Illinois, USA
‡ Department of Human Genetics, The University of Chicago, Chicago, Illinois, USA

Current Topics in Developmental Biology, Volume 96
ISSN 0070-2153, DOI: 10.1016/B978-0-12-385940-2.00008-5

data supports a broader role for dysferlin in intracellular vesicular movement, a process also important for resealing. Myoferlin is highly expressed in myoblasts that undergoing fusion, and the absence of myoferlin leads to impaired myoblast fusion. Myoferlin also regulates intracellular trafficking events, including endocytic recycling, a process where internalized vesicles are returned to the plasma membrane. The trafficking role of ferlin proteins is reviewed herein with a specific focus as to how this machinery alters myogenesis and muscle growth.

Abbreviations

EHD	Eps homology domain
FCM	fusion-competent myoblast
Fdx	fluorescein dextran
IGF	insulin-like growth factor
LGMD	limb girdle muscular dystrophy
TRIM	tripartite motif

1. Introduction

Muscle development and regeneration require the fusion of myoblast to the multinucleate syncytium. In many different genetic forms of muscular dystrophy, muscle damage or dysfunction is caused by a specific gene mutation. In the face of enhanced degeneration, there is often a concomitant increase in muscle regeneration, although typically the increase in regeneration is insufficient to match the pace of degeneration. With the imbalance favoring muscle degeneration, there is a loss-of-functioning myofibers and replacement of the muscle by fibrosis and fat. This replacement process, referred to as dystrophy, also creates an environment that is thought to further reduce effective myogenesis since the normal matrix cues that promote myoblast differentiation and support myoblast fusion are destroyed or otherwise rendered defective.

Limb girdle muscular dystrophy (LGMD) type 2B is caused by mutations in the gene encoding dysferlin. Loss-of-function mutations in the dysferlin gene lead to this autosomal recessive form of LGMD as well as the milder disorder Miyoshi Myopathy (Bashir *et al.*, 1998; Liu *et al.*, 1998). The identification of dysferlin as a disease gene has led to the identification of the ferlin family of proteins and their role in muscle repair and regeneration. In the mammalian genome, there are six ferlin proteins. To date, three have been characterized as protein products. Dysferlin is a membrane-associated protein and like all members of the ferlin family, dysferlin contains multiple C2 domains. The ferlin family is unique in that members have as few as four and as many as seven C2 domains, whereas all other C2 domain-containing proteins have one or two of these domains. C2 domains are found in at least 100 membrane-associated proteins where they mediate protein–lipid, protein–calcium, and protein–protein interactions. Ferlin proteins are found in nonmuscle cell types and have been associated with additional disease phenotypes. For example, otoferlin mutations cause non-syndromic deafness suggesting that the ferlin proteins are critical for basic functions such as intracellular trafficking (Yasunaga *et al.*, 1999). Dysferlin and myoferlin are two members of the ferlin proteins that are highly expressed in myoblasts and myofibers. This review will focus on what is known about ferlin proteins in muscle regeneration and myogenesis.

1.1. *Caenorhabditis elegans Fer-1*, the ferlin prototype

The ferlin family of proteins is defined by their sequence similarity to the *C. elegans Fer-1*, and *Fer-1* is considered the prototypical ferlin proteins. *Fer-1* was originally identified from genetic screens for fertility defects in *C. elegans*. In the normal maturation process of the spermatozoa, multiple large membranous organelle must fuse to the plasma membrane. *Fer-1* mutants were infertile and the maturing spermatozoa displayed a characteristic abnormal intracellular retention of membranous organelles (Ward *et al.*, 1981). This intracellular accumulation impaired normal motility of the spermatids leading to infertility (Achanzar and Ward, 1997). The resultant sperm cannot adhere to uterine walls and are defective in oocyte fertilization. Ten different mutations have been cataloged within *Fer-1*; five of these fall within C2 domains and two generate premature stop codons (Achanzar and Ward, 1997; Washington and Ward, 2006). These mutations disrupt calcium sensitive fusion. Fer-1 protein localizes to the membranous organelle. It was initially believed that *Fer-1* was expressed only in the testis; however recent reports counter the original observation and suggest that *Fer-1* is also expressed in *C. elegans muscle* where it alters muscle gene expression (Krajacic *et al.*, 2009).

The *Fer-1* gene encodes a 235-kDa protein that contains a carboxy-terminal transmembrane domain similar to viral fusion proteins and at least four C2 domains. C2 domains are independently folding domains formed from approximately 130 amino acids. The domains assemble as eight parallel β-strands that fold into a β sandwich. At one end is a cluster of aspartate residues formed from several different loops to generate a Ca^{2+} binding domain (Sudhof and Rizo, 1996). C2 domains mediate lipid-binding and protein–protein interactions in Ca^{2+}-dependent and Ca^{2+}-independent fashion (Davletov and Sudhof, 1993). Among the best studied C2-domain-containing proteins are the synaptotagmins; synaptotagmins are mediators of Ca^{2+}-dependent membrane fusion including the fast exocytosis that occurs at nerve terminals (Brose *et al.*, 1992; Sudhof and Rizo, 1996). By analogy, it can be expected that exocytosis events such as those related to vesicle fusion and membranous organelle fusion are mediated by ferlin proteins.

1.2. C2 domain-containing proteins, including the ferlins, are involved in lipid binding and fusion

The synaptotagmins are a family of membrane-trafficking proteins that contain a transmembrane domain and two carboxy-terminal C2 domains, C2A and C2B. Synaptotagmin I facilitates synaptic vesicle membrane fusion with the presynaptic membrane, a function that shares striking similarity to Fer-1 function (Brose *et al.*, 1992). Synaptotagmin I is located in the synaptic vesicles and interacts with syntaxin, found on the plasma membrane (Chapman *et al.*, 1995). In the presence of calcium, the C2A domain of synaptotagmin I binds syntaxin with a high affinity and also binds negatively charged phospholipids with increased affinity (Brose *et al.*, 1992; Perin *et al.*, 1990). The Ca^{2+} binding loops of the C2A and C2B domains of synaptotagmin I insert into the lipid bilayer (Bai *et al.*, 2002). Synaptotagmin I has the spatial opportunity to mediate vesicular–plasma membrane fusion while bound to the vesicular membrane and interacting with syntaxin in the presence of calcium. Thus, synaptotagmin I is thought to regulate timing of membrane fusion through its calcium-sensing ability and its ability to promote membrane fusion through Ca^{2+} loop insertion into the membrane, causing a local disruption (Peuvot *et al.*, 1999). One theory of membrane coalescence suggests that the disruption of the lipid bilayer by the angled insertion of calcium-sensing proteins induces membrane fusion (Peuvot *et al.*, 1999).

Ca^{2+} binding by C2 domains is cooperative and synaptotagmin C2 domains bind at least three Ca^{2+} molecules. A similar series of *in vitro* analyses were conducted with dysferlin and myoferlin C2 domains (Davis *et al.*, 2002; Therrien *et al.*, 2009). In these studies, the individual C2 domains were purified from bacteria as fusion proteins. The C2A domains of both myoferlin and dysferlin bind negatively charged phospholipids, particularly phosphotidylserine, but only in the presence of Ca^{2+} (Davis *et al.*, 2002). Under these same conditions, the remaining C2 domains of myoferlin were not found to bind any of the phospholipid combinations tested. A similar approach demonstrated that only the first C2 domain of dysferlin could bind phosphatidylinositol 4-phosphate and phosphatidylinositol 4,5-bis-phosphate in response to Ca^{2+} (Therrien *et al.*, 2009). The concentration of phosphotidylserine required for binding was rather high at 50%, but such concentrations may be important in the inner surface of the membrane where myoblast fusion occurs (Davis *et al.*, 2002). The Ca^{2+}

concentration required to elicit lipid binding differed slightly for myoferlin C2A compared to dysferlin C2A. The half maximal lipid binding for myoferlin C2A was 1 μM while it was 4.5 μM for dysferlin C2A (Davis *et al.*, 2002). Although synaptotagmin C2 domains bind three Ca^{2+} ions, the ferlin C2A domains may bind more. Localized Ca^{2+} concentrations and the specific composition at the inner plasma membrane surface are likely to be critical for triggering appropriate and regulated fusion events (Corbin *et al.*, 2007). A V67D mutation in the C2A domain of dysferlin that causes both LGMD type 2B and Miyoshi Myopathy (Illarioshkin *et al.*, 2000) specifically disrupts Ca^{2+}-induced lipid binding and results in the loss of cooperativity of binding (Davis *et al.*, 2002). A similar mutation, I67D, in the C2A domain of myoferlin also completely abolishes Ca^{2+}-dependent phospholipid binding (Doherty *et al.*, 2005).

1.3. The mammalian ferlin family

Humans and mice have six Fer-1-like genes that form the ferlin family: dysferlin (Fer1L1), otoferlin (Fer1L2), myoferlin (Fer1L3), Fer1L4, Fer1L5, and Fer1L6 (Achanzar and Ward, 1997; Bashir *et al.*, 1998; Britton *et al.*, 2000; Jimenez and Bashir, 2007; Liu *et al.*, 1998; Smith and Wakimoto, 2007; Yasunaga *et al.*, 1999; Fig. 8.1). The ferlins share similarity in protein structure. In common to all family members is a single pass transmembrane domain localized at the carboxy-terminus (Jimenez and Bashir, 2007; Perin *et al.*, 1990). The ferlins are type II transmembrane proteins containing a series of positively charged residues at the beginning of the transmembrane domain, and these residues are thought to be critical for anchoring the transmembrane domain within the membrane. This topology, whether at the plasma membrane or embedded in cytoplasmic vesicles, places the carboxy-terminus anchored in the membrane and the remainder of the protein, containing the C2 domains within the cytoplasm. The ferlin C2 domains are named in order from amino- to carboxy-terminus as C2A-C2F. Across ferlin family members, specific C2 domains are most homologous to each other. That is, C2A of dysferlin is most like C2A of myoferlin and less like the remaining C2 domains within dysferlin and myoferlin. Only the first C2 domain, C2A, binds to lipids and the position of the C2A domain at the very amino terminus may allow this domain to stretch some physical distance from the membrane surface itself where it could be positioned to mediate fusion of opposing membranes. These opposing membranes could be between two vesicles or between a vesicle and the plasma membrane.

1.3.1. Dysferlin

Dysferlin was originally identified through positional cloning as the gene responsible for both LGMD2B and Miyoshi Myopathy (Bashir *et al.*, 1998; Liu *et al.*, 1998). The LGMDs are a collection of genetically diverse disorders

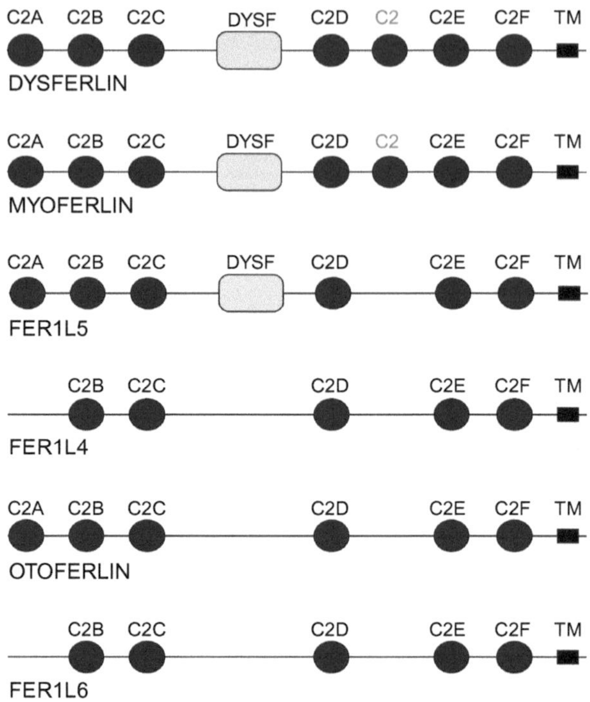

Figure 8.1 Schematic of the mammalian ferlin family members. The ferlins share similar domain architecture: a carboxy-terminal transmembrane domain and multiple amino-terminal C2 domains, including a C2A domain that has been shown to bind negatively charged phospholipids in dysferlin and myoferlin. The DYSF domain is the putative binding site of caveolin-3 and is found within dysferlin, myoferlin, and fer1L5.

that include both dominant and recessive gene mutations (Urtizberea *et al.*, 2008). The clinical presentation of LGMD2B is characterized by onset in the second decade, variable progression, and predominant wasting of the proximal muscle groups surrounding the pelvic girdle with mild shoulder girdle weakening overtime. LGMD2B is inherited autosomal recessively. Miyoshi Myopathy also has an onset in the late teens and early adulthood but predominantly affects the distal muscle groups, particularly the gastrocnemius muscle. Weakness can occur in the proximal muscles overtime. Miyoshi Myopathy is also inherited autosomally recessive. Linkage analysis showed that both LGMD2B and Miyoshi Myopathy were associated with a region on chromosome 2 and two groups independently identified dysferlin as the gene defect for these disorders (Bashir *et al.*, 1998; Liu *et al.*, 1998).

The dysferlin gene encodes a 230-kDa protein with at least six C2 domains. Some algorithms predict seven C2 domains and with an additional domain positioned between C2E and C2F. Because of this the nomenclature for C2E, C2F, and C2G may be shifted (Therrien *et al.*, 2009; Washington and Ward, 2006). Antibodies raised to dysferlin were used to show that dysferlin is highly expressed in adult skeletal muscle as well as in the heart. The predominant pattern for dysferlin immunoreactivity in adult skeletal muscle is at the cell membrane and this is well visualized in cross-sectional analysis of muscle (Anderson *et al.*, 1999). In cell culture, dysferlin is highly expressed in differentiated myotubes, although it is still detectable in myoblasts (Davis *et al.*, 2002; Doherty *et al.*, 2005; Fig. 8.2B). Muscle biopsies from humans with dysferlin mutations reveal a prominent accumulation of small subsarcolemmal vesicles and empty, swollen cisternae in the Golgi apparatus (Cenacchi *et al.*, 2005; Piccolo *et al.*, 2000; Selcen *et al.*, 2001). The finding of accumulation of intracellular vesicles is reminiscent of what was seen with *C. elegans Fer-1* mutations.

Dysferlin and *C. elegans Fer-1* each contains a DYSF domain. In dysferlin, this domain is found positioned between C2C and C2D. In *C. elegans Fer-1*, the DYSF domain is similarly positioned, but this ferlin lacks C2A and C2B. This domain is found in other proteins, notably in the yeast peroxisomal proteins Pex30p and Pex31p, and is the hypothesized binding site of caveolin-3 within dysferlin (Patel *et al.*, 2008; Yan *et al.*, 2008).

1.3.2. Otoferlin

Otoferlin is responsible for the autosomal recessive deafness disorder DFNB9 (Yasunaga *et al.*, 1999). Although otoferlin is expressed in a wide range of tissues, including skeletal muscle, it is highly expressed in the

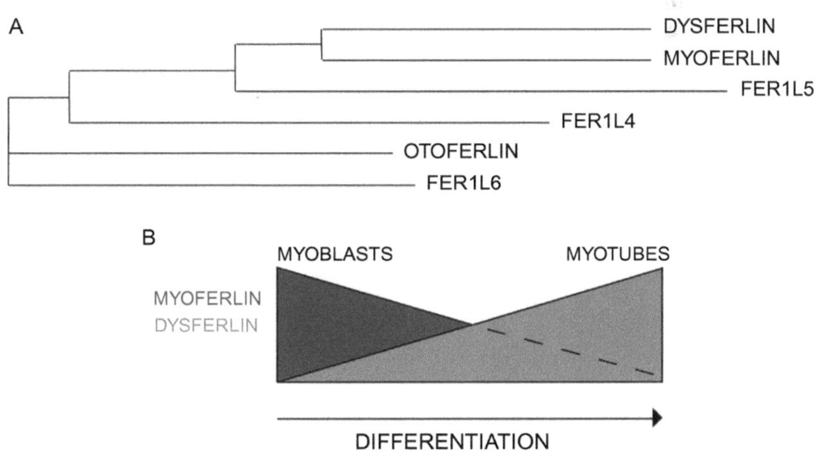

Figure 8.2 (A) Phylogram of the mammalian ferlin family. Dysferlin and myoferlin are closely related. Fer1L5 also shares close relation to dysferlin and myoferlin, while fer1L4, otoferlin, and fer1L6 are distantly related. (B) Throughout myoblast differentiation, myoferlin is highly expressed in proliferating myoblasts and nascent myotubes, while dysferlin is highly expressed in mature myotubes.

cochlea of the ear. Otoferlin has been implicated in vesicle trafficking and secretion through interaction with the endosomal protein EEA1, the Golgi marker GM130, and the GTPase Rab8a (Heidrych *et al.*, 2008; Schug *et al.*, 2006). Recent evidence suggests that otoferlin interacts with myosin VI at the inner hair cell synapse and that the interaction is involved in the recycling of synaptic vesicles (Roux *et al.*, 2009). Structurally, otoferlin is 226 kDa in size and contains either five or six C2 domains, depending on the algorithm used. Otoferlin is most like Fer1L4 and Fer1L6 and is less related to dysferlin, myoferlin, and Fer1L5. Of interest, otoferlin lacks the C2A domain suggesting that this protein may mediate different events given the lack of a lipid-binding domain.

1.3.3. Myoferlin

Myoferlin has more similarity to dysferlin than any other ferlin protein (Davis *et al.*, 2000; Fig. 8.2A). Myoferlin is also a 230-kDa protein that contains one carboxy-terminal transmembrane domain and seven C2 domains. Like dysferlin, myoferlin has a DYSF domain. Myoferlin is highly expressed in myoblasts, especially those myoblasts that have begun to differentiate (Davis *et al.*, 2000; Doherty *et al.*, 2005; Fig. 8.2B). In their proliferative stage, myoblasts can appear flattened and more fibroblast-like in appearance. When induced to differentiate, myoblasts undergo a series of morphological changes in that they become more refractile and undergo a cell shape change. Just before fusion, myoblasts are no longer fibroblast shaped but rather are elongated with a centrally position nucleus and a long axis. These prefusion myoblasts express the highest levels of myoferlin. After myoblasts fuse to form myotubes in culture, myoferlin expression is reduced. The mature myofiber expresses some myoferlin and, like dysferlin, it is found at the plasma membrane (Davis *et al.*, 2000). There is also a population of myoferlin seen at the nuclear membrane, but the role of this perinuclear myoferlin is not known. It may represent myoferlin undergoing trafficking to the plasma membrane or to intracellular vesicles. During fusion, myoferlin is enriched at sites where cells are contacted (Doherty *et al.*, 2005). Upon muscle injury, myoferlin levels are dramatically increased (Doherty *et al.*, 2005). A recent characterization of the myoferlin promoter shows that it is regulated by the NFAT transcription factors (Demonbreun *et al.*, 2010a). With muscle injury, myoferlin expression is increased in both myoblast and myotubes (Demonbreun *et al.*, 2010a).

1.3.4. Fer1L4, Fer1L5, Fer1L6

Based on database analysis, there are three other ferlin genes in the mammalian genome. The three additional ferlins family members are: Fer1L4, Fer1L5, and Fer1L6. Fer1L4 and Fer1L6 are 201 and 209 kDa, respectively, and both contain five C2 domains. Fer1L5 is a 241-kDa protein that contains six C2 domains, including the C2A domain which Fer1L4 and

Fer1L6 lack. Sequence analysis suggests that Fer1L5 is highly similar to myoferlin, Fer1L6 is highly similar to otoferlin, and Fer1L4 is similar to otoferlin but to a lesser degree than Fer1L6. A detailed characterization of the proteins produced from these genes is lacking. It is worthwhile to note that cDNAs in the electronic databases suggest that alternative splicing may occur in these mRNAs leading to a range of proteins. The role of Fer1L4, Fer1L5, and Fer1L6 in myogenesis, if any, remains to be determined.

2. Ferlin Expression and Localization and Interacting Proteins in Muscle

Dysferlin is strongly expressed in heart, brain, placenta, and skeletal muscle (Liu *et al.*, 1998). In mature skeletal muscle, dysferlin is normally localized to the plasma membrane (Anderson *et al.*, 1999). In C2C12 myotubes, dysferlin colocalizes with the marker Bin1 in T-tubules and translocates to the plasma membrane upon damage to the myotubes (Klinge *et al.*, 2007). In a rat model of muscle regeneration, dysferlin is localized to the T-tubules and dysferlin-deficient muscle shows abnormal T-tubule development suggesting that dysferlin is required for T-tubules formation (Klinge *et al.*, 2010). Also in C2C12 myotubes, dysferlin colocalizes to microtubules and immunoprecipitates with tubulin suggesting that dysferlin-containing intracellular vesicles may translocate to the plasma membrane via microtubules (Azakir *et al.*, 2010).

Dysferlin immunoprecipitates with membrane specific proteins, including caveolin-3, from normal human skeletal muscle (Matsuda *et al.*, 2001). Caveolin-3 is a small molecular mass protein that anchors into membranes leaving its amino and carboxy domains in the cytoplasm. Mutations in the gene encoding caveolin-3 have been described in patients with inherited forms of muscular dystrophy and clinically milder disorders associated with elevated serum creatine kinase and comparatively less muscle weakness. LGMD1C is a dominantly inherited muscular dystrophy that affects the proximal muscle groups and it is associated with caveolin-3 gene mutations. Patients with LGMD1C exhibit abnormally localized dysferlin. Similarly, dysferlin is aberrantly localized due to caveolin-3 mutations in a dominant negative fashion (Cai *et al.*, 2009). Dysferlin also interacts with mitsugumin 53 (MG53), a muscle-specific TRIM (tripartite motif, consisting of a RING finger domain, a B-box zinc finger domain, and a coiled-coil region) protein involved in intracellular vesicle trafficking during muscle repair. Along with MG53, dysferlin and caveolin-3 represent an essential muscle membrane repair complex and disrupting of either component affects the localization and membrane repair function of the other complex components.

Other dysferlin-interacting proteins have been identified by coimmunoprecipitation and mass spectrometry. One of these proteins includes AHNAK, a large (~700 kDa) protein primarily located near the nucleus that translocates to the plasma membrane and associates with annexin 2 and actin for stability during the membrane repair process (Benaud *et al.*, 2004; Hohaus *et al.*, 2002; Huang *et al.*, 2007). AHNAK interacts with the C2A domain of dysferlin, the same domain responsible for phospholipid binding. Downregulation of AHNAK with small-interfering RNA was found to disrupt cortical actin organization, suggesting that AHNAK participates in cytoskeletal rearrangements (Benaud *et al.*, 2004). Dysferlin also interacts with annexin A1 and A2, Ca^{2+} and phospholipid-binding proteins involved in the aggregation of lipid rafts at the plasma membrane (Lennon *et al.*, 2003). Annexin A1 and A2 localization is disrupted in dysferlin mutant muscle. Also present in the dysferlin protein complex is calpain-3, encoded by *CAPN3*. Calpain-3 is a skeletal-muscle specific nonlysosomal, Ca^{2+}-dependent cysteine protease involved in sarcomere reorganization (Huang *et al.*, 2005). Mutations in *CAPN3* cause a recessive form of muscular dystrophy called LGMD2A. Calpain-3 cleaves AHNAK into smaller fragments, and this cleavage specifically disrupts AHNAK binding to dysferlin (Huang *et al.*, 2008). Muscle from *CAPN3* mutant patients exhibits increased levels of AHNAK. These results highlight a shared property of defective sarcomeric and subsarcolemmal reorganization associated with mutations that affect either calpain-3 or dysferlin.

Myoferlin is expressed in heart and skeletal muscle, as well as at low levels in the lung and most other tissues and cell types (Davis *et al.*, 2000). In C2C12 myoblasts undergoing fusion, myoferlin is found in discrete structures at the cell periphery (Doherty *et al.*, 2008). During myoblast fusion, myoferlin is enriched at sites on the plasma membrane of cell–cell merger (Doherty *et al.*, 2005). Myoferlin directly interacts with the endocytic recycling protein EH domain 2 (EHD2), a member of the EHD-containing family of proteins, arguably members of the dynamin superfamily; EHD2 binds an asparagine-proline-phenylalanine (NPF) motif in the myoferlin C2B domain (Doherty *et al.*, 2008). Similarly, the insulin growth factor receptor-1 (IGFR-1) immunoprecipitates and colocalizes with myoferlin in proliferating myoblasts (Demonbreun *et al.*, 2010b). Myoferlin, like dysferlin, also interacts with AHNAK through its C2A domain (Benaud *et al.*, 2004).

Otoferlin is expressed in the cochlear sensory hair cells (Yasunaga *et al.*, 1999). Otoferlin colocalizes with the endosomal markers EEA1 and Golgi proteins like GM130. Also, otoferlin immunoprecipitates with the Rab8b GTPase as well as myosin VI (Heidrych *et al.*, 2008, 2009). Myosin VI and otoferlin interact at the inner hair cell ribbon synapse, providing a possible mechanism for the transport of intracellular vesicles from the *trans*-Golgi network and the recycling endosome to the plasma membrane (Roux *et al.*,

2009). These interactions may also be present with other ferlins. For instance, the Rab8a GTPase interacts with myosin Vb and localizes to endosomal tubules containing EHD2 homologs EHD1 and EHD3 (Roland *et al.*, 2007). The Rab family of interacting proteins (Rab-FIPs) contains NPF motifs, similar to the NPF motif found in myoferlin, suggesting possible binding sites for EHD interaction. Thus, the homologs of Rab GTPase and myosins may vary by tissue and cell type, but it is plausible that this interaction links the recycling endosome to intracellular trafficking via microtubules to the plasma membrane.

While dysferlin and myoferlin are strongly expressed in heart and muscle tissue, they are also expressed in other cell types. Dysferlin is present in secretory and plasma membrane vesicles of polymorphic neutrophils (Jethwaney *et al.*, 2007). In fact, neutrophil recruitment is attenuated in dysferlinopathy, suggesting that dysferlin plays a role in the immune response to muscle damage (Chiu *et al.*, 2009). This aspect of dysferlin function in the regeneration of muscle damage will be explored below. Like dysferlin, myoferlin is also expressed outside of myoblasts and myofibers. Myoferlin is also expressed in infiltrating immune cells, including Mac-1 antigen-positive cells, in regenerating muscle (Demonbreun *et al.*, 2010a). In endothelial cells, myoferlin silencing decreases clathrin- and caveolae-dependent endocytosis and overexpression of myoferlin increases endocytosis (Bernatchez *et al.*, 2009). Additionally, myoferlin has been shown to regulate the recycling of vascular endothelial growth factor receptor-2 (VEGFR-2) and exists in a complex with VEGFR-2, caveolin-1, and dynamin-2 which functions similar to EHD2 (Bernatchez *et al.*, 2007, 2009).

In common to all the ferlin family members described to date is a role in protein interactions important for trafficking vesicles within cells. For *C. elegans* Fer-1, this trafficking is critical for mobilizing the membranous organelles required for normal spermatogenesis but more recently it has been suggested to also be important for muscle function in the worm (Krajacic *et al.*, 2009). In otoferlin, intracellular vesicle movement and docking is critical for the fusion of synapse vesicles at the inner hair cell ribbon synapse. In muscle function, vesicle docking and fusion are required to repair disrupted membranes. In myogenesis, vesicle trafficking is not only important for providing membrane lipid components to a fusing myoblast but also for the protein cargo it carries to the sites of fusion and potentially for intercellular protein movement. We suggest that the ferlins share a general role in the trafficking of intracellular vesicles in multiple cell types for the exocytosis of intracompartmental contents and the membrane replacement of recycled signaling receptors. A role in endocytosis cannot be excluded, but this process is emerging as one that is an essential step in myoblast fusion, and myoblast fusion is tightly linked to differentiation.

3. Cytoskeletal Rearrangement in Myogenesis

Throughout myogenesis, singly nucleated myoblasts fuse to other singly nucleated myoblasts to form syncytial multinucleated myofibers. These fusion events are important during embryogenesis for muscle development and also during adulthood as damaged muscle undergoes regeneration. The mechanism of cell–cell fusion has previously been poorly understood, however, genetic and cell biology studies in *Drosophila* and mammalian cell culture have provided *in vivo* and *in vitro* evidence for the involvement of several molecules in myoblast fusion and revealed information about the sequential events of the fusion process.

The first step of myoblast–myoblast fusion is the recognition of opposing myoblasts. In *Drosophila*, two separate populations of myoblasts exist in the somatic mesoderm: muscle founder cells and fusion-competent cells. Founder cells are the anchors for future myofibers and are specified through Notch-mediated lateral inhibition (Haralalka and Abmayr, 2010). Fusion-compentent myoblasts (FCMs) derive from the same population of somatic mesoderm myoblasts but are not included in the founder cell selection. FCMs are attracted to and fuse to the founder cells (Chen and Olson, 2005). This attraction is thought to occur through either the random extensions of FCM filopodia or the directed extentions of FCM filopodia toward a concentrated founder cell attractant (Chen and Olson, 2004). Such an attractant has yet to be described.

Founder cells and FCMs express immunoglobulin (Ig)-domain-containing transmembrane receptors on the cell surface that are responsible for cell adhesion (Haralalka and Abmayr, 2010). In founder cells, those receptors are Dumbfounded (*Duf*) and its redundant paralog Roughest (*Rst*), while in FCMs, they are Sticks and stones (*Sns*) and its paralog Hibris (*Hbs*; Bour *et al.*, 2000; Dworak *et al.*, 2001; Ruiz-Gomez *et al.*, 2000; Strunkelnberg *et al.*, 2001). Deletion of *Duf* and *Rst* causes a complete block in myoblast fusion, but the fusion defect can be rescued with one copy of either paralog (Strunkelnberg *et al.*, 2001). Loss of *Sns* also results in a lack of fusion, while only minor fusion defects are observed with the loss of *Hbs* (Artero *et al.*, 2001). Ectopic expression of *Duf* and *Sns* in *Drosophila* S2 cells demonstrates a direct interaction between the two molecules as *Duf*-expressing cells aggregate with *Sns*-expressing cells (Galletta *et al.*, 2004). Coincidently, *Duf* and *Sns* display a specific subcellular localization and colocalize with F-actin foci at sites of myoblast fusion (Kesper *et al.*, 2007). Taken together, these results suggest that the Ig cell-adhesion molecules *Duf*/*Rst* and *Sns*/*Hbs* are responsible for the attachment, and possibly attraction, of founder cells and FCMs.

The second step of myoblast fusion, where the fusion machinery is recruited and there is cytoskeletal rearrangement to accommodate fusion, is less well understood. Electron micrographs of several steps preceding myoblast fusion have provided insight into the organelle arrangement associated with myogenesis (Doberstein *et al.*, 1997; Kalderon *et al.*, 1977). One of the early steps after cell–cell adhesion is the juxtaposition of "prefusion complexes" which consists of dense core vesicles aligned on the cytoplasmic side of both cell membranes (Doberstein *et al.*, 1997). These prefusion vesicles are observed to derive from a subcellular location near the Golgi, associate with microtubules, and are transported to actin-rich foci at the plasma membrane prior to myoblast fusion (Kim *et al.*, 2007). Thus, the membrane fusion step of myoblast fusion involves the intracellular signaling that transmits the cell–cell attachment signal to the molecules responsible for actin cytoskeleton rearrangement as well as actin foci enrichment that allows for translocation of the prefusion vesicles from exocytic origins to the plasma membrane (Fig. 8.3).

The primary signal is transmitted from the cell-adhesion molecules *Duf* and *Sns*. Truncation of the intracellular domain of *Duf* severely compromises myoblast fusion efficiency (Bulchand *et al.*, 2010). The cytoplasmic region of *Sns* is also essential for myoblast fusion (Galletta *et al.*, 2004). In founder cells, the intracellular domain of *Duf* interacts with two secondary messengers: Antisocial/Rolling pebbles (*Ants/Rols7*) and Loner (Bulchand *et al.*, 2010). *Ants/Rols7* and Loner are localized in specific subcellular foci and can be independently recruited to fusion sites at the plasma membrane

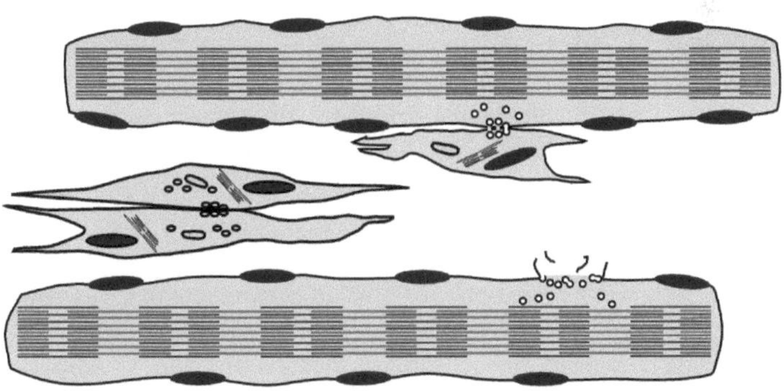

Figure 8.3 Membrane fusion events in muscle repair. Shown is a site of muscle injury. At the top is myoblast to myofiber fusion. The middle portion reflects myoblast to myoblast fusion. The lower portion of the picture depicts plasma membrane disruption undergoing resealing. Ferlin family members, particularly, dysferlin and myoferlin, are poised to mediate these events. (See Color Insert.)

by *Duf* and *Rst*. Subcellular localization of *Ants/Rols7* and Loner is *Duf/Rst*-dependent; *duf rst* double mutants show *Ants/Rols7* and Loner distributed throughout the cytoplasm (Chen and Olson, 2001; Chen *et al.*, 2003).

After receiving signals from *Duf/Rst*, *Ants/Rols7* physically interacts with a member of the CDM (CED-5, Dock180, Myoblast city) family of proteins, Myoblast city (Mbc; Chen and Olson, 2001). Mbc is the *Drosophila* homolog of the mammalian Dock180, which is responsible for modulation of the small GTPase Rac (Nolan *et al.*, 1998). Mbc forms a bipartite guanine nucleotide exchange factor with dELMO, and when overexpressed together, produce a phenotypic defect reminiscent of Drac1 overexpression, suggesting that Mbc and dELMO act in concert to activate Drac1 (Geisbrecht *et al.*, 2008). Drac1, Drac2, and Mtl are Rac GTPases with overlapping function (Hakeda-Suzuki *et al.*, 2002). Mutants of each Rac exhibit a myoblast fusion defect and a dominant negative Drac1 demonstrates unregulated actin polymerization (Hakeda-Suzuki *et al.*, 2002; Luo *et al.*, 1994). In a murine model, Rac1 mutant myoblasts fail to recruit actin fibers to myoblast contact sites (Vasyutina *et al.*, 2009). Rac1 is responsible for activation of the actin nucleation factor Arp2/3, which induces actin cytoskeletal rearrangements (Ten Klooster *et al.*, 2006). Taken together, these results suggest that Rac1 is involved in dynamic actin reorganization.

The other secondary messenger, Loner, displays guanine nucleotide exchange activity on the GTPase ARF6 (Chen *et al.*, 2003). Dominant negative ARF6 mutations exhibit a myoblast fusion defect in the *Drosophila* embryo as well as in mammalian cell cultures, similar to the defect observed for loss-of-function *loner* mutations. ARF6 has also been implicated in the control of the subcellular localization of Rac1 (Radhakrishna *et al.*, 1999). In mammalian cells, ARF6 associates in a multiprotein complex with M-cadherin, Rac1, and the guanine nucleotide exchange factor Trio, and ARF6 activates phospholipase D and phosphatidylinositol 4,5-bis-phosphate production (Bach *et al.*, 2010). Silencing of ARF6 inhibits Trio and Rac1 association with M-cadherin. ARF6 may provide a link between Loner and regulation of cytoskeletal rearrangements.

Other actin-binding and polymerizing factors with roles in myoblast fusion have also been identified. WASp is a nucleation-promoting factor that activates Arp2/3 in restricted developmental contexts (Zallen *et al.*, 2002). WASp is required for myoblast fusion and the WASp-interacting protein (WIP) Solitary (*Sltr*) recruits WASp to sites of fusion occupied by *Duf* and *Sns*. The WASp/Stlr/Arp2/3 complex is responsible for actin polymerization at fusion sites, necessary for the recruitment of the prefusion vesicles (Kim *et al.*, 2007). It remains unclear when actin polymerization would preferably proceed from downstream signaling of *Ants/Rols7* or Loner.

The last step of myoblast fusion was also observed through electron micrographs. After the alignment of paired vesicles at the plasma membrane, the paired vesicles fuse to the plasma membrane and a resultant electron-dense plaque remains (Doberstein *et al.*, 1997). The *Drosophila* gene Singles Bar (*Sing*) is required for progression past the prefusion complexes; *sing* mutant myoblasts contain more prefusion complexes than wild-type myoblasts *Sing* may be required for the fusion of the vesicles to the plasma membrane (Estrada *et al.*, 2007). The function of *Sing* is not well understood. After the prefusion complexes coalesce with the plasma membrane, fusion pores develop between the juxtaposed membranes allowing union of cytoplasmic material. Transmission electron micrographs of WASp and WIP mutants demonstrate that WASp and WIP are required at a late state of myoblast fusion, shortly after the initial fusion pores form (Massarwa *et al.*, 2007). Finally, the detached membrane forms vesicles sacs of the previous extracellular contents as the nascent bilayer fusion is resolved (Kalderon and Gilula, 1979). Efficient rearrangement of the actin cytoskeleton is also critical during this step of myoblast fusion in order to allow the paired vesicle–plasma membrane fusion and the vesiculation of detached plasma membrane.

Drosophila have one ferlin gene, *misfire*. As with the initial reports of the *C. elegans* Fer-1, misfire has been reported to be involved in spermatogenesis and not expressed in the musculature of the organism (Smith and Wakimoto, 2007). However, the characterization of misfire did not include a developmental analysis, as only adult flies were observed. A general role for ferlin involvement in intracellular trafficking is emerging and it is quite plausible that this role is conserved in *Drosophila*.

Drosophila has provided a wealth of information about the myoblast fusion process through the use of genetic models. The field has not advanced as fast in mammalian systems. Some mammalian homologs of *Drosophila myogenesis* genes are not expressed in the developing mesoderm, such as the cell-adhesion receptors *Duf*, *Rst*, *Sns*, and Hbs, and an ortholog of the secondary messenger *Ants/Rols7*, Tanc1. Cell-adhesion in mammalian cells may involve N-cadherin and the Ig receptor CDO (Krauss, 2010). Downstream activation of Rac1 can be modulated by M-cadherin. While some molecular functions are not conserved between mammals and insects, others are conserved. ARF6 can modulate Rac1 activity (Chen *et al.*, 2003). Dock180 interacts with ELMO to activate Rac1 as well (Komander *et al.*, 2008). Because of the importance of actin polymerization throughout the developing embryo for cell motility, cell–cell fusion, and organelle transport, developing genetic models is difficult since many mutants are lethal. The cre recombinase system could potentially provide an excellent way to study the effects of these genes in myoblast fusion, but many of the cre models that currently exist to study muscle are not expressed specifically in the myoblast or not expressed in the myoblast at all.

4. Ferlin Proteins in Cytoskeletal Rearrangements During Myogenesis

Myoferlin is highly expressed in proliferating and early differentiating myoblasts and at sites of myoblast–myoblast or myoblast–myotube fusion (Davis *et al.*, 2002; Doherty *et al.*, 2005). Myoferlin directly interacts with the endosomal-recycling protein EHD2, and, based on homology, it is conceivable that myoferlin may interact with other members of the EHD family (Doherty *et al.*, 2008). This interaction suggests that myoferlin participates in the intracellular trafficking of recycling vesicles (Fig. 8.4). Loss of myoferlin results is known to produce defective endocytic recycling in primary myoblasts (Doherty *et al.*, 2008). Like dysferlin-null muscle biopsies, myoferlin-null muscle shows an accumulation of small subsarcolemmal vesicles, suggesting that myoferlin is also required for fusion of these vesicles to the plasma membrane (Demonbreun *et al.*, 2010b). The primary assay for recycling relies on internalization and recycling of transferrin. It has been observed that more transferrin accumulates in myoferlin-null

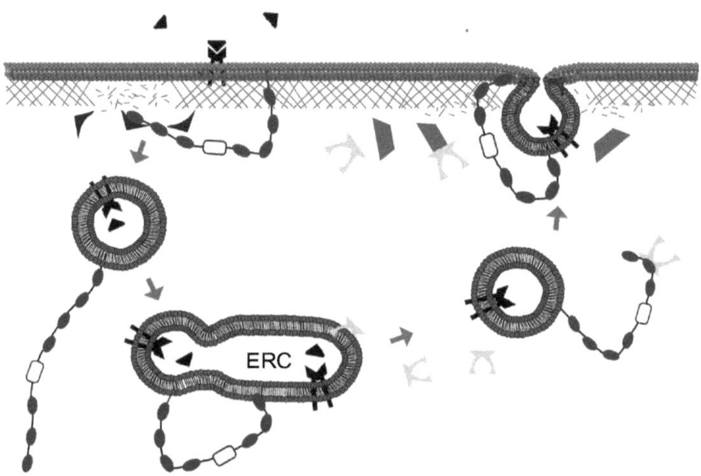

Figure 8.4 Model of ferlin-mediated endocytic recycling that contributes to myoblast–myoblast fusion, myoblast–myotube fusion, and membrane damage repair. Ferlin-interacting actin-binding proteins (ferlins—string of beads; AHNAK—boomerang shape) remodel the actin cytoskeleton, allowing endocytosis of ligand-bound receptors. Ferlins mediate fusion of endocytosed vesicles with the endocytic recycling compartment (ERC) where the ligand is removed. EHD proteins (wishbone shape) participate in the scission of recycling vesicles from the ERC. EHD proteins interact with the ferlins for transport back to the plasma membrane. EHD proteins also interact with EHBP1 (trapezoid) for actin cytoskeletal rearrangement, and ferlin proteins assist in fusion of recycling vesicles to the sarcolemma. (See Color Insert.)

myoblasts than wild-type myoblasts at baseline, and using pulse-chase experiments that the recycling of the accumulated transferrin is also delayed (Doherty *et al.*, 2008).

The EHD family of proteins consists of four highly homologous members, each containing an amino-terminal nucleotide-binding domain, a carboxy-terminal EH domain, and a central coiled-coil domain. EHDs have been implicated in the trafficking of many cell signaling and metabolic molecules, including IGF1R, EGFR, GLUT4, LDL receptor, and the transferrin receptor (Austin *et al.*, 2010; Demonbreun *et al.*, 2010b; Naslavsky *et al.*, 2007; Rotem-Yehudar *et al.*, 2001). A direct interaction between EHD1 and SNAP29, a SNARE protein, suggests that EHD1 may have a role in membrane fusion (Rotem-Yehudar *et al.*, 2001). The *C. elegans* EHD homolog, RME-1, was identified in a genetic screen for mutants defective in receptor-mediated endocytosis (Grant and Hirsh, 1999). *rme-1* mutants accumulated RME-2 yolk receptors within recycling endosomes and not on the cell surface, indicating that RME-1 functions in the endocytic recycling and not internalization (Grant *et al.*, 2001). Dominant negative substitutions within RME-1, G81R, and G429R, failed to recycle yolk receptors as well. The *Drosophila* EHD homolog, Past1, has also been implicated in endocytosis (Olswang-Kutz *et al.*, 2009). Garland cells from *Past1* mutants demonstrated a reduced ability to endocytose fluorescently labeled avidin. A genetic interaction between Past1 and the Notch signaling pathway was inferred from a wingvein phenotype, suggesting that Past1 may be responsible for the recycling of Notch receptor.

While the EHDs are highly homologous molecules, they are not completely redundant. siRNA reduction of EHD1-4 in mammalian cells demonstrated that loss of EHD1 and EHD3 function severely reduces transferrin exit from the endocytic recycling compartment (George *et al.*, 2007). However, transferrin recycling was only mildly affected by loss of EHD4, and EHD2 reduction had no affect on transferrin recycling. Deletion of the EH domain in EHD1 and EHD3 resulted in a perinuclear clustering of the cytoplasmic Rab11, while EHD2 and EHD4 deletions had no effect. Reduction of EHD protein levels demonstrated that EHD1 and EHD2 function in the exit from the recycling endosome, while EHD3 and EHD4 function in the transport from the early endosome to the endosomal-recycling compartment. EHD1 and EHD4 null mice both exhibit defects in spermatogenesis and decreased male fertility (George *et al.*, 2010; Rainey *et al.*, 2010). siRNA reduction of EHD3 in cardiomyocytes indicated that EHD3 has a role in the trafficking of the cardiac ankyrin-B membrane protein (Gudmundsson *et al.*, 2010). In addition, EHD3 and EHD4 were sufficiently elevated following myocardial infarction, suggesting that these molecules may have a role in cardiomyocyte damage response. The necessity for these molecules throughout myogenesis has yet to be determined. Yet if we posit that those prefusion complexes that align themselves adjacent to the apposed membranes at sites of fusion are of endocytic recycling origin, the EHDs are candidates for the transport from the exit of the recycling endosome to the plasma membrane. This argument is strengthened by the interaction of EHDs and the EHD-binding protein, EHBP1, which contains a novel amino-terminal C2 domain version coupled with a calponin-homology domain (Zhang and Aravind, 2010). Loss of EHBP1 disrupts recycling of GLUT4 in cultured adipocytes, and overexpression of EHBP1 disrupts transferring uptake and induces actin rearrangements, suggesting that EHBP1 regulates actin polymerization (Guilherme *et al.*, 2004a,b).

Defective endocytic recycling prevents signaling receptors from being properly returned to the plasma membrane for subsequent signaling. In myoferlin-null primary myoblasts, the IGFR-1 is abnormally internalized and is shuttled for lysosomal degradation (Demonbreun *et al.*, 2010b). IGF1 mediates cell growth and is responsible for skeletal muscle hypertrophy (Barton, 2006). Mice lacking IGF1 are smaller in size, while mice overexpressing IGF1 have increased skeletal muscle fiber number and increased fiber cross-sectional area (Liao *et al.*, 2006; Liu *et al.*, 1993; Powell-Braxton *et al.*, 1993). While wild-type myoblasts hypertrophy upon IGF stimulation, myoferlin-null myoblasts do not respond to stimulation (Demonbreun *et al.*, 2010b). Although IGF1 signaling has been shown to be important for myogenesis, there are other signaling pathways required as well. Myoferlin may be required for the intracellular trafficking of these receptors also. In addition to those receptors required for myogenesis, myoferlin also participates in the endocytosis of the transferrin receptor, the cholera toxin-B receptor, and the VEGFR-2; Bernatchez *et al.*, 2007, 2009). Myoferlin deficiency results in a reduction in VEGFR-2 protein levels; myoferlin prevents the ubiquination and degradation of VEGFR-2 (Bernatchez *et al.*, 2007).

As a result of myoferlin-null defective endocytic recycling, myoferlin-null mice have decreased body mass and muscle mass, and *in vivo* muscle analysis reveals myoferlin-null muscle fibers are decreased in area and size, similar to IGF-null mice (Doherty *et al.*, 2005; Liu *et al.*, 1993; Powell-Braxton *et al.*, 1993). Primary myoblasts from myoferlin-null mice form smaller myotubes due to defective myoblast fusion. Similarly, primary myoblasts from transgenic mice overexpressing a dominant negative IGF-1R also demonstrate impaired myoblast fusion (Heron-Milhavet *et al.*, 2010).

Defective myoblast fusion is observed with dysferlin deficiency. Primary myoblasts from dysferlin mutant patients begin to fuse later than control myoblasts and demonstrate an incomplete myotube differentiation, as majority of myotubes are binucleated (de Luna *et al.*, 2006). Myogenin is decreased in dysferlin mutant myotubes, suggesting that dysferlin may influence myogenesis by increasing *myogenin* gene expression. Correspondingly, dysferlin-deficient C2C12 cells demonstrate a reduction in long myotubes and a reduction of muscle differentiation proteins, including myosin heavy chain (MHC; Belanto *et al.*, 2010).

5. Ferlin Proteins Participate in Muscle Damage Repair

There is evidence that the ferlins participate in the repair of damaged adult muscle fibers. Dysferlin-deficient muscle demonstrates a pathological progressive pattern of fiber necrosis, fibrotic, and immune infiltrate, and increased sarcolemmal membrane disruption and regenerating fibers (Bansal *et al.*, 2003). Loss of dysferlin is also characterized by a poor repair response to muscle damage. Threefold overexpression of a skeletal-muscle specific dysferlin transgene rescues all histological defects as well as increases the recovery of muscle function after contraction injury, suggesting that the defects observed in dysferlin deficiency are initially caused by the loss of dysferlin in skeletal muscle (Millay *et al.*, 2009).

Bansal *et al.* (2003) provided the quintessential study demonstrating that dysferlin was involved in the repair of damaged myofibers. Control and dysferlin-deficient myofibers were injured using a laser in the presence of a membrane impermeable fluorescent dye. In the absence of Ca^{2+}, neither wild-type nor dysferlin-deficient myofibers were able to repair the membrane disruptions induced by the laser, demonstrating the Ca^{2+}-dependence of muscle membrane resealing. However, in the presence of Ca^{2+}, wild-type fibers efficiently prevented dye entry by repairing the damaged membrane while dysferlin-null myofibers were unable to do so. Electron micrographs of wild-type and dysferlin-null skeletal muscle also showed an accumulation of subsarcolemmal vesicles in dysferlin-null but not wild-type muscle (Piccolo *et al.*, 2000; Selcen *et al.*, 2001). Recent studies showed that antisense dysferlin morpholinos inhibit intercellular wound-triggered Ca^{2+} signaling in neighboring cells of sea urchin embryos, but not Ca^{2+} spikes in the wounded cell itself (Covian-Nares *et al.*, 2010). Taken together, these results suggest that dysferlin responds to the Ca^{2+} influx after damage and influences the fusion of intracellular and exocytic vesicles to the plasma membrane. These exocytic vesicles provide additional phospholipids to seal the damaged area through vesicular–plasma membrane fusion and also release molecules into the extracellular environment for cell–cell signaling.

Roche *et al.* (2008) injected wild-type and dysferlin-null mice intraperitoneally with fluorescein dextran (FDx) and induced large-strain injury in the ankle dorsifelxors. FDx is a fluorescent dye, similar to that used in the Bansal *et al.* laser-wounding study that diffuses into the myofiber through the damaged sarcolemmal membrane and is trapped inside the cell as the membrane is resealed. Wild-type tibialis anterior muscle retained the FDx 14 days postinjury, dysferlin-deficient TA muscle retains FDx 3 days post-injury but demonstrated a significant reduction in FDx retention 7 days postinjury (Roche *et al.*, 2008, 2010). Three days postinjury, there was a significant increase in mononuclear cells, mostly CD68+ macrophages, in dysferlin-null TA muscle, but not wild-type TA muscle (Roche *et al.*, 2010). Dysferlin-null TA muscle also showed a significant increase in centrally nucleated fibers and developmental myosin heavy chain positive (dMHC+) fibers 14 days postinjury. Central nuclei and dMHC expression are markers of regenerating myofibers. Interestingly, inhibition of satellite cell activation by hindlimb irradiation prior to injury prevented the increase in centrally nucleated and dMHC+ fibers in dysferlin-null muscle, which suggests that central nuclei and dMHC expression rely on satellite cell activation and myogenesis. The single localized 25 Gy dose of radiation did not inhibit macrophage infiltration; instead, inflammation was increased in both wild-type and dysferlin-null muscle following a longer large-strain injury protocol. These results are interesting because they demonstrate that satellite cell activation and myogenesis is increased with dysferlin deficiency. However, wild-type muscle did not show an increase in central nuclei and only showed a modest increase in dMHC+ fibers, indicating that the repair of the sarcolemmal membrane may occur independent of myogenesis. Supporting these findings, dysferlin is expressed in a subset of c-met+ satellite cells in normal muscle and satellite cells in pathological muscles (De Luna *et al.*, 2004). All activated (MyoD+) satellite cells also express dysferlin. Thus, dysferlin has an apparent role in satellite cells, possibly through the regulation of satellite cell activation.

Myoferlin is normally expressed throughout myogenesis in myoblasts and nascent myotubes (Davis *et al.*, 2000; Doherty *et al.*, 2005). Myoferlin protein levels are normally low in adult skeletal muscle and nearly absent in healthy myofibers. One plausible mechanism for the increase in myogenesis in dysferlin-null muscle is that an increase in satellite cell activation leads to an accumulation of mononuclear myoferlin-positive myoblasts that anxiously wait to repair damaged myofibers; irradiation could destroy these myoblasts and prevent the myogenesis observed by Roche *et al.* Conversely, an alternate mechanism is that an increase in myoferlin expression, due to dysferlin deficiency, could directly stimulate satellite cell activation. Studies have demonstrated that alterations in dysferlin expression lead to changes in the expression levels of myogenic transcription factors. Increased induction of dysferlin with dexamethasone treatment in C2C12 muscle cells increases the expression of muscle differentiation proteins, including MHC (Belanto *et al.*, 2010). Dysferlin deficiency causes a sharp reduction in myogenin, a myogenic transcription factor expressed in activated satellite cells, so this alternative mechanism is plausible (de Luna *et al.*, 2006).

In disease and injury models, myoferlin appears to have a role in the repair of the damaged muscle. Myoferlin mRNA is upregulated 7.3-fold in quadriceps biopsies from Duchenne Muscular Dystrophy patients (Haslett *et al.*, 2002). Myoferlin mRNA, as well as IGF-1 mRNA, was significantly upregulated in the hindlimb muscles of resistance exercise trained rats (Adams *et al.*, 2007). At the protein level, myoferlin is focally upregulated at the membrane in *mdx* skeletal muscle, suggesting that it is important during muscle regeneration (Davis *et al.*, 2002; Demonbreun *et al.*, 2010a). When *Sgcg*-null mice are injected intraperitoneally with Evan's blue dye (EBD), a marker sequestered in myofibers with disrupted sarcolemma, myoferlin is abundantly expressed in dye-positive but not dye-negative fibers (Demonbreun *et al.*, 2010a). *In vitro*, myoferlin is increased in myotubes after cardiotoxin-induced injury, while nearly myoferlin is absent in uninjured myotubes. A GFP reporter for the myoferlin promoter replicates these findings. The myoferlin GFP reporter is normally expressed in myoblasts but not mature myotubes. However, upon injury, when endogenous myoferlin is normally upregulated, GFP is expressed. These GFP-positive myofibers accumulate EBD as well as increased intracellular Ca^{2+}, suggesting that myoferlin is upregulated in myofibers with damaged membranes. Myoferlin-null mice also demonstrate delayed muscle repair after injury, highlighting that myoblast fusion is not only required during embryogenesis but also during damage repair and muscle regeneration (Demonbreun *et al.*, 2010b; Doherty *et al.*, 2005).

6. Concluding Remarks

The discovery of the ferlin family of proteins, specifically dysferlin and myoferlin, has led to a new appreciation for the importance of vesicle trafficking and its importance to normal muscle growth and repair. The ferlin proteins harbor the capacity to bind directly to negatively charged phospholipids and additionally scaffold a number of distinct proteins via their C2 domains. Future studies will focus on identifying binding partners and understanding the dynamic intercellular changes that occur for these proteins. The role of dysferlin in disease likely arises from both defects in the mature myofiber and the myoblast, reinforcing the importance of vesicular trafficking for these processes. The ferlins interact with other trafficking proteins during damage repair to transport internal membrane structures to the site of damage and seal the breach in sarcolemma. These interacting molecules include those covered earlier with roles in exocytic vesicle transportation, membrane docking, and actin cytoskeletal rearrangement.

ACKNOWLEDGMENTS

Supported by NIH R01 NS47726, T32 HL 7381, and the Muscular Dystrophy Association.

REFERENCES

Achanzar, W. E., and Ward, S. (1997). A nematode gene required for sperm vesicle fusion. *J. Cell Sci.* **110**(Pt 9), 1073–1081.

Adams, G. R., Haddad, F., Bodell, P. W., Tran, P. D., and Baldwin, K. M. (2007). Combined isometric, concentric, and eccentric resistance exercise prevents unloading-induced muscle atrophy in rats. *J. Appl. Physiol.* **103,** 1644–1654.

Anderson, L. V., Davison, K., Moss, J. A., Young, C., Cullen, M. J., Walsh, J., Johnson, M. A., Bashir, R., Britton, S., Keers, S., Argov, Z., Mahjneh, I., *et al.* (1999). Dysferlin is a plasma membrane protein and is expressed early in human development. *Hum. Mol. Genet.* **8,** 855–861.

Artero, R. D., Castanon, I., and Baylies, M. K. (2001). The immunoglobulin-like protein Hibris functions as a dose-dependent regulator of myoblast fusion and is differentially controlled by Ras and Notch signaling. *Development* **128,** 4251–4264.

Austin, P., Heller, M., Williams, D. E., McIntosh, L. P., Vogl, A. W., Foster, L. J., Andersen, R. J., Roberge, M., and Roskelley, C. D. (2010). Release of membrane-bound vesicles and inhibition of tumor cell adhesion by the peptide Neopetrosiamide A. *PLoS ONE* **5,** e10836.

Azakir, B. A., Di Fulvio, S., Therrien, C., and Sinnreich, M. (2010). Dysferlin interacts with tubulin and microtubules in mouse skeletal muscle. *PLoS ONE* **5,** e10122.

Bach, A. S., Enjalbert, S., Comunale, F., Bodin, S., Vitale, N., Charrasse, S., and Gauthier-Rouviere, C. (2010). ADP-ribosylation factor 6 regulates mammalian myoblast fusion through phospholipase D1 and phosphatidylinositol 4, 5-bisphosphate signaling pathways. *Mol. Biol. Cell* **21,** 2412–2424.

Bai, J., Wang, P., and Chapman, E. R. (2002). C2A activates a cryptic Ca(2+)-triggered membrane penetration activity within the C2B domain of synaptotagmin I. *Proc. Natl. Acad. Sci. USA* **99,** 1665–1670.

Bansal, D., Miyake, K., Vogel, S. S., Groh, S., Chen, C. C., Williamson, R., McNeil, P. L., and Campbell, K. P. (2003). Defective membrane repair in dysferlin-deficient muscular dystrophy. *Nature* **423,** 168–172.

Barton, E. R. (2006). The ABCs of IGF-I isoforms: Impact on muscle hypertrophy and implications for repair. *Appl. Physiol. Nutr. Metab.* **31,** 791–797.

Bashir, R., Britton, S., Strachan, T., Keers, S., Vafiadaki, E., Lako, M., Richard, I., Marchand, S., Bourg, N., Argov, Z., Sadeh, M., Mahjneh, I., *et al.* (1998). A gene related to *Caenorhabditis elegans* spermatogenesis factor fer-1 is mutated in limb-girdle muscular dystrophy type 2B. *Nat. Genet.* **20,** 37–42.

Belanto, J. J., Diaz-Perez, S. V., Magyar, C. E., Maxwell, M. M., Yilmaz, Y., Topp, K., Boso, G., Jamieson, C. H., Cacalano, N. A., and Jamieson, C. A. (2010). Dexamethasone induces dysferlin in myoblasts and enhances their myogenic differentiation. *Neuromuscul. Disord.* **20,** 111–121.

Benaud, C., Gentil, B. J., Assard, N., Court, M., Garin, J., Delphin, C., and Baudier, J. (2004). AHNAK interaction with the annexin 2/S100A10 complex regulates cell membrane cytoarchitecture. *J. Cell Biol.* **164,** 133–144.

Bernatchez, P. N., Acevedo, L., Fernandez-Hernando, C., Murata, T., Chalouni, C., Kim, J., Erdjument-Bromage, H., Shah, V., Gratton, J. P., McNally, E. M., Tempst, P., and Sessa, W. C. (2007). Myoferlin regulates vascular endothelial growth factor receptor-2 stability and function. *J. Biol. Chem.* **282,** 30745–30753.

Bernatchez, P. N., Sharma, A., Kodaman, P., and Sessa, W. C. (2009). Myoferlin is critical for endocytosis in endothelial cells. *Am. J. Physiol. Cell Physiol.* **297,** C484–C492.

Bour, B. A., Chakravarti, M., West, J. M., and Abmayr, S. M. (2000). Drosophila SNS, a member of the immunoglobulin superfamily that is essential for myoblast fusion. *Genes Dev.* **14,** 1498–1511.

Britton, S., Freeman, T., Vafiadaki, E., Keers, S., Harrison, R., Bushby, K., and Bashir, R. (2000). The third human FER-1-like protein is highly similar to dysferlin. *Genomics* **68,** 313–321.

Brose, N., Petrenko, A. G., Sudhof, T. C., and Jahn, R. (1992). Synaptotagmin: A calcium sensor on the synaptic vesicle surface. *Science* **256,** 1021–1025.

Bulchand, S., Menon, S. D., George, S. E., and Chia, W. (2010). The intracellular domain of Dumbfounded affects myoblast fusion efficiency and interacts with Rolling pebbles and Loner. *PLoS ONE* **5,** e9374.

Cai, C., Weisleder, N., Ko, J. K., Komazaki, S., Sunada, Y., Nishi, M., Takeshima, H., and Ma, J. (2009). Membrane repair defects in muscular dystrophy are linked to altered interaction between MG53, caveolin-3, and dysferlin. *J. Biol. Chem.* **284,** 15894–15902.

Cenacchi, G., Fanin, M., De Giorgi, L. B., and Angelini, C. (2005). Ultrastructural changes in dysferlinopathy support defective membrane repair mechanism. *J. Clin. Pathol.* **58,** 190–195.

Chapman, E. R., Hanson, P. I., An, S., and Jahn, R. (1995). Ca2+ regulates the interaction between synaptotagmin and syntaxin 1. *J. Biol. Chem.* **270,** 23667–23671.

Chen, E. H., and Olson, E. N. (2001). Antisocial, an intracellular adaptor protein, is required for myoblast fusion in Drosophila. *Dev. Cell* **1,** 705–715.

Chen, E. H., and Olson, E. N. (2004). Towards a molecular pathway for myoblast fusion in Drosophila. *Trends Cell Biol.* **14,** 452–460.

Chen, E. H., and Olson, E. N. (2005). Unveiling the mechanisms of cell–cell fusion. *Science* **308,** 369–373.

Chen, E. H., Pryce, B. A., Tzeng, J. A., Gonzalez, G. A., and Olson, E. N. (2003). Control of myoblast fusion by a guanine nucleotide exchange factor, loner, and its effector ARF6. *Cell* **114,** 751–762.

Chiu, Y. H., Hornsey, M. A., Klinge, L., Jorgensen, L. H., Laval, S. H., Charlton, R., Barresi, R., Straub, V., Lochmuller, H., and Bushby, K. (2009). Attenuated muscle regeneration is a key factor in dysferlin-deficient muscular dystrophy. *Hum. Mol. Genet.* **18,** 1976–1989.

Corbin, J. A., Evans, J. H., Landgraf, K. E., and Falke, J. J. (2007). Mechanism of specific membrane targeting by C2 domains: Localized pools of target lipids enhance Ca2+ affinity. *Biochemistry* **46,** 4322–4336.

Covian-Nares, J. F., Koushik, S. V., Puhl, H. L., 3rd, and Vogel, S. S. (2010). Membrane wounding triggers ATP release and dysferlin-mediated intercellular calcium signaling. *J. Cell Sci.* **123,** 1884–1893.

Davis, D. B., Delmonte, A. J., Ly, C. T., and McNally, E. M. (2000). Myoferlin, a candidate gene and potential modifier of muscular dystrophy. *Hum. Mol. Genet.* **9,** 217–226.

Davis, D. B., Doherty, K. R., Delmonte, A. J., and McNally, E. M. (2002). Calcium-sensitive phospholipid binding properties of normal and mutant ferlin C2 domains. *J. Biol. Chem.* **277,** 22883–22888.

Davletov, B. A., and Sudhof, T. C. (1993). A single C2 domain from synaptotagmin I is sufficient for high affinity Ca2+/phospholipid binding. *J. Biol. Chem.* **268,** 26386–26390.

De Luna, N., Gallardo, E., and Illa, I. (2004). In vivo and in vitro dysferlin expression in human muscle satellite cells. *J. Neuropathol. Exp. Neurol.* **63,** 1104–1113.

de Luna, N., Gallardo, E., Soriano, M., Dominguez-Perles, R., de la Torre, C., Rojas-Garcia, R., Garcia-Verdugo, J. M., and Illa, I. (2006). Absence of dysferlin alters

myogenin expression and delays human muscle differentiation "in vitro" *J. Biol. Chem.* **281,** 17092–17098.
Demonbreun, A. R., Lapidos, K. A., Heretis, K., Levin, S., Dale, R., Pytel, P., Svensson, E. C., and McNally, E. M. (2010a). Myoferlin regulation by NFAT in muscle injury, regeneration and repair. *J. Cell Sci.* **123,** 2413–2422.
Demonbreun, A. R., Posey, A. D., Heretis, K., Swaggart, K. A., Earley, J. U., Pytel, P., and McNally, E. M. (2010b). Myoferlin is required for insulin-like growth factor response and muscle growth. *FASEB J.* **24,** 1284–1295.
Doberstein, S. K., Fetter, R. D., Mehta, A. Y., and Goodman, C. S. (1997). Genetic analysis of myoblast fusion: Blown fuse is required for progression beyond the prefusion complex. *J. Cell Biol.* **136,** 1249–1261.
Doherty, K. R., Cave, A., Davis, D. B., Delmonte, A. J., Posey, A., Earley, J. U., Hadhazy, M., and McNally, E. M. (2005). Normal myoblast fusion requires myoferlin. *Development* **132,** 5565–5575.
Doherty, K. R., Demonbreun, A. R., Wallace, G. Q., Cave, A., Posey, A. D., Heretis, K., Pytel, P., and McNally, E. M. (2008). The endocytic recycling protein EHD2 interacts with myoferlin to regulate myoblast fusion. *J. Biol. Chem.* **283,** 20252–20260.
Dworak, H. A., Charles, M. A., Pellerano, L. B., and Sink, H. (2001). Characterization of Drosophila hibris, a gene related to human nephrin. *Development* **128,** 4265–4276.
Estrada, B., Maeland, A. D., Gisselbrecht, S. S., Bloor, J. W., Brown, N. H., and Michelson, A. M. (2007). The MARVEL domain protein, Singles Bar, is required for progression past the pre-fusion complex stage of myoblast fusion. *Dev. Biol.* **307,** 328–339.
Galletta, B. J., Chakravarti, M., Banerjee, R., and Abmayr, S. M. (2004). SNS: Adhesive properties, localization requirements and ectodomain dependence in S2 cells and embryonic myoblasts. *Mech. Dev.* **121,** 1455–1468.
Geisbrecht, E. R., Haralalka, S., Swanson, S. K., Florens, L., Washburn, M. P., and Abmayr, S. M. (2008). Drosophila ELMO/CED-12 interacts with Myoblast city to direct myoblast fusion and ommatidial organization. *Dev. Biol.* **314,** 137–149.
George, M., Ying, G., Rainey, M. A., Solomon, A., Parikh, P. T., Gao, Q., Band, V., and Band, H. (2007). Shared as well as distinct roles of EHD proteins revealed by biochemical and functional comparisons in mammalian cells and *C. elegans*. *BMC Cell Biol.* **8,** 3.
George, M., Rainey, M. A., Naramura, M., Ying, G., Harms, D. W., Vitaterna, M. H., Doglio, L., Crawford, S. E., Hess, R. A., Band, V., and Band, H. (2010). Ehd4 is required to attain normal prepubertal testis size but dispensable for fertility in male mice. *Genesis* **48,** 328–342.
Grant, B., and Hirsh, D. (1999). Receptor-mediated endocytosis in the *Caenorhabditis elegans* oocyte. *Mol. Biol. Cell* **10,** 4311–4326.
Grant, B., Zhang, Y., Paupard, M. C., Lin, S. X., Hall, D. H., and Hirsh, D. (2001). Evidence that RME-1, a conserved *C. elegans* EH-domain protein, functions in endocytic recycling. *Nat. Cell Biol.* **3,** 573–579.
Gudmundsson, H., Hund, T. J., Wright, P. J., Kline, C. F., Snyder, J. S., Qian, L., Koval, O. M., Cunha, S. R., George, M., Rainey, M. A., Kashef, F. E., Dun, W., *et al.* (2010). EH domain proteins regulate cardiac membrane protein targeting. *Circ. Res.* **107,** 84–95.
Guilherme, A., Soriano, N. A., Bose, S., Holik, J., Bose, A., Pomerleau, D. P., Furcinitti, P., Leszyk, J., Corvera, S., and Czech, M. P. (2004a). EHD2 and the novel EH domain binding protein EHBP1 couple endocytosis to the actin cytoskeleton. *J. Biol. Chem.* **279,** 10593–10605.
Guilherme, A., Soriano, N. A., Furcinitti, P. S., and Czech, M. P. (2004b). Role of EHD1 and EHBP1 in perinuclear sorting and insulin-regulated GLUT4 recycling in 3T3-L1 adipocytes. *J. Biol. Chem.* **279,** 40062–40075.

Hakeda-Suzuki, S., Ng, J., Tzu, J., Dietzl, G., Sun, Y., Harms, M., Nardine, T., Luo, L., and Dickson, B. J. (2002). Rac function and regulation during Drosophila development. *Nature* **416,** 438–442.

Haralalka, S., and Abmayr, S. M. (2010). Myoblast fusion in Drosophila. *Exp. Cell Res.* **316,** 3007–3013.

Haslett, J. N., Sanoudou, D., Kho, A. T., Bennett, R. R., Greenberg, S. A., Kohane, I. S., Beggs, A. H., and Kunkel, L. M. (2002). Gene expression comparison of biopsies from Duchenne muscular dystrophy (DMD) and normal skeletal muscle. *Proc. Natl. Acad. Sci. USA* **99,** 15000–15005.

Heidrych, P., Zimmermann, U., Bress, A., Pusch, C. M., Ruth, P., Pfister, M., Knipper, M., and Blin, N. (2008). Rab8b GTPase, a protein transport regulator, is an interacting partner of otoferlin, defective in a human autosomal recessive deafness form. *Hum. Mol. Genet.* **17,** 3814–3821.

Heidrych, P., Zimmermann, U., Kuhn, S., Franz, C., Engel, J., Duncker, S. V., Hirt, B., Pusch, C. M., Ruth, P., Pfister, M., Marcotti, W., Blin, N., *et al.* (2009). Otoferlin interacts with myosin VI: Implications for maintenance of the basolateral synaptic structure of the inner hair cell. *Hum. Mol. Genet.* **18,** 2779–2790.

Heron-Milhavet, L., Mamaeva, D., LeRoith, D., Lamb, N. J., and Fernandez, A. (2010). Impaired muscle regeneration and myoblast differentiation in mice with a muscle-specific KO of IGF-IR. *J. Cell. Physiol.* **225,** 1–6.

Hohaus, A., Person, V., Behlke, J., Schaper, J., Morano, I., and Haase, H. (2002). The carboxyl-terminal region of ahnak provides a link between cardiac L-type Ca2+ channels and the actin-based cytoskeleton. *FASEB J.* **16,** 1205–1216.

Huang, Y., Verheesen, P., Roussis, A., Frankhuizen, W., Ginjaar, I., Haldane, F., Laval, S., Anderson, L. V., Verrips, T., Frants, R. R., de Haard, H., Bushby, K., *et al.* (2005). Protein studies in dysferlinopathy patients using llama-derived antibody fragments selected by phage display. *Eur. J. Hum. Genet.* **13,** 721–730.

Huang, Y., Laval, S. H., van Remoortere, A., Baudier, J., Benaud, C., Anderson, L. V., Straub, V., Deelder, A., Frants, R. R., den Dunnen, J. T., Bushby, K., and van der Maarel, S. M. (2007). AHNAK, a novel component of the dysferlin protein complex, redistributes to the cytoplasm with dysferlin during skeletal muscle regeneration. *FASEB J.* **21,** 732–742.

Huang, Y., de Morree, A., van Remoortere, A., Bushby, K., Frants, R. R., Dunnen, J. T., and van der Maarel, S. M. (2008). Calpain 3 is a modulator of the dysferlin protein complex in skeletal muscle. *Hum. Mol. Genet.* **17,** 1855–1866.

Illarioshkin, S. N., Ivanova-Smolenskaya, I. A., Greenberg, C. R., Nylen, E., Sukhorukov, V. S., Poleshchuk, V. V., Markova, E. D., and Wrogemann, K. (2000). Identical dysferlin mutation in limb-girdle muscular dystrophy type 2B and distal myopathy. *Neurology* **55,** 1931–1933.

Jethwaney, D., Islam, M. R., Leidal, K. G., de Bernabe, D. B., Campbell, K. P., Nauseef, W. M., and Gibson, B. W. (2007). Proteomic analysis of plasma membrane and secretory vesicles from human neutrophils. *Proteome Sci.* **5,** 12.

Jimenez, J. L., and Bashir, R. (2007). In silico functional and structural characterisation of ferlin proteins by mapping disease-causing mutations and evolutionary information onto three-dimensional models of their C2 domains. *J. Neurol. Sci.* **260,** 114–123.

Kalderon, N., and Gilula, N. B. (1979). Membrane events involved in myoblast fusion. *J. Cell Biol.* **81,** 411–425.

Kalderon, N., Epstein, M. L., and Gilula, N. B. (1977). Cell-to-cell communication and myogenesis. *J. Cell Biol.* **75,** 788–806.

Kesper, D. A., Stute, C., Buttgereit, D., Kreiskother, N., Vishnu, S., Fischbach, K. F., and Renkawitz-Pohl, R. (2007). Myoblast fusion in *Drosophila melanogaster* is mediated through a fusion-restricted myogenic-adhesive structure (FuRMAS). *Dev. Dyn.* **236,** 404–415.

Kim, S., Shilagardi, K., Zhang, S., Hong, S. N., Sens, K. L., Bo, J., Gonzalez, G. A., and Chen, E. H. (2007). A critical function for the actin cytoskeleton in targeted exocytosis of prefusion vesicles during myoblast fusion. *Dev. Cell* **12,** 571–586.

Klinge, L., Laval, S., Keers, S., Haldane, F., Straub, V., Barresi, R., and Bushby, K. (2007). From T-tubule to sarcolemma: Damage-induced dysferlin translocation in early myogenesis. *FASEB J.* **21,** 1768–1776.

Klinge, L., Harris, J., Sewry, C., Charlton, R., Anderson, L., Laval, S., Chiu, Y. H., Hornsey, M., Straub, V., Barresi, R., Lochmuller, H., and Bushby, K. (2010). Dysferlin associates with the developing T-tubule system in rodent and human skeletal muscle. *Muscle Nerve* **41,** 166–173.

Komander, D., Patel, M., Laurin, M., Fradet, N., Pelletier, A., Barford, D., and Cote, J. F. (2008). An alpha-helical extension of the ELMO1 pleckstrin homology domain mediates direct interaction to DOCK180 and is critical in Rac signaling. *Mol. Biol. Cell* **19,** 4837–4851.

Krajacic, P., Hermanowski, J., Lozynska, O., Khurana, T. S., and Lamitina, T. (2009). *C. elegans* dysferlin homolog fer-1 is expressed in muscle, and fer-1 mutations initiate altered gene expression of muscle enriched genes. *Physiol. Genomics* **40,** 8–14.

Krauss, R. S. (2010). Regulation of promyogenic signal transduction by cell–cell contact and adhesion. *Exp. Cell Res.* **316,** 3042–3049.

Lennon, N. J., Kho, A., Bacskai, B. J., Perlmutter, S. L., Hyman, B. T., and Brown, R. H., Jr. (2003). Dysferlin interacts with annexins A1 and A2 and mediates sarcolemmal wound-healing. *J. Biol. Chem.* **278,** 50466–50473.

Liao, L., Dearth, R. K., Zhou, S., Britton, O. L., Lee, A. V., and Xu, J. (2006). Liver-specific overexpression of the insulin-like growth factor-I enhances somatic growth and partially prevents the effects of growth hormone deficiency. *Endocrinology* **147,** 3877–3888.

Liu, J. P., Baker, J., Perkins, A. S., Robertson, E. J., and Efstratiadis, A. (1993). Mice carrying null mutations of the genes encoding insulin-like growth factor I (Igf-1) and type 1 IGF receptor (Igf1r). *Cell* **75,** 59–72.

Liu, J., Aoki, M., Illa, I., Wu, C., Fardeau, M., Angelini, C., Serrano, C., Urtizberea, J. A., Hentati, F., Hamida, M. B., Bohlega, S., Culper, E. J., *et al.* (1998). Dysferlin, a novel skeletal muscle gene, is mutated in Miyoshi myopathy and limb girdle muscular dystrophy. *Nat. Genet.* **20,** 31–36.

Luo, L., Liao, Y. J., Jan, L. Y., and Jan, Y. N. (1994). Distinct morphogenetic functions of similar small GTPases: *Drosophila* Drac1 is involved in axonal outgrowth and myoblast fusion. *Genes Dev.* **8,** 1787–1802.

Massarwa, R., Carmon, S., Shilo, B. Z., and Schejter, E. D. (2007). WIP/WASp-based actin-polymerization machinery is essential for myoblast fusion in *Drosophila*. *Dev. Cell* **12,** 557–569.

Matsuda, C., Hayashi, Y. K., Ogawa, M., Aoki, M., Murayama, K., Nishino, I., Nonaka, I., Arahata, K., and Brown, R. H., Jr. (2001). The sarcolemmal proteins dysferlin and caveolin-3 interact in skeletal muscle. *Hum. Mol. Genet.* **10,** 1761–1766.

Millay, D. P., Maillet, M., Roche, J. A., Sargent, M. A., McNally, E. M., Bloch, R. J., and Molkentin, J. D. (2009). Genetic manipulation of dysferlin expression in skeletal muscle: Novel insights into muscular dystrophy. *Am. J. Pathol.* **175,** 1817–1823.

Naslavsky, N., Rahajeng, J., Rapaport, D., Horowitz, M., and Caplan, S. (2007). EHD1 regulates cholesterol homeostasis and lipid droplet storage. *Biochem. Biophys. Res. Commun.* **357,** 792–799.

Nolan, K. M., Barrett, K., Lu, Y., Hu, K. Q., Vincent, S., and Settleman, J. (1998). Myoblast city, the *Drosophila* homolog of DOCK180/CED-5, is required in a Rac signaling pathway utilized for multiple developmental processes. *Genes Dev.* **12,** 3337–3342.

Olswang-Kutz, Y., Gertel, Y., Benjamin, S., Sela, O., Pekar, O., Arama, E., Steller, H., Horowitz, M., and Segal, D. (2009). *Drosophila* Past1 is involved in endocytosis and is required for germline development and survival of the adult fly. *J. Cell Sci.* **122,** 471–480.

Patel, P., Harris, R., Geddes, S. M., Strehle, E. M., Watson, J. D., Bashir, R., Bushby, K., Driscoll, P. C., and Keep, N. H. (2008). Solution structure of the inner DysF domain of myoferlin and implications for limb girdle muscular dystrophy type 2b. *J. Mol. Biol.* **379,** 981–990.

Perin, M. S., Fried, V. A., Mignery, G. A., Jahn, R., and Sudhof, T. C. (1990). Phospholipid binding by a synaptic vesicle protein homologous to the regulatory region of protein kinase C. *Nature* **345,** 260–263.

Peuvot, J., Schanck, A., Lins, L., and Brasseur, R. (1999). Are the fusion processes involved in birth, life and death of the cell depending on tilted insertion of peptides into membranes? *J. Theor. Biol.* **198,** 173–181.

Piccolo, F., Moore, S. A., Ford, G. C., and Campbell, K. P. (2000). Intracellular accumulation and reduced sarcolemmal expression of dysferlin in limb-girdle muscular dystrophies. *Ann. Neurol.* **48,** 902–912.

Powell-Braxton, L., Hollingshead, P., Warburton, C., Dowd, M., Pitts-Meek, S., Dalton, D., Gillett, N., and Stewart, T. A. (1993). IGF-I is required for normal embryonic growth in mice. *Genes Dev.* **7,** 2609–2617.

Radhakrishna, H., Al-Awar, O., Khachikian, Z., and Donaldson, J. G. (1999). ARF6 requirement for Rac ruffling suggests a role for membrane trafficking in cortical actin rearrangements. *J. Cell Sci.* **112**(Pt 6), 855–866.

Rainey, M. A., George, M., Ying, G., Akakura, R., Burgess, D. J., Siefker, E., Bargar, T., Doglio, L., Crawford, S. E., Todd, G. L., Govindarajan, V., Hess, R. A., *et al.* (2010). The endocytic recycling regulator EHD1 is essential for spermatogenesis and male fertility in mice. *BMC Dev. Biol.* **10,** 37.

Roche, J. A., Lovering, R. M., and Bloch, R. J. (2008). Impaired recovery of dysferlin-null skeletal muscle after contraction-induced injury in vivo. *NeuroReport* **19,** 1579–1584.

Roche, J. A., Lovering, R. M., Roche, R., Ru, L. W., Reed, P. W., and Bloch, R. J. (2010). Extensive mononuclear infiltration and myogenesis characterize recovery of dysferlin-null skeletal muscle from contraction-induced injuries. *Am. J. Physiol. Cell Physiol.* **298,** C298–C312.

Roland, J. T., Kenworthy, A. K., Peranen, J., Caplan, S., and Goldenring, J. R. (2007). Myosin Vb interacts with Rab8a on a tubular network containing EHD1 and EHD3. *Mol. Biol. Cell* **18,** 2828–2837.

Rotem-Yehudar, R., Galperin, E., and Horowitz, M. (2001). Association of insulin-like growth factor 1 receptor with EHD1 and SNAP29. *J. Biol. Chem.* **276,** 33054–33060.

Roux, I., Hosie, S., Johnson, S. L., Bahloul, A., Cayet, N., Nouaille, S., Kros, C. J., Petit, C., and Safieddine, S. (2009). Myosin VI is required for the proper maturation and function of inner hair cell ribbon synapses. *Hum. Mol. Genet.* **18,** 4615–4628.

Ruiz-Gomez, M., Coutts, N., Price, A., Taylor, M. V., and Bate, M. (2000). *Drosophila* dumbfounded: A myoblast attractant essential for fusion. *Cell* **102,** 189–198.

Schug, N., Braig, C., Zimmermann, U., Engel, J., Winter, H., Ruth, P., Blin, N., Pfister, M., Kalbacher, H., and Knipper, M. (2006). Differential expression of otoferlin in brain, vestibular system, immature and mature cochlea of the rat. *Eur. J. Neurosci.* **24,** 3372–3380.

Selcen, D., Stilling, G., and Engel, A. G. (2001). The earliest pathologic alterations in dysferlinopathy. *Neurology* **56,** 1472–1481.

Smith, M. K., and Wakimoto, B. T. (2007). Complex regulation and multiple developmental functions of misfire, the *Drosophila melanogaster* ferlin gene. *BMC Dev. Biol.* **7,** 21.

Strunkelnberg, M., Bonengel, B., Moda, L. M., Hertenstein, A., de Couet, H. G., Ramos, R. G., and Fischbach, K. F. (2001). rst and its paralogue kirre act redundantly during embryonic muscle development in *Drosophila. Development* **128,** 4229–4239.

Sudhof, T. C., and Rizo, J. (1996). Synaptotagmins: C2-domain proteins that regulate membrane traffic. *Neuron* **17,** 379–388.

Ten Klooster, J. P., Evers, E. E., Janssen, L., Machesky, L. M., Michiels, F., Hordijk, P., and Collard, J. G. (2006). Interaction between Tiam1 and the Arp2/3 complex links activation of Rac to actin polymerization. *Biochem. J.* **397,** 39–45.

Therrien, C., Di Fulvio, S., Pickles, S., and Sinnreich, M. (2009). Characterization of lipid binding specificities of dysferlin C2 domains reveals novel interactions with phosphoinositides. *Biochemistry* **48,** 2377–2384.

Urtizberea, J. A., Bassez, G., Leturcq, F., Nguyen, K., Krahn, M., and Levy, N. (2008). Dysferlinopathies. *Neurol. India* **56,** 289–297.

Vasyutina, E., Martarelli, B., Brakebusch, C., Wende, H., and Birchmeier, C. (2009). The small G-proteins Rac1 and Cdc42 are essential for myoblast fusion in the mouse. *Proc. Natl. Acad. Sci. USA* **106,** 8935–8940.

Ward, S., Argon, Y., and Nelson, G. A. (1981). Sperm morphogenesis in wild-type and fertilization-defective mutants of *Caenorhabditis elegans. J. Cell Biol.* **91,** 26–44.

Washington, N. L., and Ward, S. (2006). FER-1 regulates Ca2+ -mediated membrane fusion during *C. elegans* spermatogenesis. *J. Cell Sci.* **119,** 2552–2562.

Yan, M., Rachubinski, D. A., Joshi, S., Rachubinski, R. A., and Subramani, S. (2008). Dysferlin domain-containing proteins, Pex30p and Pex31p, localized to two compartments, control the number and size of oleate-induced peroxisomes in *Pichia pastoris. Mol. Biol. Cell* **19,** 885–898.

Yasunaga, S., Grati, M., Cohen-Salmon, M., El-Amraoui, A., Mustapha, M., Salem, N., El-Zir, E., Loiselet, J., and Petit, C. (1999). A mutation in OTOF, encoding otoferlin, a FER-1-like protein, causes DFNB9, a nonsyndromic form of deafness. *Nat. Genet.* **21,** 363–369.

Zallen, J. A., Cohen, Y., Hudson, A. M., Cooley, L., Wieschaus, E., and Schejter, E. D. (2002). SCAR is a primary regulator of Arp2/3-dependent morphological events in *Drosophila. J. Cell Biol.* **156,** 689–701.

Zhang, D., and Aravind, L. (2010). Identification of novel families and classification of the C2 domain superfamily elucidate the origin and evolution of membrane targeting activities in eukaryotes. *Gene* **469,** 18–30.

CHAPTER NINE

Circadian Rhythms, the Molecular Clock, and Skeletal Muscle

Mellani Lefta,[1] Gretchen Wolff,[1] *and* Karyn A. Esser

Contents

Abstract

Almost all organisms ranging from single cell bacteria to humans exhibit a variety of behavioral, physiological, and biochemical rhythms. In mammals, circadian rhythms control the timing of many physiological processes over a 24-h period, including sleep-wake cycles, body temperature, feeding, and

Center for Muscle Biology, Department of Physiology, College of Medicine, University of Kentucky, Lexington, Kentucky, USA

[1] These authors contributed equally to the review.

Current Topics in Developmental Biology, Volume 96

ISSN 0070-2153, DOI: 10.1016/B978-0-12-385940-2.00009-7

hormone production. This body of research has led to defined characteristics of circadian rhythms based on period length, phase, and amplitude. Underlying circadian behaviors is a molecular clock mechanism found in most, if not all, cell types including skeletal muscle. The mammalian molecular clock is a complex of multiple oscillating networks that are regulated through transcriptional mechanisms, timed protein turnover, and input from small molecules. At this time, very little is known about circadian aspects of skeletal muscle function/metabolism but some progress has been made on understanding the molecular clock in skeletal muscle. The goal of this chapter is to provide the basic terminology and concepts of circadian rhythms with a more detailed review of the current state of knowledge of the molecular clock, with reference to what is known in skeletal muscle. Research has demonstrated that the molecular clock is active in skeletal muscles and that the muscle-specific transcription factor, *MyoD*, is a direct target of the molecular clock. Skeletal muscle of clock-compromised mice, $Bmal1^{-/-}$ and $Clock^{\Delta 19}$ mice, are weak and exhibit significant disruptions in expression of many genes required for adult muscle structure and metabolism. We suggest that the interaction between the molecular clock, *MyoD*, and metabolic factors, such as *PGC-1*, provide a potential system of feedback loops that may be critical for both maintenance and adaptation of skeletal muscle.

1. Introduction

The term Circadian comes from the Latin *circa*, "around," and *diem*, "day," meaning "about a day." Almost all organisms ranging from single cell bacteria to humans exhibit a variety of behavioral, physiological, and biochemical circadian rhythms (Albrecht and Oster, 2001; Hastings *et al.*, 2008; Merrow *et al.*, 2005). The presence of a molecular clock within a cell and/or organism provides the necessary timekeeping for anticipation of daily changes in environmental/external conditions (Albrecht, 2002; Gekakis *et al.*, 1998; Hastings *et al.*, 2008; Holzberg and Albrecht, 2003; Schibler, 2009; Takahashi *et al.*, 2008; Zhang and Kay, 2010). Synchronizing the molecular clock and intracellular physiology with external day–night cycles represents an evolutionary survival advantage for organisms (Albrecht and Oster, 2001; Holzberg and Albrecht, 2003; Oster *et al.*, 2002). While much has been learned about circadian rhythms and the molecular clock, there is still very little known about its regulation and function in skeletal muscle (Almon *et al.*, 2008; Andrews *et al.*, 2010; McCarthy *et al.*, 2007; Miller *et al.*, 2007; Zhang *et al.*, 2009). Thus, the goal of this chapter is twofold. First, we provide some fundamental background in circadian rhythms with introductions to terminology and concepts in circadian research. Second, we review what is known about the molecular clock, and when possible, incorporate research in skeletal muscle.

At this stage, this is a very new and open field of research, so much is yet to be done to mechanistically link the function of the molecular clock in skeletal muscle to known biochemical and physiological outcomes. Additionally, much is still to be done to understand the mechanisms in place to coordinate and synchronize the clocks among the diverse groups of skeletal muscle throughout the body with the central clock and other peripheral tissues.

2. Characteristics of Oscillating Systems

The term "circadian" was first used in the late 1950s, to describe a rhythm with a period length of ~24 h by Franz Halberg. This was the start of the standardization of terms for the growing research in chronobiology (Halberg *et al.*, 1959). Although the understanding and definition of many circadian terms have developed as discoveries have been made, the notion that organisms operate in a cyclic manner close to 24 h remains (Halberg, 1969). Many of the terms that will be discussed in this chapter are listed in Table 9.1.

2.1. Tau

The term tau refers to the period of a rhythm and is the length, in time, of one complete cycle. One circadian cycle is classically found to be 20–28 h in length and is based on the photoperiod or exposure to light (Halberg *et al.*, 1977). Period length can be measured for any variable that has a changing pattern, for example, oscillation or rhythm. The tau, or period, for a cycling variable is calculated by selecting one point in the cycle and measuring how long it takes to get to that same point in the next cycle. Period length is a principal measurement in chronobiology, as it illustrates a fundamental feature of time keeping, for example, how long is the cycle? Period can be measured in behaviors that cycle such as sleeping and wakefulness, locomotor activity, or eating/drinking. Many physiological, biochemical, and molecular variables are known to oscillate including body temperature, circulating cortisol levels, and tissue gene expression (mRNA/protein levels).

To identify different circadian rhythms across species of mammals, researchers commonly evaluate behaviors under constant conditions, such as total darkness. For example, strains of inbred mice demonstrate a shorter period of locomotor activity in darkness (<24 h), while humans have a longer *tau* (Czeisler *et al.*, 1980). Rats and hamsters also display a period length close to 24 h that tends to decrease as the animals age (Rosenberg *et al.*, 1991). While there are some variations in tau across animals, it is also clear that period length of cycling behaviors across many species are

Table 9.1 Terms used in circadian rhythms research

Term	Description
Amplitude	The difference between the mean of the rhythm and the peak or the trough
Circadian Time (CT)	Time as it relates to the endogenous rhythm of an organism measured under constant conditions. CT0 is defined as the onset of activity for a diurnal animal while CT12 is the onset of activity for nocturnal animals.
DD	Constant darkness
Dampening	Persistent decrease in the amplitude of a rhythm
Entrainment	When a stable phase relationship is established between an endogenous, self-sustained oscillator and an external time giver ("Zeitgeber")
Free-running	A state when there are no capable entrainment cues; the self-sustained rhythm of the organism is observed
LD	Light cycle followed by a dark cycle; 12:12 LD- 12 h of light then 12 h of dark repeating every 24 h
LL	Constant light
Masking	A change in the rhythm that does not reflect the true phase or period of the rhythm
Period	The time it takes to complete 1 cycle
Phase	The time of any point in a rhythm
Phase shift	A shift in the time of any point in a rhythm to earlier or later (phase delay, phase advance)
PRC	Phase response curve; A graphical representation of a stimuli's temporal effect on the phase of a rhythm
Zeitgeber Time (ZT)	Time as it relates to an external Zeitgeber ("Zeit = time, geber = giver"). ZT0 refers to lights on for a diumal organism, such as humans, and ZT12 defines lights off for a nocturnal organism, such as rodents

remarkably close to 24 h even after weeks without exposure to light, or in the case of blind animals, an entire lifetime (Lowe *et al.*, 1967). As will be discussed in more detail in later sections, there is much research currently on understanding the tau of the molecular clock within single cells, coupled cells, or whole tissues. In these studies, period can be measured by rhythms generated by molecular oscillators. The PERIOD2::LUCIFERASE (Per2:: Luc) mouse is an animal model used to study endogenous molecular rhythms. *mPer2*, the mammalian *Period2* gene is a core component of the molecular clock in mammals (see Section 5). Yoo *et al.* (2004) created a mouse in which the luciferase cDNA was knocked into the endogenous *mPer2* locus (Yoo *et al.*, 2004). The result is a chimeric protein that can be used to measure bioluminescence that is directly related to Period2 protein

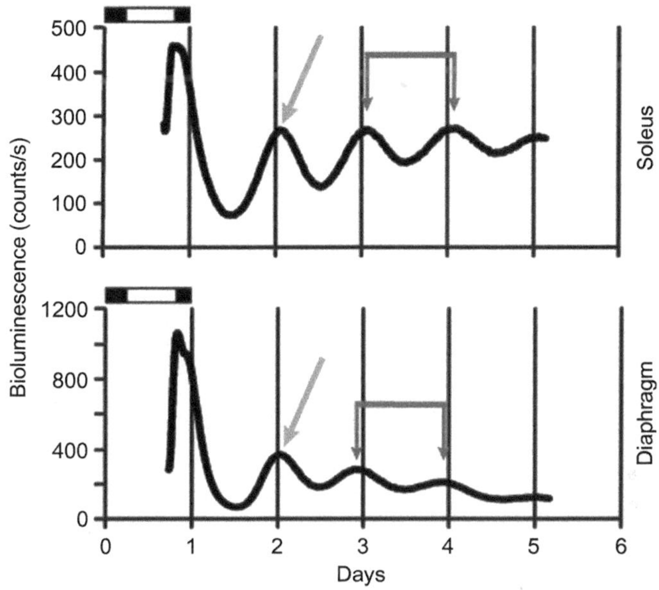

Figure 9.1 Real-time bioluminescence data from soleus and diaphragm explants using PERIOD2::LUCIFERASE mice demonstrate that different muscles within the same animal are in phase. Soleus muscle (top) and diaphragm (bottom); bioluminescence on the *y*-axis measured in (photon) counts/second for 5 days (days on *x*-axis). Dark gray double arrows indicate peak bioluminescence on days 3 and 4 for each muscle; the time between these arrows represents the period length and is 23.92 ± 0.01 (avg ± Sd dev). The lighter gray single arrows point to the first peak following 24 h in culture; this peak reflects the phase of the molecular clock and for these samples = 24.10 ± 0.80 for the soleus muscle and 23.43 ± 0.65 for the diaphragm. (Wolff and Esser, unpublished data).

accumulation and degradation in the cell. An illustration of period measured using data from the diaphragm and soleus muscles taken from Per2::Luc mice is provided in Fig. 9.1. As can be seen in this figure, the tau is the time between two peaks in luminescence as noted by the two red arrows. As can be seen in Fig. 9.1, the molecular clock in two different skeletal muscles, diaphragm and soleus, exhibits period lengths that are very similar to 23.92 ± 0.01 h (mean ± sd dev). In normal, healthy conditions, the tau of the molecular clock will be the same, or very close to the tau for behavioral rhythms.

2.2. Phase

Another principal characteristic of circadian rhythms is phase. Phase describes the timing of a particular point within a rhythm. Most of the time, researchers use the peak (highest point), trough (lowest point), or another point in the cycle for reference when discussing phase. When using phase to describe a rhythm, it is necessary to identify the point of interest, such as the onset of light or dark or the peak cortisol level. The phase of a rhythm is often used to describe its relationship to another oscillator, such as the phase of the peak of gene expression relative to the phase of activity onset or light onset. The phase of gene expression (mRNA and protein) in molecular oscillators usually maintains a stable relationship, across tissues in an animal. For example, the peak in *Per2* gene expression in the central clock in the suprachiasmatic nucleus (SCN) occurs about 4–6 h prior to the peak in *Per2* gene expression in skeletal muscles and other peripheral tissues (Yamazaki *et al.*, 2000; Yoo *et al.*, 2004). In Fig. 9.1, the phase of luminescence (blue arrow) was measured at the first peak after 24 h in culture. We have also found that the phase between different skeletal muscles is very similar, with the phase of the soleus being 24.10 $\pm$ 0.80 h versus 23.43 $\pm$ 0.65 in the diaphragm. This means that the first peak of Per2::Luc bioluminescence occurs right around midnight in both muscles, $\sim$5–6 h after the peak in the SCN. The fact that the molecular clocks between different muscles are in phase is not surprising as the molecular clocks across diverse tissues are synchronized under normal conditions. However, this is the first report comparing the phase between two different muscles of different function and developmental origin. The synchronization of this phase relationship will be discussed later (Sections 3.2. and 4).

2.3. Amplitude

Amplitude is a third principal descriptor of circadian rhythms. Amplitude is defined as a measure of the difference between the calculated mean value of a rhythm and the peak or trough (highest and lowest points, respectively). Amplitude is often used to describe the robustness of the oscillator or rhythm. Within the field of chronobiology, a robust rhythm is sometimes viewed as more responsive or more stable and this is in contrast to a weak rhythm or one that is dampened (Aschoff, 1998). Changes in amplitude have been observed in aged animals, where many investigators have reported alterations in the timing and responsiveness of endogenous and behavioral rhythms (Turek *et al.*, 1995; Wise *et al.*, 1988; Wyse and Coogan, 2010). One piece of evidence for a destabilized clock in aged animals was demonstrated by Davidson *et al.* (2006) when they used a jet-lag experimental paradigm to demonstrate that persistent advances in the onset of light could produce accelerated mortality in aged animals compared with a younger cohort (Davidson *et al.*, 2006). There are no data available on the amplitude of circadian rhythms in skeletal muscle at this time. Discussions of jet-lag and phase shifting will be provided in Section 3.3.

3. Environmental Influence on Endogenous Oscillators

The ability to synchronize an endogenous rhythm with an environmental time cue provides the animal with a biological advantage when performing daily activities (Feillet *et al.*, 2006; Halberg and Cornelissen, 1993). Environmental stimuli can affect how long the clock is running (period), what time the clock is set to in relation to other clocks (phase), and the stability of the clock (amplitude). Although it has been demonstrated that clock-driven rhythms continue to run in the experimentally controlled absence of cues from the environment (Aschoff, 1990; Gibbs, 1976; Wilson *et al.*, 1976), exogenous stimuli can profoundly impact the rhythm of endogenous oscillators, and can be used to understand how the organisms maintain a synchronous relationship with their surroundings.

3.1. Entrainment

An animal is considered entrained when its rhythm is aligned with a known external cue, such as light (Aschoff *et al.*, 1975). The rhythm can be behavioral or molecular; it is endogenous and self-sustained (Aschoff, 1967; Aschoff and Wever, 1976). Behavioral measurements such as locomotor activity or drinking activity can be used to monitor entrainment (Aschoff, 1978). In an entrained animal, the period length (average time between activity onset from one day to the next) will match the period of the environmental cue or zeitgeber ("time giver," zeitgebers will be discussed in Section 3.2). For example, under experimental conditions an animal may be housed within a 24-h light–dark (LD) cycle with 12 h of light and 12 h of darkness (LD 12:12); the lights come on every 24 h. In an entrained state, the period length of the animal's behavior will also be 24 h. The phase of the behavior will also be synchronized with the LD cycle. Phase is often measured in hours or sometimes degrees. Under entraining conditions, the phase relationship between the animal's rhythm and the LD cycle is considered stable. That means that there is no shift in the timing of the rhythm relative to that of the entraining cycle, so there are 0 h (or 0°) of difference between the phase of the rhythm and the entraining cue (Aschoff, 1960; Mrosovsky *et al.*, 1992). In the presence of a zeitgeber, such as light, the time is measured relative to the zeitgeber, thus zeitgeber time (ZT). ZT is usually measured in 24:00 h with ZT12 designated as the time of lights-off for a nocturnal animal, such as a mouse. In contrast to the entrained state, free running is defined as rhythmic behavior of an animal in the absence of a competent zeitgeber (Aschoff, 1978, 1998, 1999). Free running describes an animal's activity when the endogenous self-sustaining rhythm is in control of behavior and physiology. As in the example above of a 24-h LD cycle, if the same animal is released into constant darkness, the endogenous rhythm of the animal will persist and the period for that animal will likely not measure exactly 24 h, as described earlier in Section 2.1 (Schwartz and Zimmerman, 1990). Endogenous period length differs among species and strains (Holmes and Mistlberger, 2000; Mistlberger, 1991; Mrosovsky, 1999; Stephan, 1983). When animals are housed under conditions with constant light or darkness, the "time" is often measured relative to the endogenous circadian rhythm. Under these conditions, researchers cannot use ZT to define time since there is no set time cue for reference so they developed the terminology circadian time (CT). CT denotes the time relative to a point during the animal's behavioral rhythm. The most common point of reference for using CT is the onset of animal activity. In these conditions, activity onset is defined as CT12 for a nocturnal animal and times before or after activity onset are reported relative to CT12.

3.2. Zeitgebers

Environmental stimuli that are able to shift the timing of behavioral and molecular rhythms are critically important to understanding how oscillators can be synchronized to the environment and to each other. Figure 9.2 illustrates some of the major input and output pathways within the circadian system and the relationships that have been established or that are being evaluated by ongoing research. By far, the most comprehensively studied circadian zeitgeber is the photic time cue, light. Synchronization of the molecular clock in the SCN with the LD cycle via the retinohypothalamic tract (RHT) has been well established (Johnson *et al.*, 1988). This major input pathway (light to SCN) synchronizes behavioral rhythms as well as molecular rhythms in peripheral oscillators as demonstrated in Fig. 9.2 by the thick arrows going from the light to SCN and on to the peripheral tissues (Guo *et al.*, 2005). While the LD cycle has been studied vigorously in many species (Devlin and Kay, 2001) nonphotic zeitgebers, such as feeding and physical activity/exercise, have been shown to influence behavior and molecular rhythms (Edgar and Dement, 1991; Mistlberger, 1991; Mrosovsky, 1996).

As depicted by the thin solid lines in Fig. 9.2, changing the time of food presentation has measurable effects on the molecular clock in the liver and lung and there is evidence that specific brain regions are involved when access to food is restricted (Damiola *et al.*, 2000; Honma *et al.*, 1983; Mieda *et al.*, 2006). Several groups have demonstrated that the presentation of food at almost any time of day, for as little as 4 h or as many as 12 h, alters the rhythm of animal behavior and the molecular clock in tissues (Davidson *et al.*, 2002; Diaz-Munoz *et al.*, 2000; Escobar *et al.*, 1998; Satoh *et al.*, 2006; Stokkan *et al.*, 2001). When the animal is presented with food for a

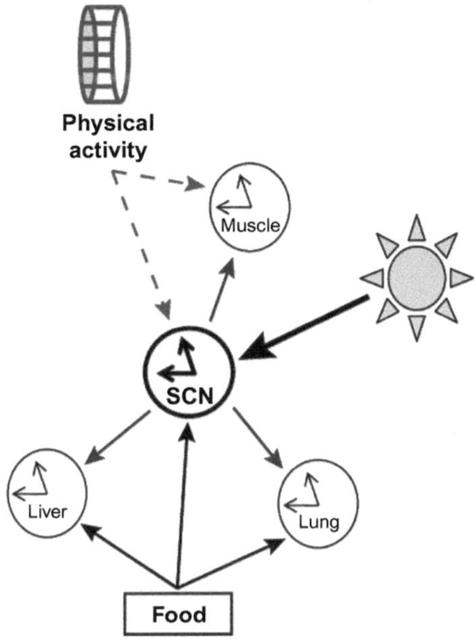

Figure 9.2 Molecular clocks within the central and peripheral tissues respond to both photic and nonphotic timing cues. This cartoon depicts the proposed interplay among time cues, the central clock, SCN, and peripheral clocks (liver/muscle/lung). The central clock, SCN, is synchronized by light (Takahashi *et al.*, 1984), and the peripheral tissues (primarily liver) have been shown to be synchronized by time of feeding (Escobar *et al.*, 1998; Honma *et al.*, 1983). The SCN is also possibly affected by time of feeding (Angeles-Castellanos *et al.*, 2007). Additionally, we suggest that physical activity, in the form of a running, may also be a time cue that acts to synchronize clocks in skeletal muscle (Yamanaka *et al.*, 2008).

restricted time, locomotor activity levels in the hours preceding the meal are greatly increased. This activity is termed food-anticipatory activity or FAA (Mistlberger, 1994; Stephan *et al.*, 1979). Evidence of FAA in the absence of the SCN suggests the existence of a central oscillator located in another region of the brain outside the SCN (Angeles-Castellanos *et al.* 2005; Hara *et al.*, 2001). The likelihood that the brain could harbor such a structure has made it a target of ongoing research with respect to restricted feeding.

Feeding is the only nonphotic time cue known in mammals but there are other nonphotic cues, such as scheduled activity and repeated treatment with methamphetamine, that have been shown to alter behavioral and molecular rhythms in animals. Scheduled bouts of activity can entrain behavioral rhythms in rodents and enhance reentrainment following phase shifts in rodents and humans (Edgar and Dement, 1991; Mrosovsky, 1996; Mrosovsky *et al.*, 1992; Yagita *et al.*, 2010; Yamanaka *et al.*, 2008). The results of some studies suggest that the molecular clock in skeletal muscle is responsive to exercise and may be entrained by scheduled bouts of physical activity; in Fig. 9.2, this is represented by the dashed line between physical activity and muscle (Yamanaka *et al.*, 2008; Zambon *et al.*, 2003). In 2008, Yamanaka *et al.* advanced the time of lights on by 8 h for 4 consecutive days while providing running wheels to a subset of mice. The mice that received the wheel synchronized their behavior with the new LD cycle more quickly than the control group. Additionally, the molecular clock in skeletal muscle was rapidly shifted to the new LD cycle (Yamanaka *et al.*, 2008). This result suggests that the molecular clock in skeletal muscle could be synchronized by a cue(s) from physical activity. The effect of repeated methamphetamine treatment on circadian rhythms is also an area of ongoing research. In this work, the most profound interaction between methamphetamines and circadian behavior was demonstrated by synchronization of locomotor rhythms in an arrhythmic animal treated with methamphetamines (Honma *et al.*, 1987; Iijima *et al.*, 2002). Animals with genetic mutations as well as SCN lesion have a similar behavioral response to methamphetamine introduction via drinking water (Hiroshige *et al.*, 1991; Masubuchi *et al.*, 2001). They show consolidated running behavior with a period length greater than 24 h (Mohawk *et al.*, 2009). Still, a methamphetamine-sensitive oscillator has not been identified, nor the method by which the drug synchronizes rhythms.

3.3. Phase response curves, phase shifts, and jet-lag

The early studies on phase were focused on the relationship between external zeitgebers (time cues) and their effect on endogenous rhythms, because if the clock (timing) is important, then understanding what synchronizes it is essential (Aschoff, 1967; Haus and Halberg, 1959). Phase shifting occurs when a particular point in the rhythm is advanced or delayed. The phase of the rhythm often refers to the time of the peak in a rhythm or the onset of a behavioral rhythm. A phase advance is a shift in the timing of the phase to occur earlier. A phase delay is a shift to a later time. In order to study the effect of different environmental stimuli on phase, the phase response curve (PRC) has been used to illustrate the behavioral response of an animal to an external time cue (Mrosovsky, 1996; Mrosovsky *et al.*, 1992). Many studies have shown that the response of the circadian system to an external cue is very complex. For example, light cues given at different times and/or durations can have alternate effects on phase of the rhythmic activity. One example of a PRC that could reflect the effect of light pulses on circadian behavior is drawn in Fig. 9.3. In this graph, time of day is along the x-axis and the amount of the phase shift, either an advance (+) or delay (−), in hours are along the y-axis. A PRC will have one peak and one nadir within a 24-h period.

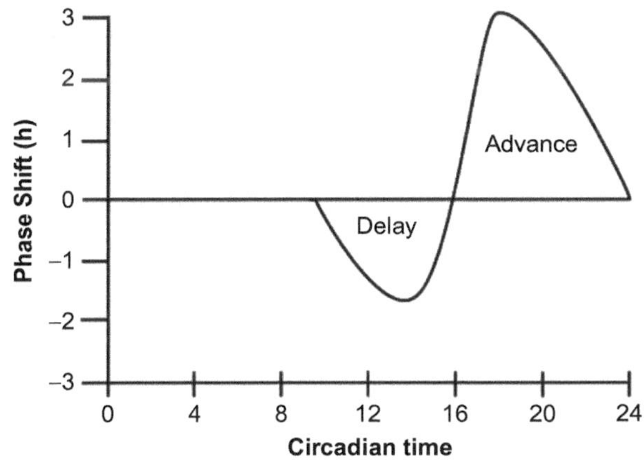

Figure 9.3 Phase response curve for a nocturnal animal in response to light pulses given at different times of day. This is an example of a phase response curve, where CT0–CT12 indicates the subjective day (light) for a nocturnal animal and CT12–CT24 is subjective night (dark). During the first 12 h (CT0–CT12) administering light pulses have no effect on phase of locomotor behavior in a nocturnal animal. In contrast, a light pulse given in the early subjective night (CT12–CT16) will produce phase delays in the onset of locomotor activity. Whereas, the same light pulse given between CT16–CT22 induces phase advances in the onset of locomotor activity. This example just provides repeatable behavioral evidence in mice demonstrating a time of day effect for the response to a defined light cue.

PRCs, such as the example in Fig. 9.3, are created when the animal is in constant conditions (DD) so that CT0–CT12 is the subjective day and CT12–CT24 is the subjective night; CT12 is the time of activity onset for a nocturnal animal. As illustrated in Fig. 9.3, light pulses have no effect on the phase of activity onset during the subjective day (through CT12), which is the time when the animal would normally be exposed to daylight. However, light pulses during the subjective night cause phase delays early on (CT12–CT16) and phase advances during the latter portion of the subjective night (CT16–CT22). The subjective night is when the nocturnal animal would normally be in darkness and active. Unlike the PRC created by photic stimuli, exposure to a running wheel (nonphotic cue) at any time during the subjective day produces a significant phase advance while the same wheel has no effect on phase during the subjective night. While there is a lot of complexity when evaluating the PRCs from different animals with different stimuli, what is clear is that a single environmental stimulus can cause either a phase advance or a phase delay depending on the time at which the animal was exposed to the stimulus. This time of day effect is a fundamental observation in circadian biology and illustrates the dynamic nature of the molecular system that underlies the behavioral biology.

It has been established that light pulses do shift phase and continual shifting produces a phenomenon known as jet-lag. When an animal's environmental cues are shifting chronically, the clock has difficulty catching up and studies have shown that over time, or under compromised conditions, the consequences of continued shifting of the clock are significant (Davidson *et al.*, 2006; Filipski *et al.*, 2002). In an article in *Current Biology*, Davidson *et al.* (2006) demonstrated that aged mice exposed to chronic phase advances of 6 h every week for 8 weeks had increased mortality compared with nonshifted controls. The effects of jet-lag are also significant when examining the clinical problems associated with working a shifted schedule (Harrington *et al.* 1990; Knutsson, 2003). Therefore, the discussion is growing around what sets behavioral and molecular rhythms and the presence of nonphotic timing cues that may be able to offset the negative effects of shift work (Angeles-Castellanos *et al.*, 2005; Mieda *et al.*, 2006).

4. Organization of the Circadian System

The presence of circadian rhythms that persist in constant conditions were the observations that led to the suggestion that circadian behaviors are driven by an organized endogenous timing system within the animal (Aschoff, 1960, 1967; Halberg *et al.*, 1965). The SCN of the hypothalamus was first suggested as the circadian pacemaker when its surgical ablation caused arrhymic behavior in rats (Stephan and Zucker, 1972). A second demonstration that the SCN was responsible for behavioral rhythms was revealed when transplantation of cells from the SCN of intact animals restored rhythms to arrhythmic SCN-lesioned animals. Further, the endogenous period length of the donor was restored in the recipient, suggesting that period length is an intrinsic property of the SCN (DeCoursey and Buggy, 1989; Lehman *et al.*, 1987; Ralph *et al.*, 1990). Additionally, it was discovered that the SCN was directly linked to the RHT and was thereby receiving timing information in the form of light, from the LD cycle (Johnson *et al.*, 1988).

A major focus of circadian research has been on the SCN and neurological control of behavioral rhythms. However, the introduction of the *Luciferase* reporter system in mice has provided scientists with the tools to ask more questions about peripheral oscillators and how the network of tissues/organs is synchronized in the animal. Firefly luciferase has been widely used in animals, mainly rodents, to measure rhythms in real-time from explanted tissues (Davidson *et al.*, 2002; Yamazaki and Takahashi, 2005; Yamazaki *et al.*, 2000; Yoo *et al.*, 2004). Using these tools has allowed for the discovery of oscillators at the cellular level in nonneuronal tissue (Izumo *et al.*, 2003; Welsh *et al.*, 2004). Once the existence of self-sustaining oscillators in peripheral tissues was established, the hypothesis rose that somehow these oscillators must be synchronized to the LD cycle through the SCN or perhaps another pacemaker. In 2005, Guo *et al.* utilized the SCN-lesioned mice with a parabiosis approach to ask whether there is some blood-borne factor that facilitates the synchronization of peripheral tissues to the pacemaker in the SCN. SCN-lesioned mice were surgically united with SCN intact mice and the shared circulation provided for a humoral exchange between the two mice. If the SCN was controlling the secretion of a synchronizing agent, the molecular rhythms in the arrhythmic mice would be restored. They found that some peripheral tissues (liver, kidney) were resynchronized in the SCN-lesioned animals, but others were not (heart, skeletal muscle; Guo *et al.*, 2005). This data suggests that there must be other nonhumoral pathways, such as innervation, by which the peripheral oscillators are synchronized within the animal to the environmental time cues.

The presence of circadian oscillators in all cell types within the body has lead to a divergence of interesting work within the field. The reality that SCN lesions produce behavioral and molecular arrhythmicity, suggested its role as dominant pacemaker and has maintained the current hierarchical view of a master/slave relationship between the pacemaker in the SCN and the peripheral oscillators which it controls. However, there is growing evidence that peripheral clocks can be synchronized independently of the SCN and that nonphotic timing cues may alternatively entrain peripheral clocks through pathways unknown at this time (Castillo *et al.*, 2004; Hara *et al.*, 2001; Tahara *et al.* 2010).

5. The Organization of the Mammalian Molecular Clock

The molecular circadian clock is a genetically based mechanism inherent to each mammalian cell type, including skeletal muscle cells. The circadian clock generates cell-autonomous and self-sustaining rhythms, which prepare the cell to anticipate and adapt to exogenous stimuli (Grundschober *et al.*, 2001; Panda *et al.*, 2002; Storch *et al.*, 2002; Yamazaki *et al.*, 2000; Zambon *et al.*, 2003). The generation of circadian rhythms is driven by a series of interconnected positive and negative transcriptional/translational feedback loops, sometimes referred to as TTFL (Zhang and Kay). Interwoven with these autoregulatory feedback loops are posttranslational modifications of core clock proteins, which govern protein stability and degradation and modulate the length of the circadian period (tau; Lee *et al.*, 2001). The circadian clock is a stable and redundant network; multiple loops work in concert to generate rhythms;

within each loop, core clock members have functional homologues that complement each other; and several pathways of transcriptional, translational, and posttranslational control are in place to compensate for each other and ensure the stability and accuracy of the molecular clock (Lowrey and Takahashi, 2004; Shearman *et al.*, 2000). While individual components are different, the molecular clock mechanism is evolutionary conserved from cyanobacteria to humans, confirming its importance in normal physiology and behavior (Grundschober *et al.*, 2001; Harmer *et al.*, 2001; Panda *et al.*, 2002; Rosbash, 2009; Storch *et al.*, 2002).

5.1. The core molecular clock

Using a forward genetics approach, the *Clock* gene (Circadian Locomotor Output Cycles Kaput) was the first component of the molecular clock to be identified in mammals. Mice homozygous for the *Clock* mutation (*Clock*$^{\Delta 19}$) have significantly longer circadian periods (tau ~27.3 h) and become behaviorally arrhythmic when housed in constant dark conditions (Lee *et al.*, 2001; Vitaterna *et al.*, 1994). The CLOCK protein plays a fundamental role in circadian rhythm generation; it defines not only tau but also rhythm persistence under constant conditions. The CLOCK$^{\Delta 19}$ protein lacks the transactivation domain and acts in a dominant negative fashion, as the circadian phenotype of the heterozygous mutant is less severe than that of a hemizygous (one mutated copy of *Clock*, one null allele; King *et al.*, 1997).

The *Clock*$^{\Delta 19}$ phenotype was rescued by transgenic expression of a bacterial artificial chromosome harboring the wild-type *Clock* gene, further supporting a functional role of *Clock* as an integral component of the molecular clock. In addition, CLOCK was found to be a member of the bHLH-PAS (basic Helix Loop Helix-Period Arnt Single-minded) family of transcription factors. These factors work as dimers through binding via their HLH-PAS domains, suggesting that CLOCK binds to another bHLH-PAS protein to perform its function as a regulator of circadian rhythms (Antoch *et al.*, 1997; Kewley *et al.*, 2004).

The bHLH-PAS protein BMAL1 (Brain and Muscle ARNT Like protein 1) was identified in a yeast two hybrid screen as a binding partner for CLOCK. The *Bmal1* transcript is coexpressed with *Clock* in the SCN and retina, implicating *Bmal1* as another component of the molecular clock (Gekakis *et al.*, 1998). The knockout of *Bmal1* was generated in mice by targeted deletion of the bHLH domain. The *Bmal1*$^{-/-}$ mice were arrhythmic in behavior and these findings confirmed that BMAL1 is a central and nonredundant component of the molecular clock (Bunger *et al.*, 2000).

Dimerization of CLOCK to BMAL1 via their HLH-PAS domains results in translocation of the heterodimer to the nucleus, where it binds to an E-box sequence (CACGTG) in the promoter regions of the other components of the core clock, such as *Period* (*Per1* and *Per2*) and *Cryptochrome* (*Cry 1, Cry 2,* and *Cry 3*) genes. CLOCK and BMAL1 activate transcription of *Per* and *Cry* and are referred to as the positive arm of the molecular clock (Gekakis *et al.*, 1998; Hogenesch *et al.*, 1998; Kume *et al.*, 1999).

BMAL1 availability is the rate-limiting factor for the CLOCK:BMAL1 complex formation, nuclear translocation, and transcriptional activation. The BMAL1 protein accumulates with a circadian profile in both the SCN and peripheral tissues (Lee *et al.*, 2001). Although total CLOCK protein levels do not oscillate in the SCN (Gekakis *et al.*, 1998), there is a clear circadian oscillation in nuclear to cytoplasmic distribution of CLOCK both *in vitro* and *in vivo*. This is disrupted in *Bmal1*$^{-/-}$ mice, suggesting that BMAL1 is necessary for rhythmic accumulation of the CLOCK protein in the nucleus and hence the rhythmicity of CLOCK:BMAL1-dependent transcriptional activity (Kondratov *et al.*, 2003).

CLOCK:BMAL1-mediated transcription of *Per* and *Cry* leads to their protein accumulation in the cytoplasm. PER and CRY proteins form multimeric complexes with each other and with casein kinase 1ε and translocate to the nucleus to inhibit the transcriptional activity of the CLOCK:BMAL1 heterodimer (Eide *et al.*, 2002; Kume *et al.*, 1999; Lee *et al.*, 2001; Sangoram *et al.*, 1998). PER is responsible for regulation of nuclear entry of the complex, as shown by lack of nuclear CRY accumulation in *Per1*$^{-/-}$*Per2*$^{-/-}$ mutant mice (Lee *et al.*, 2001). CRY is primarily responsible for the inhibitory actions of the complex. CRY inhibits the histone acetyltransferase p300, leading to a decrease in CLOCK:BMAL1-mediated transcription (Etchegaray *et al.*, 2003). The CRY:PER-mediated repression of the CLOCK:BMAL1 activity constitutes the negative feedback component of the molecular clock loop.

The CLOCK:BMAL1-CRY:PER loop is referred to as the central autoregulatory feedback loop due to the strong inhibition that PER and CRY impose on their own transcription. Additional transcriptional/translational feedback loops exist to ensure the robustness and fidelity of the molecular clock. One such loop involves members of the retinoic acid-related orphan nuclear receptor family of transcription factors, RevErbα and RORα. Transcription of RevErbα and RORα is enhanced by CLOCK:BMAL1-binding at E-box sequences in their promoter regions (Preitner *et al.*, 2002; Sato *et al.*, 2004). In turn, accumulation of RevErbα protein inhibits the expression of *Bmal1* (Preitner *et al.*, 2002). RORα on the other hand, activates the expression of *Bmal1* (Sato *et al.*, 2004). Both RevErbα and RORα exert their function by binding to retinoic acid-related orphan nuclear receptor response elements (RORE sequences) in the promoter region of *Bmal1*. Other members of the ROR and RevErb families have been found to have similar functions in the molecular clock (Guillaumond *et al.*, 2005).

These autoregulatory transcriptional/translational feedback loops drive the rhythmic expression of core clock components, with the positive regulators oscillating antiphase to the negative regulators. Figure 9.4 shows a simplified version of the temporal events of circadian rhythm generation in rodent peripheral tissues. There is a 0–6-h delay between the mRNA expression and protein abundance rhythms, therefore transcription and translation alone cannot account for the generation of 24-h circadian rhythms. Delays are imposed on the system by posttranslational modifications controlling protein–protein interactions, nuclear localization, and protein degradation, as discussed in the following section (Harmer *et al.*, 2001; Lee *et al.*, 2001; Lowrey and Takahashi, 2004; Panda *et al.*, 2002; Shearman *et al.*, 2000).

5.2. Posttranslational modifications as modulators of the circadian period

Transcription and translation of core clock components plays a critical role in rhythm generation, whereas delays imposed by posttranslational modifications (PTMs) are important for determining the period (tau) of circadian rhythms (Dey *et al.*, 2005; Godinho *et al.*, 2007; Iitaka *et al.*, 2005; Lee *et al.*, 2001; Lowrey *et al.*, 2000; Nakahata *et al.*, 2008; Sanada *et al.*, 2004; Siepka *et al.*, 2007; Vanselow *et al.*, 2006). PTMs play essential roles in regulating nuclear localization, protein stability, and protein degradation (Eide *et al.*, 2002, 2005; Lowrey *et al.*, 2000; Sahar *et al.*, 2010; Shirogane *et al.*, 2005; Vielhaber *et al.*, 2000; Yin *et al.*, 2010). PTMs are carried out by enzymes that not only respond to environmental stimuli such as nutrient availability but also control a wide range of cellular processes outside the molecular clock (Etchegaray *et al.*, 2003; Lee *et al.*, 2001; Nakahata *et al.*, 2008; Rayasam *et al.*, 2009; Sanada *et al.*, 2002). On one hand, this indicates that the molecular clock tau can potentially be influenced by the environment, as modulated in part by the actions of PTM enzymes. On the other hand, while the environment might modulate tau, the significant overlap in the system of PTMs, is an indicator that several mechanisms are operating to insure the fidelity of the molecular clock and that multiple signals are needed to disturb molecular clock function. In this section, we will briefly discuss what is currently known about three types of posttranslational modifications imposed on core clock proteins: phosphorylation, acetylation, and ubiquitination and their effect on the molecular clock function.

5.2.1. Phosphorylation of clock proteins

Several kinases work together to phosphorylate molecular clock components. Phosphorylation has different effects on the circadian period depending on the time of day, the kinase involved, and the proteins being phosphorylated. Phosphorylation modulates the period of circadian rhythms by controlling the start, duration, and termination of both the

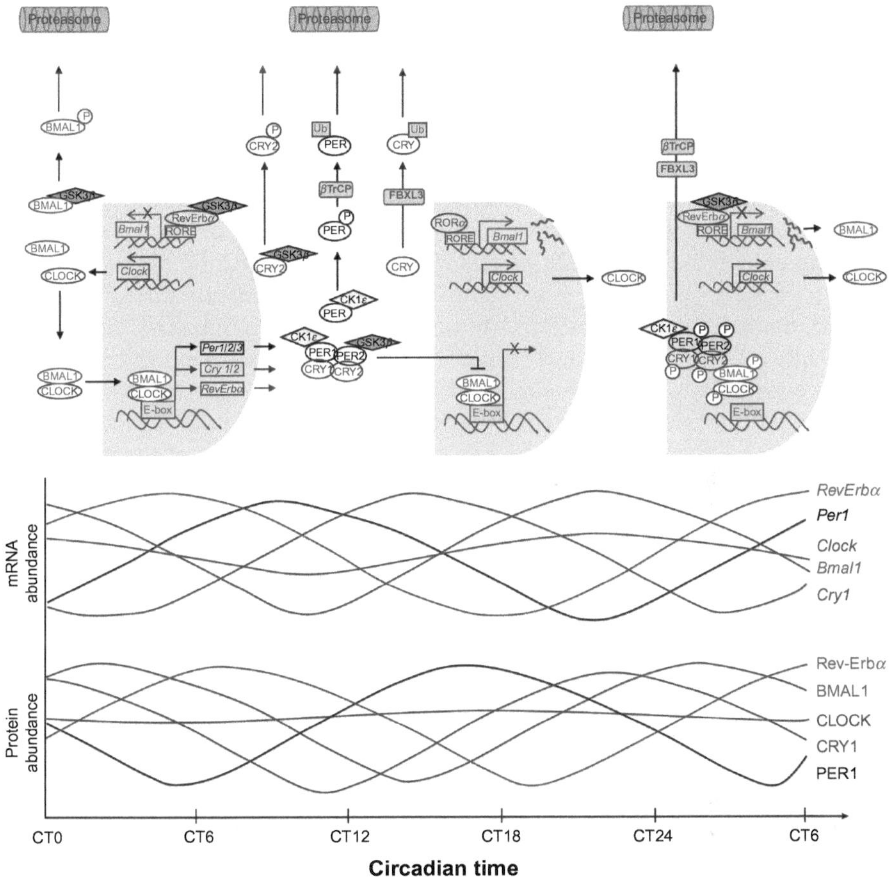

Figure 9.4 Mammalian molecular clock and oscillation of mRNA and protein in peripheral tissues. The molecular clock is regulated by interacting transcription–translation feedback loops. As depicted in this figure, the expression of components of the molecular clock are determined by the balance of synthetic versus degradative processes. Specifically, the circadian cycle starts with the accumulation of the BMAL1 protein in the cytoplasm (CT15–CT3). As BMAL1 accumulates to sufficient levels in the cytoplasm, it forms heterodimers with CLOCK, which is constitutively present over the circadian cycle. Nondimerized BMAL1 is targeted by the kinase GSK3β, for phosphorylation and subsequent degradation. Upon dimerization, the CLOCK: BMAL1 heterodimer translocates to the nucleus where it binds to E-box sequences in the promoter region of the negative regulator genes (*Per, Cry, RevErba*) and induces their transcription. This leads to peaks in the mRNA levels of *Per1, Per2, Per3, Cry1, Cry2,* and *RevErba* between CT2 and CT17 followed by peaks in their protein levels 4–6 h later. In the cytoplasm, the negative regulators form multimeric complexes that translocate to the nucleus to inhibit the transcriptional activity of the CLOCK:BMAL1 heterodimer. The kinetics of this interaction depends on the availability of the negative regulators. Free CRY and PER are targeted by GSK3β and CK1ε respectively for phosphorylation and subsequent degradation via the proteasomal pathway. At around CT15–CT20, RORα binds to RORE elements in the promoter region of *Bmal1* and

activating and repressing phases of the molecular clock mechanism. During the early subjective day in peripheral tissues, phosphorylation of the positive components leads to their degradation via the 26S proteasomal pathway, delaying the onset of positive transcriptional activation (Sahar *et al.*, 2010; Spengler *et al.*, 2009; Yin *et al.*, 2006). Toward the end of the subjective day, phosphorylation of the negative regulators inhibits their nuclear entry and primes them for degradation. This keeps the negative regulators out of the nucleus and unable to perform their repressive function, imposing a delay in the molecular clock (Akashi *et al.*, 2002; Eide *et al.*, 2005; Harada *et al.*, 2005; Iitaka *et al.*, 2005; Shirogane *et al.*, 2005; Vielhaber *et al.*, 2000). Accumulation of the kinases in the nucleus at the beginning of the subjective night phosphorylates and enhances the repressor function of the inhibitory complex (Eide *et al.*, 2002; Sanada *et al.*, 2002, 2004). Hyperphosphorylation of the inhibitory complex at the end of the circadian night leads to degradation of the complex and termination of the repressive state (Lee *et al.*, 2001). Here, we will review the major effects of three important kinases on the molecular clock: casein kinase 1 epsilon, glycogen synthase kinase 3 beta, and mitogen-activating protein kinase.

5.2.1.1. Casein kinase 1 epsilon Casein kinase 1 epsilon (CK1ε) is a member of the serine/threonine family of protein kinases, known to phosphorylate a broad range of substrates (Chergui *et al.*, 2005; Klimowski *et al.*, 2006). CK1ε was first implicated in circadian rhythms when a point mutation (Arg178Cys) in its catalytic region was found to be the basis of the *tau* mutation in the short circadian period hamster (Lowrey *et al.*, 2000; Ralph and Menaker, 1988).

CK1ε does not show circadian rhythms in abundance or activity *in vivo* (Lee *et al.*, 2001). CK1ε targets members of both the positive and negative limbs of the molecular clock and primarily controls the duration of the repressor state. At the end of the circadian day, CK1ε binds to and phosphorylates PER1, inducing a conformational change hindering the nuclear localization signal. This inhibits nuclear entry of PER1 and leads to its cytoplasmic accumulation and subsequent degradation. By keeping PER1 out of the nucleus, CK1ε delays the PER1-mediated repression of the CLOCK:BMAL1-dependent transcription. CK1ε controls PER1

induces *Bmal1* transcription. This allows for the accumulation of *Bmal1* mRNA with a peak at around CT21. Meanwhile, CK1ε phosphorylates all members of the PER/CRY complex as well as CLOCK and BMAL1. This hyperphosphorylated complex is shuttled to the proteasome for degradation between CT21 and CT24, freeing the E-box site for another round of CLOCK:BMAL1 binding. The delay imposed by transcription, translation, and posttranslational modifications is what defines the period of circadian rhythms (Lee *et al.*, 2001; Lowrey and Takahashi, 2004; Sahar *et al.*, 2010; Shearman *et al.*, 2000; Ueda *et al.*, 2002). (See Color Insert.)

availability in the cytoplasm, which is a rate-limiting step for inhibitory complex formation (Vielhaber *et al.*, 2000). In addition, CK1ε phosphorylates the other members of the PER family (PER2 and PER3) and primes them for ubiquitination and degradation (Akashi *et al.*, 2002; Eide *et al.*, 2005; Shirogane *et al.*, 2005). At the beginning of the circadian night, CRY1 protein levels begin to rise; CRY1 binds to the PER:CK1ε complex in the cytoplasm, it promotes nuclear entry of the CRY1: PER:CK1ε complex and increases nuclear localization of CK1ε (Eide *et al.*, 2002). Once in the nucleus, the CRY1: PER:CK1ε complex binds to and inhibits the transcriptional activity of the CLOCK:BMAL1 heterodimer. Following the negative regulation, the CLOCK:BMAL1: PER:CRY complex becomes hyperphosphorylated by CK1ε and the entire complex is shuttled to degradation pathways at the end of the circadian night. This ends the repressive state, opens up the E-box for another round of CLOCK: BMAL1-binding, and signals the beginning of a new circadian cycle (Lee *et al.*, 2001). Through this series of events, CK1ε controls the beginning, duration, and termination of the repressor state of the molecular clock.

Further, nuclear CK1ε increases BMAL1 phosphorylation, which in turn enhances CLOCK:BMAL1-mediated transcription *in vitro*. Phosphorylation of BMAL1 by CK1ε is not dependent on the presence of the PER protein, suggesting that CK1ε phosphorylates BMAL1 at a time when the inhibitory complex is not bound to the CLOCK:BMAL1 heterodimer (Eide *et al.*, 2002). Hence, CK1ε also controls the length of the transcriptional activation state of the molecular clock.

Phosphorylation of core clock proteins by CK1ε controls their stability, subcellular localization, and protein turnover and has an effect on the period of circadian rhythms. This is supported by studies in both hamster and humans in which the mutations of either CK1ε or PER2 affect period. As mentioned above, in the *tau* mutant hamster, CK1ε is unable to phosphorylate the PER proteins. Hypophosphorylated PER is translocated to the nucleus, where its accumulation leads to earlier repression of the CLOCK: BMAL1-mediated transcription, shortening the circadian period to 20 h (Lowrey *et al.*, 2000). Although the circadian expression profile of *Per1* and *Per2* is not altered in the SCN of *tau* mutants, the nuclear PER protein levels decline more rapidly. The authors suggest that the repressive phase of the cycle is shorter, leading to a shorter circadian cycle (Dey *et al.*, 2005). In humans, familial advance sleep phase syndrome (FASPS) is a condition in which people go to sleep earlier and earlier each day. Studies of these individuals determined that they have a very short endogenous period of $\sim$20 h versus $>$24 h in most humans. This is due to a mutation in PER2 that affects a phosphorylation site (target of CK1ε). This leads to a decrease in PER2 protein stability, premature nuclear clearance of PER2 containing complexes, early termination of transcriptional repression, and a subsequent shortening of the circadian period (Vanselow *et al.*, 2006).

5.2.1.2. Glycogen synthase kinase 3 beta Glycogen synthase kinase 3 beta (GSK-3β), is another member of the serine/threonine family of protein kinases, known to phosphorylate components of a wide variety of processes, such as glycogen synthesis, cell proliferation, embryogenesis, axon growth, and cardiomyocyte hypertrophy (Rayasam *et al.*, 2009).

GSK-3β phosphorylation (which is indirectly correlated to its activity level) oscillates with a circadian profile in both SCN and liver. In the liver, GSK-3β is active throughout the circadian cycle, with peak activity at the end of circadian night and beginning of circadian day. During these times, GSK-3β is responsible for phosphorylation of the positive loop members, determining the onset of transcriptional activation (Sahar *et al.*, 2010; Yin *et al.*, 2006). During the late circadian day and early circadian night, GSK-3β phosphorylates members of the negative loop and modulates the start of transcriptional repression (Harada *et al.*, 2005; Iitaka *et al.*, 2005).

During the early circadian day, GSK-3β targets BMAL1 for phosphorylation, followed by subsequent ubiquitination and degradation via the proteasomal pathway. A mutated GSK-3β increases BMAL1 protein levels and dampens BMAL1 protein oscillations. This suggests that GSK-3β plays a critical role in the regulation of cytoplasmic BMAL1 abundance, which is a rate-limiting step in the CLOCK:BMAL1 heterodimer formation, nuclear translocation, and transcriptional activation (Sahar *et al.*, 2010). Further, GSK-3β phosphorylates CLOCK and primes it for degradation. GSK-3β can only phosphorylate CLOCK after a BMAL1-dependent priming phosphorylation event at a different residue on CLOCK. A CLOCK mutant lacking the phosphorylation site reduces CLOCK degradation and enhances the CLOCK:BMAL1 transcriptional activity, leading to a delay in the peak of *Per1* transcript and a lengthening of the circadian period (Spengler *et al.*, 2009). At the same time, GSK3β phosphorylates and stabilizes the RevErbα protein by preventing its degradation by the proteasome. Stabilized RevErbα inhibits *Bmal1* transcription at a time when BMAL1 protein is accumulating in the cytoplasm (Yin *et al.*, 2006). GSK-3β imposes a necessary delay step in the clock, between BMAL1 protein accumulation at the end of the circadian night and CLOCK:BMAL1 transcriptional activation during early-mid circadian day.

During the late circadian day–early circadian night, GSK-3β phosphorylates members of the negative limb, but has opposing effects on their activity, depending on the member involved. GSK-3β interacts with PER2 in the PER:CRY:CK1ε complex, enhances its nuclear entry, and triggers the start of transcriptional repression. This is supported by the finding that GSK-3β inhibition through pharmacological agents leads to the cytoplasmic accumulation of PER2, a delay in the start of the transcriptional repression state and hence a lengthening of the circadian period *in vitro* (Iitaka *et al.*, 2005). In addition, GSK-3β phosphorylates CRY2 and targets it for proteasomal degradation *in vivo*. Phosphorylation of CRY2 by GSK-3β requires prior phosphorylation by MAPK at a different serine residue showing that both kinases must work in concert to mediate CRY degradation (Harada *et al.*, 2005; Sanada *et al.*, 2004). By mediating CRY2 degradation, GSK-3β delays inhibitory complex formation and the start of the repressive state, lengthening the circadian period. This finding might explain the shortening of the circadian period length observed *in vitro* when GSK-3β is inhibited by siRNA interference or by small molecules (Hirota *et al.*, 2008). Depending on the availability of PER2 and CRY2, GSK-3β can either delay or speed up the molecular clock by regulating the time of onset of transcriptional repression. By regulating the onset of both transcriptional activation and transcriptional repression, GSK-3β can modulate the period length of circadian rhythms.

5.2.1.3. ***Mitogen activated protein kinase*** Mitogen-activated protein kinase (MAPK) is the largest subfamily of the serine/threonine family of protein kinases. They participate in many signal transduction pathways with roles in protein turnover, cell growth, transcription factor activation, chromatin modification, and gene expression (Cuadrado and Nebreda, 2010).

MAPK exhibits circadian rhythms of activity *in vivo* (Obrietan *et al.*, 1998) and has been shown to phosphorylate members of both the positive and negative loops of the molecular clock. This activity contributes to regulation of the termination of transcriptional activation as well as the duration of transcriptional repression. MAPK binds to and phosphorylates BMAL1 in the nucleus and negatively regulates CLOCK:BMAL1-mediated transcription (Sanada *et al.*, 2002). This leads to the termination of the CLOCK:BMAL1-mediated transcriptional activation phase during the late circadian day–early circadian night.

During the circadian night, MAPK phosphorylates both CRY1 and CRY2 and enhances their repression of the CLOCK:BMAL1 heterodimer, lengthening the period of circadian rhythms. This is supported by the finding that mutations in the MAPK phosphorylation sites of CRY lead to a decrease in the CRY repressor function on the CLOCK:BMAL1-mediated transcription. This could be due to the reduced affinity of CRY for the CLOCK: BMAL1 dimer. The authors propose that MAPK functions to lengthen the period of negative regulation during the subjective night. This is a time when protein levels of the negative regulators are declining, yet their mRNA levels are still low. By regulating both the positive and negative arms of the molecular clock, it is suggested that MAPK imposes the appropriate time lag for the generation of 24-h rhythms (Sanada *et al.*, 2002, 2004).

5.2.2. Acetylation/deacetylation of clock proteins

Acetylation and deacetylation are important enzymatic reactions that control gene expression via chromatin remodeling. Acetylation of histones unfolds chromatin to expose promoter regions to the transcription machinery and is associated with activation of gene expression, whereas deacetylation leads to silencing of gene expression (Etchegaray *et al.*, 2003; Nakahata *et al.*, 2008). In addition to histones, histone acetyltransferases (HAT) and histone deacetylases (HDAC) target members of the molecular clock, modulate their expression, and have an effect on the circadian period length (Asher *et al.*, 2008; Doi *et al.*, 2006; Etchegaray *et al.*, 2003; Nakahata *et al.*, 2008).

The histone acetyltransferase, p300, associates with CLOCK in a circadian manner, suggesting that p300 is a member of the transcription activating complex. Consistent with this, H3 histone acetylation at the promoter region of *Per1* and *Per2* shows robust circadian rhythms in phase with the *Per1*/*Per2* mRNA and p300: CLOCK complex formation. Further, p300 is inhibited by the CRY protein leading to a decrease in CLOCK:BMAL1-mediated transcription at the *Per1* promoter, suggesting that the repressor action of CRY is mediated in part by its actions on chromatin structure (Etchegaray *et al.*, 2003). The histone deacetylase *SirtT1* targets histone H3, leading to chromatin condensation, hindrance of the promoter sites, and gene silencing. SirtT1 deacetylase activity shows a robust circadian rhythm antiphase to the rhythm of histone H3 acetylation (Nakahata *et al.*, 2008). This antiphase oscillation of HAT and HDAC activity implies that acetylation/deacetylation are important mechanisms in controlling clock gene expression and that they have roles in the initiation, duration, and termination of both the activating and repressing phases of the circadian cycle.

In addition to its role as a transcription factor, it has been shown that CLOCK protein can function as a histone acetylase (Doi *et al.*, 2006) capable of acetylating its binding partner, BMAL1. Dimerization of CLOCK to BMAL1 is critical for the acetylation of BMAL1 at a specific lysine residue in the carboxy terminal. BMAL1 acetylation by CLOCK is rhythmic over the circadian cycle in mouse liver and increases CRY binding to and repression of the CLOCK:BMAL1 heterodimer (Hirayama *et al.*, 2007). Thus acetylation is important for determining the time of transcriptional repression initiation. SirT1 also binds to the CLOCK:BMAL1 dimer in the promoter regions of circadian genes and deacetylates BMAL1, inducing gene expression silencing. This is consistent with the finding that $SirT1^{-/-}$ mice show increases in the amplitude and length of circadian gene rhythms in the liver. Further, SirT1 deacetylates PER2 protein in the PER:CLOCK:BMAL1 complex and contributes to its degradation (Asher *et al.*, 2008; Nakahata *et al.*, 2008).

Recent work has placed SirT1 in the crossroad between the molecular clock and cellular metabolism. This role of SirT1 will be explored further in Section 6.2. The interplay between acetylases and deacetylases is of critical importance, assuring that molecular clock components are activated/silenced appropriately in a time of day-dependent manner.

5.2.3. Ubiquitination of clock proteins

As mentioned above, phosphorylation of most clock components primes them for degradation via the 26S proteasomal pathway. An intermediate and necessary step for degradation via the proteasome is the labeling of proteins with ubiquitins, mediated by the action of ubiquitin ligases. Ubiquitin ligases preferentially target members of the negative limb of the molecular clock and control their protein turnover by promoting their degradation (Busino *et al.*, 2007; Eide *et al.*, 2005; Yin *et al.*, 2010). Ubiquitin ligases increase the transcriptional activity of the CLOCK: BMAL1 heterodimer and regulate the initiation and termination of both the transcriptional activation and transcriptional repression states and hence the period of circadian rhythms (Godinho *et al.*, 2007; Shirogane *et al.*, 2005; Siepka *et al.*, 2007). Here, we will briefly discuss what is currently known about the roles of four main ubiquitin ligases in the circadian system: β-TrCP, Fbxl3, Arf-bp1, and Pam.

β-TrCP is a ubiquitin ligase adapter protein, member of the SCF (Skp1/cullin/F-box) E3 ubiquitin ligase complex. β-TrCP recognizes the phosphorylated form of PER2. Phosphorylation of PER2 by CK1ε exposes a binding site for β-TrCP on PER2. This is followed by polyubiquitination and subsequent degradation (Eide *et al.*, 2005). The two isoforms of β-TrCP (β-TrCP1 and β-TrCP2) bind to phosphorylated PER1 on a region different from the CK1ε binding site and lead to its degradation. Knockdown of β-TrCP leads to stabilization of the PER1 protein and to a decrease in CLOCK:BMAL1-mediated transcription (Shirogane *et al.*, 2005).

The orphan F-box protein, Fbxl3, is another member of the SCF E3 ubiquitin ligases, which is found to bind to CRY1 and CRY2 and decrease their half-life. Knocking down Fbxl3 or silencing it by small hairpin RNAs, stabilizes the CRY proteins and increases their half-lives. This leads to repression of CLOCK:BMAL1 dependent transcription as shown by a decrease in oscillation and expression of *Per1*, *Per2*, and *Cry1*. Fbxl3 polyubiquitinates the CRY proteins and shuttles them to the proteasome (Busino *et al.*, 2007). Consistent with this, a mutation in Fbxl3 which decreases CRY2 binding increases the circadian period to 27 h (Godinho *et al.*, 2007). A second Fbxl3 mutation leading to its loss of function in a murine model is associated with decreased CRY1 degradation, increased CRY1 protein stability, longer repression of CLOCK:BMAL1-dependent gene transcription, and a longer circadian period of over 26 h (Siepka *et al.*, 2007). Fbxl3 thus plays a role in regulating circadian period length by controlling the start and end times of both transcriptional activation and transcriptional repression.

Two additional members of the E3 ligase family, Arf-bp1 and Pam, were recently found to associate with RevErbα and mediate its ubiqitination and degradation. Inhibiting Arf-bp1 and Pam via small interfering RNAs leads to stabilization of RevErbα protein. Knocking down Arf-bp1 and Pam altered the amplitude of RevErbα oscillations and suppressed both *Bmal1* and CLOCK:BMAL1-dependent gene expression. This suggests that Arf-bp1 and Pam can modulate the molecular clock by regulating RevErbα protein stability and thus the duration of the inhibitory actions of RevErbα on *Bmal1* transcription (Yin *et al.*, 2010).

Ubiquitin ligases control the protein stability of the negative regulators of the molecular clock. By promoting PER and CRY degradation at the end of the circadian day, ubiquitin ligases lengthen the transcriptional activation state and delay the start of the transcriptional repression state. Ubiquitination of PER and CRY at the end of the circadian night facilitates inhibitory complex degradation and signals the termination of the repressing state, allowing CLOCK:BMAL1 to start a new transcription cycle (Busino *et al.*, 2007; Eide *et al.*, 2005; Yin *et al.*, 2010). Thus, ubiquitin ligases serve as a balance between the activating and repressing states of the molecular clock oscillator.

In summary, posttranslational modifications of core clock proteins are important for introducing necessary delays in the clock mechanism to ensure 24-h oscillations. PTMs control protein stability, nuclear localization, and protein degradation, which in turn affect protein activity and the timing of molecular clock transcriptional activation and transcriptional repression. PTMs contribute to the fidelity of the molecular clock and provide fine tuning of the core clock mechanism.

6. The Molecular Clock in Skeletal Muscle

Several studies have documented that the molecular clock is present and functional in skeletal muscle, however, not much is known about its function in skeletal muscle physiology and pathology (Yamazaki *et al.*, 2000; Zambon *et al.*, 2003; Zylka *et al.*, 1998). Expression profiling determined that ~7% of skeletal muscle transcriptome is expressed in a circadian manner, including genes involved in protein metabolism, transcription, cytoskeletal organization, and signaling (McCarthy *et al.*, 2007). This suggests that the molecular clock is involved in some aspects of skeletal muscle physiology and is supported by the finding that the *Clock* $^{\Delta 19}$ skeletal muscle shows a significant loss or phase shifting in the circadian rhythms of genes involved in structural organization, contractile performance, and metabolism (McCarthy *et al.*, 2007).

In this section, we will discuss what is currently known about the role of the molecular clock in skeletal muscle. The molecular clock regulates the rhythmic expression of skeletal muscle-specific clock control genes, it has functions in skeletal muscle metabolism, and when aberrant, it leads to skeletal muscle pathologies (Andrews *et al.*, 2010; Bae *et al.*, 2006; Fulco *et al.*, 2003; Kennaway *et al.*, 2007; Kondratov *et al.*, 2006; Liu *et al.*, 2007; McCarthy *et al.*, 2007; McDearmon *et al.*, 2006; Vieira *et al.*, 2008).

6.1. MyoD as a clock-controlled gene in skeletal muscle

Clock-controlled genes are genes outside the core clock, whose rhythmic expression is driven by the activity of the CLOCK:BMAL1 heterodimer in their promoter regions (Lowrey and Takahashi, 2004). Recent work from our laboratory has identified *MyoD* (myogenic determination factor 1) as a skeletal muscle-specific clock controlled gene. The CLOCK:BMAL1 heterodimer binds to core enhancer (CE) sequences in the promoter region of *MyoD* and drives its transcription (Andrews *et al.*, 2010). This is consistent with expression profiling data showing that MyoD mRNA oscillates with a circadian profile in rodent skeletal muscle (McCarthy *et al.*, 2007).

MyoD is a skeletal muscle-specific bHLH transcription factor, which is activated early during myogenesis and commits undifferentiated cells to the muscle lineage. *MyoD* is known as a master regulator of the muscle-specific transcription program by controlling the expression of many structural, functional, and metabolic skeletal muscle-specific genes (Bergstrom *et al.*, 2002; Molkentin and Olson, 1996; Rudnicki *et al.*, 1993; Tapscott, 2005). The circadian oscillation of MyoD mRNA is abolished in the skeletal muscle of both $Bmal1^{-/-}$ and $CLOCK^{\Delta 19}$ mice. This is associated with a downregulation of many *MyoD* controlled genes, with functions ranging from skeletal muscle structure to metabolism. Further, skeletal muscle from both $Bmal1^{-/-}$ and $CLOCK^{\Delta 19}$ mice show disturbances in both force production and myofilament organization (discussed in more detail in Section 6.3), suggesting that the molecular clock regulates skeletal muscle structure and function by controlling *MyoD* expression (Andrews *et al.*, 2010).

6.2. The clock and metabolism in skeletal muscle

Recent work has shown that the molecular clock in skeletal muscle can be modulated by molecules that are part of the metabolic sensing machinery. These metabolic sensors respond to extracellular environmental changes, such as nutrient availability, contractile activity, energy balance, and redox status and modulate skeletal muscle physiological processes (Fulco *et al.*, 2003; Lira *et al.*, 2010; Muoio and Koves, 2007; Puigserver and Spiegelman, 2003; Vinciguerra *et al.*, 2010; Witczak *et al.*, 2008). At the same time, metabolic sensors introduce posttranslational modifications in the core molecular clock components and modulate molecular clock rhythms (Asher *et al.*, 2008; Lamia *et al.*, 2009; Liu *et al.*, 2007; Nakahata *et al.*, 2008). Not surprisingly, these metabolic sensors exhibit circadian rhythms in their activity levels, suggesting that a mutually dependent link exists between cell metabolism and circadian rhythms (Andrews *et al.*, 2010;

Lamia *et al.*, 2009; Liu *et al.*, 2007; McCarthy *et al.*, 2007). In this chapter, we will focus on the actions of AMPK, PGC1α, and SirT1and their roles as metabolic sensors and molecular clock regulators.

AMPK (adenosine monophosphate-activated protein kinase) is a well-recognized mediator of metabolic signals. In skeletal muscle, AMPK is activated by processes that increase the AMP:ATP ratio, such as exercise, oxidative stress, glucose restriction, and intracellular Ca^{2+} concentrations. In response to these stimuli, activated AMPK participates in the regulation of protein, carbohydrate, and lipid metabolic pathways in skeletal muscle (Witczak *et al.*, 2008). Transgenic mice lacking the AMPKγ subunit in skeletal muscle specifically show increased glycogen synthesis post exercise. In addition, they exhibit an increase in lipid oxidation as shown by the lack of triglyceride accumulation in skeletal muscle (Vieira *et al.*, 2008).

Activation of AMPK by low glucose in mouse embryonic fibroblasts increases the period of circadian rhythms and dampens the amplitude of *Bmal1* oscillations. AMPK phosphorylates CRY1 and primes it for degradation. Decreasing CRY1 stability leads to reduction of input from the negative feedback loop and the subsequent lengthening of the circadian period. Therefore, AMPK acts as a sensor of metabolic activity and relays nutrient/metabolic signals to circadian clocks. Recently, studies have found that AMPK itself exhibits circadian rhythms in expression levels, nuclear localization, and phosphorylation (a measure of activity) providing more evidence for the interrelationship between AMPK and the molecular clock (Lamia *et al.*, 2009). Under conditions of oxidative stress, energy depletion (exercise), or low nutrient availability, activated AMPK in skeletal muscle may be able to impose a delay in the molecular clock, by increasing the duration of transcriptional activation.

PGC1α/β (peroxisome proliferator-activated receptor gamma (PPARγ) coactivator-1 alpha/beta) is a well-known transcriptional coactivator that senses the energy balance of the organism. In skeletal muscle, PGC1 responds to nutrient stimuli and exercise and controls rate-limiting enzymes in the tricarboxylic acid cycle and oxidative phosphorylation pathways. PGC1α interacts with MEF2 (myocyte enhancer factor 2) and increases GLUT4 levels leading to increase in glucose uptake in skeletal muscle. PGC1α/β stimulates mitochondrial biogenesis, regulates fatty acid metabolism, mediates fiber type switching, and enhances exercise tolerance (Lira *et al.*, 2010; Muoio and Koves, 2007; Puigserver and Spiegelman, 2003).

PGC1α/β mRNA and protein levels in skeletal muscle oscillate with a circadian period (Andrews *et al.*, 2010; Liu *et al.*, 2007; McCarthy *et al.*, 2007). PGC1α enhances *Bmal1* transcription by activating RORα and RORγ. This interaction is mediated by the action of the p300 histone acetyltransferase recruited to the *Bmal1* promoter by PGC1α and is dependent on the levels of RevErbα and RevErbβ. This is consistent with the finding that PGC1α protein peaks at a time when *Bmal1* transcription is highest. The *Bmal1* transcription is high at a time when the CLOCK: BMAL1 dependent transcription is repressed. By controlling the time of *Bmal1* transcription, PGC1α regulates the length of circadian rhythms (Liu *et al.*, 2007).

The importance of PGC1 as a communicator between the core clock and cell metabolism is further supported by the finding that circadian rhythms of locomotor activity were disrupted in PGC1α-null mice, associated with a disruption in either amplitude or period in several circadian clock genes in skeletal muscle and liver. In addition, rate-limiting enzymes in oxidative phosphorylation and tricarboxylic acid (TCA) cycle had lost their diurnal pattern of expression in the PGC1α-null skeletal muscle and liver (Liu *et al.*, 2007). Unlike AMPK, PGC1α puts the clock in a transcriptional repressive state by promoting *Bmal1* transcription. It remains to be established, but it is possible that certain skeletal muscle processes, such as oxidative phosphorylation and the TCA cycle might preferentially occur during specific CTs.

Sirtuin 1 (*SirT1*), a member of the NAD^+-dependent protein deacetylase family, is a mediator of metabolism in response to nutritional availability and exercise. The activity of *SirT1* is influenced by the NAD^+: NADH ratio, which reflects cellular energy homeostasis and redox status. The majority of the NAD^+ is synthesized by the conversion of nicotinamide to nicotinamide mononucleotide utilizing the rate-limiting enzyme NAMPT (nicotinamide phosphoribisyl transferase; Fulco *et al.*, 2003; Ramsey *et al.*, 2009). *In vitro,* SirT1 overexpression in skeletal muscle cells blocks myocyte proliferation and differentiation by deacetylating and inhibiting *MyoD* (Fulco *et al.*, 2003). Blockage of differentiation also occurs when cultured myoblasts are grown in a medium with restricted glucose. This is due to activation of AMPK, which activates NAMPT, leading to NAD^+ accumulation and SirT1 activation (Vinciguerra *et al.*, 2010).

SirT1 exhibits circadian rhythms in both protein levels and histone deacetylase activity. As discussed in Section 5.2.2, SirT1 deacetylates and degrades PER2 and affects the amplitude of circadian clock gene expression, including *Bmal1*, *Per2*, and *Cry1*. In addition, SirT1 deacetylates and silences *Bmal1* and forms a complex with CLOCK:BMAL1 in the promoter regions of clock control genes (Asher *et al.*, 2008; Nakahata *et al.*, 2008). In turn, the CLOCK:BMAL1 dimer binds to E box sequences in the promoter region of *NAMPT* and upregulates *NAMPT* transcription, leading to circadian oscillations in *NAMPT* mRNA levels (Ramsey *et al.*, 2009). SirT1 activity puts the clock in a transcriptional repressive state and simultaneously inhibits *MyoD*, a skeletal muscle-specific transcription factor that regulates skeletal muscle growth and differentiation. That skeletal muscle physiological and metabolic processes exhibit circadian rhythms, has been documented by several groups and reviewed recently (Zhang *et al.*, 2009). SirT1 then serves as a sensor of metabolism and responds by regulating the time of the molecular clock, so that processes involved in skeletal muscle growth, differentiation, and maintenance can occur at appropriate CTs.

At this time, it is clear that there are links between the molecular clock and metabolism in all cells/organisms studied. From this work, we propose that metabolic signals will act as both inputs and outputs of the molecular clock in muscle. The circadian and metabolic oscillators can modulate each other through the interlocking actions of key players such as AMPK, PGC1α, and SirT1. A summary of the interplay between metabolism and the molecular clock is shown in Fig. 9.5.

6.3. Skeletal muscle pathologies in circadian mutants

Although expression profiling data have identified key processes in skeletal muscle physiology that may be affected or regulated by the molecular clock, not much is known regarding the role of the molecular clock in normal skeletal muscle physiology. However, certain lessons can be learned from the work with various circadian mutants that exhibit, among other things, pathologies in skeletal muscle (summarized in Table 9.2).

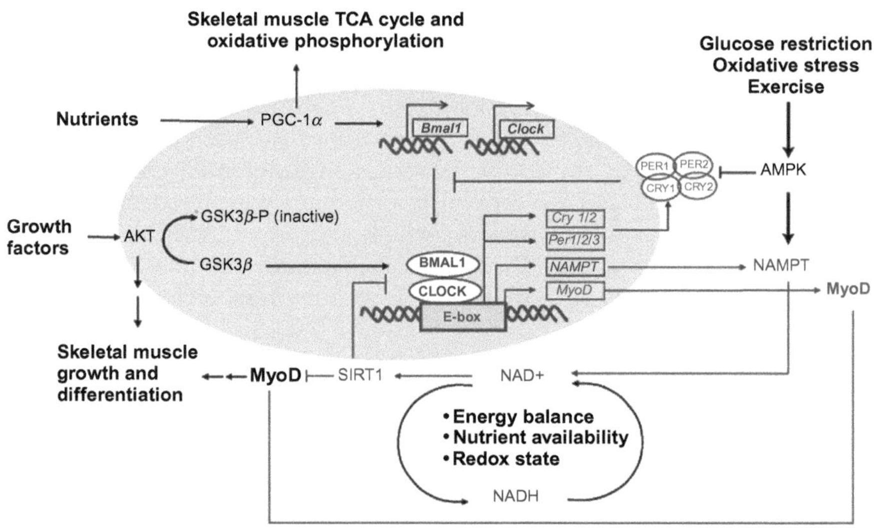

Figure 9.5 Interplay between the circadian clock and metabolism in skeletal muscle. There is a significant overlap between the factors that modulate the molecular clock and the factors critical for metabolism in skeletal muscle. Metabolic sensors, such as AMPK, SirT1, and PGC1α respond to nutrient availability, contractile activity, energy balance, and redox status, and modulate both skeletal muscle metabolism and the molecular clock. In turn, the molecular clock controls the rhythmic expression of these metabolic sensors, suggesting that there is an interdependent link between skeletal muscle metabolism and the molecular clock. (See Color Insert.)

Table 9.2 Skeletal muscle abnormalities in circadian mutants

Mutation	Circadian phenotype	Skeletal muscle pathology	Reference
$Bmal1^{-/-}$	Arrhythmic behavior and expression of clock genes in both SCN and peripheral tissues	Sarcopenia age-associated decrease in muscle fiber number and fiber diameter Reduced lifespan	Kondratov *et al.* (2006)
Muscle-rescued $Bmal1^{-/-}$	Arrhythmic wheel running behavior	Normal body weight Normal activity levels Improved longevity	McDearmon *et al.* (2006)
$Bmal1^{-/-}$ $Clock^{\Delta 19}$	Arrhythmic wheel running behavior	Decreased maximal force production Myofilament disarrangement Mitochondrial pathology	Andrews *et al.* (2010)
$Clock^{\Delta 19}$ + Mel	Liver and skeletal muscle-specific arrhythmic expression of clock genes	Decreased GLUT4 expression Possible skeletal muscle insulin resistance	Kennaway *et al.* (2007)
$Per2^{-/-}$	Short circadian period (tau = 22 h) Arrhythmic in constant darkness	Reduced forced locomotor performance without alteration in skeletal muscle contractile function	Bae *et al.* (2006)
$ROR\beta^{-/-}$	Slightly longer circadian period, (tau = 24.3 h)	Muscle weakness when young, gain strength with age "Duck Like" gait Locomotor difficulties	Andre *et al.* (1998)

The *Bmal1*$^{-/-}$ mice exhibit an arrhythmic behavior and demonstrate decreased activity, decreased body weight, increased sleep fragmentation, arthropathy, age-associated pathologies, and a subsequent decrease in lifespan (Bunger *et al.*, 2005; Kondratov *et al.*, 2006; Laposky *et al.*, 2005). Skeletal muscle pathologies include sarcopenia, or a decrease in muscle mass. Muscle mass loss is very pronounced in the *Bmal1*$^{-/-}$ mice and is only surpassed by fat mass loss. Sarcopenia is associated with a decrease in muscle fiber number and in the diameter of the remaining fibers. Moreover, this pathology starts developing later in life, as muscles from young *Bmal1*$^{-/-}$ animals (10 weeks of age) are indistinguishable from controls (Kondratov *et al.*, 2006). Rescuing *Bmal1* only in skeletal muscle was not sufficient to rescue the rhythmic behavior, but it did result in increased locomotor activity levels, restored body weight, and increased lifespan (McDearmon *et al.*, 2006). These findings suggest that an intact molecular clock within skeletal muscle is sufficient for maintenance of skeletal muscle. In addition, these findings are intriguing as they indicate that (1) the health of the skeletal muscle can influence voluntary locomotor activity levels and (2) increased muscle mass can decrease mortality.

Recent work from our laboratory is the first evidence of direct evaluation of skeletal muscle function in two phenotypically different circadian mutant mice. We studied skeletal muscle from the *Bmal1*$^{-/-}$ mice (arrhythimc) and *Clock*$^{\Delta 19}$ mice (long period length). Surprisingly, we found that specific tension was significantly reduced in skeletal muscle of both *Bmal1*$^{-/-}$ and *Clock*$^{\Delta 19}$ mice. This reduction in force was associated with a disorganization of myofilament arrangement and decreased expression of key structural genes, such as actin, myosin, and titin. Further, skeletal muscles of *Bmal1*$^{-/-}$ and *Clock*$^{\Delta 19}$ mice exhibit signs of a mitochondrial pathology, characterized by a significant decrease in mitochondrial volume and compromised function (reduced respiratory capacity ratio and increased respiration uncoupling) of the remaining mitochondria. The presence of a similar skeletal muscle phenotype in two different circadian mutants confirms the importance of the circadian clock in the maintenance of skeletal muscle structural and functional integrity (Andrews *et al.*, 2010).

Using an approach opposite to McDearmon *et al.* (2006) Kennaway and colleagues used the "CLOCK$^{\Delta 19}$ + Mel" mutant as a model of intact central rhythmicity, but aberrant rhythmicity in liver and skeletal muscle only. In addition to impaired glucose tolerance and reduced insulin secretion, with loss of rhythmicity in rate-limiting enzymes of glucose metabolism in the liver, the CLOCK$^{\Delta 19}$ + Mel mutant has reduced expression levels of Glut4 in skeletal muscle. The authors proposed that a reduction in Glut4 could lead to decreased insulin-mediated glucose uptake, suggesting that skeletal muscle could be insulin-resistant in the CLOCK$^{\Delta 19}$ + Mel mutant mouse (Kennaway *et al.*, 2007). This suggests that the peripheral clock in skeletal muscle contributes to the regulation of metabolic pathways in skeletal muscle and that this alone is sufficient to affect systemic metabolic health.

Per2$^{-/-}$ mice have a shorter period length (tau ~22 h) and become arrhythmic in constant dark conditions (Zheng *et al.*, 1999). The running endurance of *Per2*$^{-/-}$ mice is significantly lower when compared to wild-type mice, as measured by forced running distance. This is not due to changes in muscle contractile parameters or changes in contractile proteins. Glycolytic enzymes are upregulated in the *Per2*$^{-/-}$ skeletal muscle, suggesting that *Per2*$^{-/-}$ skeletal muscle depends more on glycolysis for energy production (Bae *et al.*, 2006). Components of the clock are thus involved in the regulation of both skeletal muscle performance and the shift in metabolic pathways of energy production.

ROR$\beta^{-/-}$ mice show a significant increase in the circadian period length, implicating RORβ as a component of the molecular clock. ROR$\beta^{-/-}$ mice show muscle weakness and smaller body size when young, but this ameliorates with age. The muscle weakness is assessed as an inability to walk and frequent falls sideways while attempting to support and balance body weight. As adults, ROR$\beta^{-/-}$ mice show a "duck-like" gait. These symptoms might be part of a broader syndrome, the *vacillans* phenotype. These symptoms could be a result of central nervous system problems, or could reflect intrinsic problems with the muscles (Andre *et al.*, 1998).

While there are very few studies on direct skeletal muscle evaluation in circadian mutants, disruption of the molecular clock network via genetic ablation of core clock components, has provided several lines of evidence for the importance of circadian rhythms in normal skeletal muscle physiology. The molecular clock is needed for skeletal muscle growth and maintenance, contractile performance, structural organization, glucose metabolism, and energy production. Whether these processes require central or skeletal muscle-specific molecular clock function is currently under investigation and has not been fully elucidated.

6.4. Exercise as a potential zeitgeber for the skeletal muscle molecular clock

As described in Section 3.2, exercise may be a potential zeitgeber for the skeletal muscle molecular clock. Resistance exercise has been shown to phase shift the circadian expression of clock genes in skeletal muscle in humans (Zambon *et al.*, 2003). Although this has not been demonstrated in skeletal muscle cells, work with neonatal cardiomyocytes shows that contractile activity can modulate the molecular clock through the actions of the CLOCK protein. CLOCK localizes to the Z-disk in neonatal cardiomyocytes. In response to contractile activity, CLOCK translocates to the nucleus to influence gene expression. The stretch sensing machinery of the sarcomere is composed of a multitude of structural and signaling molecules located in the Z-disk region. Localization of CLOCK in proximity to the Z-disk puts CLOCK in the right position to sense energy expenditure associated with contractile activity. Thus, contraction of the myocardium may influence the molecular clock and the molecular clock in turn can influence the contractile activity of cardiac muscle (Qi and Boateng, 2006).

7. Summary

Circadian biology has been studied across plant and animal species for many decades with much of the early work focused on animal/organism behavior. This body of research has led to defined characteristics of circadian rhythms based on period length, phase, and amplitude. These criteria can be applied at the whole body level to circadian behavior through to the molecular behavior of components of the clock. The goal of this chapter was to provide the basic terminology and concepts of circadian rhythms with a more detailed review of the current state of knowledge of the molecular clock with reference to what is known in skeletal muscle. The study of the circadian behavior and the molecular clock in skeletal muscle is in the very early stages. Research has demonstrated that the molecular clock is active in skeletal muscles and that the muscle-specific transcription factor, *MyoD*, is a direct target of the molecular clock. Skeletal muscle of clock-compromised mice, *Bmal1*$^{-/-}$ and *Clock*$^{\Delta 19}$ mice, exhibit significant disruption in normal expression of many genes required for adult muscle structure and metabolism. We suggest that the interaction between the molecular clock, MyoD, and metabolic factors, such as PGC-1 provide an interesting system of feedback loops that are critical for both maintenance and adaptation of skeletal muscle.

References

Akashi, M., Tsuchiya, Y., Yoshino, T., and Nishida, E. (2002). Control of intracellular dynamics of mammalian period proteins by casein kinase I epsilon (CKIepsilon) and CKIdelta in cultured cells. *Mol. Cell Biol.* **22,** 1693–1703.

Albrecht, U. (2002). Invited review: regulation of mammalian circadian clock genes. *J. Appl. Physiol.* **92,** 1348–1355.

Albrecht, U., and Oster, H. (2001). The circadian clock and behavior. *Behav. Brain Res.* **125,** 89–91.

Almon, R. R., Yang, E., Lai, W., Androulakis, I. P., Ghimbovschi, S., Hoffman, E. P., Jusko, W. J., and Dubois, D. C. (2008). Relationships between circadian rhythms and modulation of gene expression by glucocorticoids in skeletal muscle. *Am. J. Physiol. Regul. Integr. Comp. Physiol.* **295,** R1031–R1047.

Andre, E., Conquet, F., Steinmayr, M., Stratton, S. C., Porciatti, V., and Becker-Andre, M. (1998). Disruption of retinoid-related orphan receptor beta changes circadian behavior, causes retinal degeneration and leads to vacillans phenotype in mice. *EMBO J.* **17,** 3867–3877.

Andrews, J. L., Zhang, X., McCarthy, J. J., McDearmon, E. L., Hornberger, T. A., Russell, B., Campbell, K. S., Arbogast, S., Reid, M. B., Walker, J. R., Hogenesch, J. B., Takahashi, J. S., *et al.* (2010). CLOCK and BMAL1 regulate MyoD and are necessary for maintenance of skeletal muscle phenotype and function. *Proc. Natl. Acad. Sci. USA* **107,** 19090–19095.

Angeles-Castellanos, M., Mendoza, J., Diaz-Munoz, M., and Escobar, C. (2005). Food entrainment modifies the c-Fos expression pattern in brain stem nuclei of rats. *Am. J. Physiol. Regul. Integr. Comp. Physiol.* **288,** R678–R684.

Angeles-Castellanos, M., Mendoza, J., and Escobar, C. (2007). Restricted feeding schedules phase shift daily rhythms of c-Fos and protein Per1 immunoreactivity in corticolimbic regions in rats. *Neuroscience* **144,** 344–355.

Antoch, M. P., Song, E. J., Chang, A. M., Vitaterna, M. H., Zhao, Y., Wilsbacher, L. D., Sangoram, A. M., King, D. P., Pinto, L. H., and Takahashi, J. S. (1997). Functional identification of the mouse circadian Clock gene by transgenic BAC rescue. *Cell* **89,** 655–667.

Aschoff, J. (1960). Exogenous and endogenous components in circadian rhythms. *Cold Spring Harb. Symp. Quant. Biol.* **25,** 11–28.

Aschoff, J. (1967). Human circadian rhythms in activity, body temperature and other functions. *Life Sci. Space Res.* **5,** 159–173.

Aschoff, J. (1978). Features of circadian rhythms relevant for the design of shift schedules. *Ergonomics* **21,** 739–754.

Aschoff, J. (1990). Interdependence between locomotor activity and duration of wakefulness in humans during isolation. *Experientia* **46,** 870–871.

Aschoff, J. (1998). Circadian parameters as individual characteristics. *J. Biol. Rhythms* **13,** 123–131.

Aschoff, J. (1999). Masking and parametric effects of high-frequency light-dark cycles. *Jpn. J. Physiol.* **49,** 11–18.

Aschoff, J., Hoffmann, K., Pohl, H., and Wever, R. (1975). Re-entrainment of circadian rhythms after phase-shifts of the Zeitgeber. *Chronobiologia* **2,** 23–78.

Aschoff, J., and Wever, R. (1976). Human circadian rhythms: a multioscillatory system. *Fed. Proc.* **35,** 236–232.

Asher, G., Gatfield, D., Stratmann, M., Reinke, H., Dibner, C., Kreppel, F., Mostoslavsky, R., Alt, F. W., and Schibler, U. (2008). SIRT1 regulates circadian clock gene expression through PER2 deacetylation. *Cell* **134,** 317–328.

Bae, K., Lee, K., Seo, Y., Lee, H., Kim, D., and Choi, I. (2006). Differential effects of two period genes on the physiology and proteomic profiles of mouse anterior tibialis muscles. *Mol. Cells* **22,** 275–284.

Bergstrom, D. A., Penn, B. H., Strand, A., Perry, R. L., Rudnicki, M. A., and Tapscott, S. J. (2002). Promoter-specific regulation of MyoD binding and signal transduction cooperate to pattern gene expression. *Mol. Cell* **9,** 587–600.

Bunger, M. K., Walisser, J. A., Sullivan, R., Manley, P. A., Moran, S. M., Kalscheur, V. L., Colman, R. J., and Bradfield, C. A. (2005). Progressive arthropathy in mice with a targeted disruption of the Mop3/Bmal-1 locus. *Genesis* **41,** 122–132.

Bunger, M. K., Wilsbacher, L. D., Moran, S. M., Clendenin, C., Radcliffe, L. A., Hogenesch, J. B., Simon, M. C., Takahashi, J. S., and Bradfield, C. A. (2000). Mop3 is an essential component of the master circadian pacemaker in mammals. *Cell* **103,** 1009–1017.

Busino, L., Bassermann, F., Maiolica, A., Lee, C., Nolan, P. M., Godinho, S. I., Draetta, G. F., and Pagano, M. (2007). SCFFbxl3 controls the oscillation of the circadian clock by directing the degradation of cryptochrome proteins. *Science* **316,** 900–904.

Castillo, M. R., Hochstetler, K. J., Tavernier, R. J., Jr., Greene, D. M., and Bult-Ito, A. (2004). Entrainment of the master circadian clock by scheduled feeding. *Am. J. Physiol. Regul. Integr. Comp. Physiol.* **287,** R551–R555.

Chergui, K., Svenningsson, P., and Greengard, P. (2005). Physiological role for casein kinase 1 in glutamatergic synaptic transmission. *J. Neurosci.* **25,** 6601–6609.

Cuadrado, A., and Nebreda, A. R. (2010). Mechanisms and functions of p38 MAPK signalling. *Biochem. J.* **429,** 403–417.

Czeisler, C. A., Weitzman, E., Moore-Ede, M. C., Zimmerman, J. C., and Knauer, R. S. (1980). Human sleep: its duration and organization depend on its circadian phase. *Science* **210,** 1264–1267.

Damiola, F., Le Minh, N., Preitner, N., Kornmann, B., Fleury-Olela, F., and Schibler, U. (2000). Restricted feeding uncouples circadian oscillators in peripheral tissues from the central pacemaker in the suprachiasmatic nucleus. *Genes Dev.* **14,** 2950–2961.

Davidson, A. J., Sellix, M. T., Daniel, J., Yamazaki, S., Menaker, M., and Block, G. D. (2006). Chronic jet-lag increases mortality in aged mice. *Curr. Biol.* **16,** R914–R916.

Davidson, A. J., Stokkan, K. A., Yamazaki, S., and Menaker, M. (2002). Food-anticipatory activity and liver per1-luc activity in diabetic transgenic rats. *Physiol. Behav.* **76,** 21–26.

DeCoursey, P. J., and Buggy, J. (1989). Circadian rhythmicity after neural transplant to hamster third ventricle: specificity of suprachiasmatic nuclei. *Brain Res.* **500,** 263–275.

Devlin, P. F., and Kay, S. A. (2001). Circadian photoperception. *Annu. Rev. Physiol.* **63,** 677–694.

Dey, J., Carr, A. J., Cagampang, F. R., Semikhodskii, A. S., Loudon, A. S., Hastings, M. H., and Maywood, E. S. (2005). The tau mutation in the Syrian hamster differentially reprograms the circadian clock in the SCN and peripheral tissues. *J. Biol. Rhythms* **20,** 99–110.

Diaz-Munoz, M., Vazquez-Martinez, O., Aguilar-Roblero, R., and Escobar, C. (2000). Anticipatory changes in liver metabolism and entrainment of insulin, glucagon, and corticosterone in food-restricted rats. *Am. J. Physiol. Regul. Integr. Comp. Physiol.* **279,** R2048–R2056.

Doi, M., Hirayama, J., and Sassone-Corsi, P. (2006). Circadian regulator CLOCK is a histone acetyltransferase. *Cell* **125,** 497–508.

Edgar, D. M., and Dement, W. C. (1991). Regularly scheduled voluntary exercise synchronizes the mouse circadian clock. *Am. J. Physiol.* **261,** R928–R933.

Eide, E. J., Vielhaber, E. L., Hinz, W. A., and Virshup, D. M. (2002). The circadian regulatory proteins BMAL1 and cryptochromes are substrates of casein kinase Iepsilon. *J. Biol. Chem.* **277,** 17248–17254.

Eide, E. J., Woolf, M. F., Kang, H., Woolf, P., Hurst, W., Camacho, F., Vielhaber, E. L., Giovanni, A., and Virshup, D. M. (2005). Control of mammalian circadian rhythm by CKIepsilon-regulated proteasome-mediated PER2 degradation. *Mol. Cell Biol.* **25,** 2795–2807.

Escobar, C., Diaz-Munoz, M., Encinas, F., and Aguilar-Roblero, R. (1998). Persistence of metabolic rhythmicity during fasting and its entrainment by restricted feeding schedules in rats. *Am. J. Physiol.* **274,** R1309–R1316.

Etchegaray, J. P., Lee, C., Wade, P. A., and Reppert, S. M. (2003). Rhythmic histone acetylation underlies transcription in the mammalian circadian clock. *Nature* **421,** 177–182.

Feillet, C. A., Albrecht, U., and Challet, E. (2006). "Feeding time" for the brain: a matter of clocks. *J. Physiol. Paris* **100,** 252–260.

Filipski, E., King, V. M., Li, X., Granda, T. G., Mormont, M. C., Liu, X., Claustrat, B., Hastings, M. H., and Levi, F. (2002). Host circadian clock as a control point in tumor progression. *J. Natl. Cancer Inst.* **94,** 690–697.

Fulco, M., Schiltz, R. L., Iezzi, S., King, M. T., Zhao, P., Kashiwaya, Y., Hoffman, E., Veech, R. L., and Sartorelli, V. (2003). Sir2 regulates skeletal muscle differentiation as a potential sensor of the redox state. *Mol. Cell* **12,** 51–62.

Gekakis, N., Staknis, D., Nguyen, H. B., Davis, F. C., Wilsbacher, L. D., King, D. P., Takahashi, J. S., and Weitz, C. J. (1998). Role of the CLOCK protein in the mammalian circadian mechanism. *Science* **280,** 1564–1569.

Gibbs, F. P. (1976). Correlation of plasma corticosterone levels with running activity in the blinded rat. *Am. J. Physiol.* **231,** 817–821.
Godinho, S. I., Maywood, E. S., Shaw, L., Tucci, V., Barnard, A. R., Busino, L., Pagano, M., Kendall, R., Quwailid, M. M., Romero, M. R., O'Neill, J., Chesham, J. E., *et al.* (2007). The after-hours mutant reveals a role for Fbxl3 in determining mammalian circadian period. *Science* **316,** 897–900.
Grundschober, C., Delaunay, F., Puhlhofer, A., Triqueneaux, G., Laudet, V., Bartfai, T., and Nef, P. (2001). Circadian regulation of diverse gene products revealed by mRNA expression profiling of synchronized fibroblasts. *J. Biol. Chem.* **276,** 46751–46758.
Guillaumond, F., Dardente, H., Giguere, V., and Cermakian, N. (2005). Differential control of Bmal1 circadian transcription by REV-ERB and ROR nuclear receptors. *J. Biol. Rhythms* **20,** 391–403.
Guo, H., Brewer, J. M., Champhekar, A., Harris, R. B., and Bittman, E. L. (2005). Differential control of peripheral circadian rhythms by suprachiasmatic-dependent neural signals. *Proc. Natl. Acad. Sci. USA* **102,** 3111–3116.
Halberg, F. (1969). Chronobiology. *Annu. Rev. Physiol.* **31,** 675–725.
Halberg, F., Carandente, F., Cornelissen, G., and Katinas, G. S. (1977). Glossary of chronobiology (author's transl). *Chronobiologia* **4**(Suppl 1), 1–189.
Halberg, F., and Cornelissen, G. (1993). Rhythms and blood pressure. *Ann. Ist Super Sanita* **29,** 647–665.
Halberg, F., Engeli, M., Hamburger, C., and Hillman, D. (1965). Spectral resolution of low-frequency, small-amplitude rhythms in excreted 17-ketosteroids; probable androgen-induced circaseptan desynchronization. *Acta Endocrinol. (Copenh)* **50**(Suppl. 103), 101–154.
Halberg, F., Peterson, R. E., and Silber, R. H. (1959). Phase relations of 24-hour periodicities in blood corticosterone, mitoses in cortical adrenal parenchyma, and total body activity. *Endocrinology* **64,** 222–230.
Hara, R., Wan, K., Wakamatsu, H., Aida, R., Moriya, T., Akiyama, M., and Shibata, S. (2001). Restricted feeding entrains liver clock without participation of the suprachiasmatic nucleus. *Genes Cells* **6,** 269–278.
Harada, Y., Sakai, M., Kurabayashi, N., Hirota, T., and Fukada, Y. (2005). Ser-557-phosphorylated mCRY2 is degraded upon synergistic phosphorylation by glycogen synthase kinase-3 beta. *J. Biol. Chem.* **280,** 31714–31721.
Harmer, S. L., Panda, S., and Kay, S. A. (2001). Molecular bases of circadian rhythms. *Annu. Rev. Cell Dev. Biol.* **17,** 215–253.
Harrington, M. E., Eskes, G. A., Dickson, P., and Rusak, B. (1990). Lesions dorsal to the suprachiasmatic nuclei abolish split activity rhythms of hamsters. *Brain Res. Bull* **24,** 593–597.
Hastings, M. H., Maywood, E. S., and Reddy, A. B. (2008). Two decades of circadian time. *J. Neuroendocrinol.* **20,** 812–819.
Haus, E., and Halberg, F. (1959). 24-Hour rhythm in susceptibility of C mice to a toxic dose of ethanol. *J. Appl. Physiol.* **14,** 878–880.
Hirayama, J., Sahar, S., Grimaldi, B., Tamaru, T., Takamatsu, K., Nakahata, Y., and Sassone-Corsi, P. (2007). CLOCK-mediated acetylation of BMAL1 controls circadian function. *Nature* **450,** 1086–1090.
Hiroshige, T., Honma, K., and Honma, S. (1991). SCN-independent circadian oscillators in the rat. *Brain Res. Bull.* **27,** 441–445.
Hirota, T., Lewis, W. G., Liu, A. C., Lee, J. W., Schultz, P. G., and Kay, S. A. (2008). A chemical biology approach reveals period shortening of the mammalian circadian clock by specific inhibition of GSK-3beta. *Proc. Natl. Acad. Sci. USA* **105,** 20746–20751.

Hogenesch, J. B., Gu, Y. Z., Jain, S., and Bradfield, C. A. (1998). The basic-helix-loop-helix-PAS orphan MOP3 forms transcriptionally active complexes with circadian and hypoxia factors. *Proc. Natl. Acad. Sci. USA* **95,** 5474–5479.

Holmes, M. M., and Mistlberger, R. E. (2000). Food anticipatory activity and photic entrainment in food-restricted BALB/c mice. *Physiol. Behav.* **68,** 655–666.

Holzberg, D., and Albrecht, U. (2003). The circadian clock: a manager of biochemical processes within the organism. *J. Neuroendocrinol.* **15,** 339–343.

Honma, K., Honma, S., and Hiroshige, T. (1987). Activity rhythms in the circadian domain appear in suprachiasmatic nuclei lesioned rats given methamphetamine. *Physiol. Behav.* **40,** 767–774.

Honma, K., von Goetz, C., and Aschoff, J. (1983). Effects of restricted daily feeding on freerunning circadian rhythms in rats. *Physiol. Behav.* **30,** 905–913.

Iijima, M., Nikaido, T., Akiyama, M., Moriya, T., and Shibata, S. (2002). Methamphetamine-induced, suprachiasmatic nucleus-independent circadian rhythms of activity and mPer gene expression in the striatum of the mouse. *Eur. J. Neurosci.* **16,** 921–929.

Iitaka, C., Miyazaki, K., Akaike, T., and Ishida, N. (2005). A role for glycogen synthase kinase-3beta in the mammalian circadian clock. *J. Biol. Chem.* **280,** 29397–29402.

Izumo, M., Johnson, C. H., and Yamazaki, S. (2003). Circadian gene expression in mammalian fibroblasts revealed by real-time luminescence reporting: temperature compensation and damping. *Proc. Natl. Acad. Sci. USA* **100,** 16089–16094.

Johnson, R. F., Moore, R. Y., and Morin, L. P. (1988). Loss of entrainment and anatomical plasticity after lesions of the hamster retinohypothalamic tract. *Brain Res.* **460,** 297–313.

Kennaway, D. J., Owens, J. A., Voultsios, A., Boden, M. J., and Varcoe, T. J. (2007). Metabolic homeostasis in mice with disrupted Clock gene expression in peripheral tissues. *Am. J. Physiol. Regul. Integr. Comp. Physiol.* **293,** R1528–R1537.

Kewley, R. J., Whitelaw, M. L., and Chapman-Smith, A. (2004). The mammalian basic helix-loop-helix/PAS family of transcriptional regulators. *Int. J. Biochem. Cell Biol.* **36,** 189–204.

King, D. P., Vitaterna, M. H., Chang, A. M., Dove, W. F., Pinto, L. H., Turek, F. W., and Takahashi, J. S. (1997). The mouse Clock mutation behaves as an antimorph and maps within the W19H deletion, distal of Kit. *Genetics* **146,** 1049–1060.

Klimowski, L. K., Garcia, B. A., Shabanowitz, J., Hunt, D. F., and Virshup, D. M. (2006). Site-specific casein kinase 1epsilon-dependent phosphorylation of Dishevelled modulates beta-catenin signaling. *FEBS J.* **273,** 4594–4602.

Knutsson, A. (2003). Health disorders of shift workers. *Occup. Med. (Lond.)* **53,** 103–108.

Kondratov, R. V., Chernov, M. V., Kondratova, A. A., Gorbacheva, V. Y., Gudkov, A. V., and Antoch, M. P. (2003). BMAL1-dependent circadian oscillation of nuclear CLOCK: posttranslational events induced by dimerization of transcriptional activators of the mammalian clock system. *Genes Dev.* **17,** 1921–1932.

Kondratov, R. V., Kondratova, A. A., Gorbacheva, V. Y., Vykhovanets, O. V., and Antoch, M. P. (2006). Early aging and age-related pathologies in mice deficient in BMAL1, the core componentof the circadian clock. *Genes Dev.* **20,** 1868–1873.

Kume, K., Zylka, M. J., Sriram, S., Shearman, L. P., Weaver, D. R., Jin, X., Maywood, E. S., Hastings, M. H., and Reppert, S. M. (1999). mCRY1 and mCRY2 are essential components of the negative limb of the circadian clock feedback loop. *Cell* **98,** 193–205.

Lamia, K. A., Sachdeva, U. M., DiTacchio, L., Williams, E. C., Alvarez, J. G., Egan, D. F., Vasquez, D. S., Juguilon, H., Panda, S., Shaw, R. J., Thompson, C. B., and Evans, R. M. (2009). AMPK regulates the circadian clock by cryptochrome phosphorylation and degradation. *Science* **326,** 437–440.

Laposky, A., Easton, A., Dugovic, C., Walisser, J., Bradfield, C., and Turek, F. (2005). Deletion of the mammalian circadian clock gene BMAL1/Mop3 alters baseline sleep architecture and the response to sleep deprivation. *Sleep* **28,** 395–409.

Lee, C., Etchegaray, J. P., Cagampang, F. R., Loudon, A. S., and Reppert, S. M. (2001). Posttranslational mechanisms regulate the mammalian circadian clock. *Cell* **107,** 855–867.

Lehman, M. N., Silver, R., Gladstone, W. R., Kahn, R. M., Gibson, M., and Bittman, E. L. (1987). Circadian rhythmicity restored by neural transplant. Immunocytochemical characterization of the graft and its integration with the host brain. *J. Neurosci.* **7,** 1626–1638.

Lira, V. A., Benton, C. R., Yan, Z., and Bonen, A. (2010). PGC-1alpha regulation by exercise training and its influences on muscle function and insulin sensitivity. *Am. J. Physiol. Endocrinol. Metab.* **299,** E145–E161.

Liu, C., Li, S., Liu, T., Borjigin, J., and Lin, J. D. (2007). Transcriptional coactivator PGC-1alpha integrates the mammalian clock and energy metabolism. *Nature* **447,** 477–481.

Lowe, C. H., Hinds, D. S., Lardner, P. J., and Justice, K. E. (1967). Natural free-running period in vertebrate animal populations. *Science* **156,** 531–534.

Lowrey, P. L., Shimomura, K., Antoch, M. P., Yamazaki, S., Zemenides, P. D., Ralph, M. R., Menaker, M., and Takahashi, J. S. (2000). Positional syntenic cloning and functional characterization of the mammalian circadian mutation tau. *Science* **288,** 483–492.

Lowrey, P. L., and Takahashi, J. S. (2004). Mammalian circadian biology: elucidating genome-wide levels of temporal organization. *Annu. Rev. Genomics Hum. Genet.* **5,** 407–441.

Masubuchi, S., Honma, S., Abe, H., Nakamura, W., and Honma, K. (2001). Circadian activity rhythm in methamphetamine-treated Clock mutant mice. *Eur. J. Neurosci.* **14,** 1177–1180.

McCarthy, J. J., Andrews, J. L., McDearmon, E. L., Campbell, K. S., Barber, B. K., Miller, B. H., Walker, J. R., Hogenesch, J. B., Takahashi, J. S., and Esser, K. A. (2007). Identification of the circadian transcriptome in adult mouse skeletal muscle. *Physiol. Genomics* **31,** 86–95.

McDearmon, E. L., Patel, K. N., Ko, C. H., Walisser, J. A., Schook, A. C., Chong, J. L., Wilsbacher, L. D., Song, E. J., Hong, H. K., Bradfield, C. A., and Takahashi, J. S. (2006). Dissecting the functions of the mammalian clock protein BMAL1 by tissue-specific rescue in mice. *Science* **314,** 1304–1308.

Merrow, M., Spoelstra, K., and Roenneberg, T. (2005). The circadian cycle: daily rhythms from behaviour to genes. *EMBO Rep.* **6,** 930–935.

Mieda, M., Williams, S. C., Richardson, J. A., Tanaka, K., and Yanagisawa, M. (2006). The dorsomedial hypothalamic nucleus as a putative food-entrainable circadian pacemaker. *Proc. Natl. Acad. Sci. USA* **103,** 12150–12155.

Miller, B. H., McDearmon, E. L., Panda, S., Hayes, K. R., Zhang, J., Andrews, J. L., Antoch, M. P., Walker, J. R., Esser, K. A., Hogenesch, J. B., and Takahashi, J. S. (2007). Circadian and CLOCK-controlled regulation of the mouse transcriptome and cell proliferation. *Proc. Natl. Acad. Sci. USA* **104,** 3342–3347.

Mistlberger, R. E. (1991). Effects of daily schedules of forced activity on free-running rhythms in the rat. *J. Biol. Rhythms* **6,** 71–80.

Mistlberger, R. E. (1994). Circadian food-anticipatory activity: formal models and physiological mechanisms. *Neurosci. Biobehav. Rev.* **18,** 171–195.

Mohawk, J. A., Baer, M. L., and Menaker, M. (2009). The methamphetamine-sensitive circadian oscillator does not employ canonical clock genes. *Proc. Natl. Acad. Sci. USA* **106,** 3519–3524.

Molkentin, J. D., and Olson, E. N. (1996). Defining the regulatory networks for muscle development. *Curr. Opin. Genet. Dev.* **6,** 445–453.

Mrosovsky, N. (1996). Locomotor activity and non-photic influences on circadian clocks. *Biol. Rev. Camb. Philos. Soc.* **71,** 343–372.

Mrosovsky, N. (1999). Further experiments on the relationship between the period of circadian rhythms and locomotor activity levels in hamsters. *Physiol. Behav.* **66,** 797–801.
Mrosovsky, N., Salmon, P. A., Menaker, M., and Ralph, M. R. (1992). Nonphotic phase shifting in hamster clock mutants. *J. Biol. Rhythms* **7,** 41–49.
Muoio, D. M., and Koves, T. R. (2007). Skeletal muscle adaptation to fatty acid depends on coordinated actions of the PPARs and PGC1 alpha: implications for metabolic disease. *Appl. Physiol. Nutr. Metab.* **32,** 874–883.
Nakahata, Y., Kaluzova, M., Grimaldi, B., Sahar, S., Hirayama, J., Chen, D., Guarente, L. P., and Sassone-Corsi, P. (2008). The NAD+-dependent deacetylase SIRT1 modulates CLOCK-mediated chromatin remodeling and circadian control. *Cell* **134,** 329–340.
Obrietan, K., Impey, S., and Storm, D. R. (1998). Light and circadian rhythmicity regulate MAP kinase activation in the suprachiasmatic nuclei. *Nat. Neurosci.* **1,** 693–700.
Oster, H., Maronde, E., and Albrecht, U. (2002). The circadian clock as a molecular calendar. *Chronobiol. Int.* **19,** 507–516.
Panda, S., Antoch, M. P., Miller, B. H., Su, A. I., Schook, A. B., Straume, M., Schultz, P. G., Kay, S. A., Takahashi, J. S., and Hogenesch, J. B. (2002). Coordinated transcription of key pathways in the mouse by the circadian clock. *Cell* **109,** 307–320.
Preitner, N., Damiola, F., Lopez-Molina, L., Zakany, J., Duboule, D., Albrecht, U., and Schibler, U. (2002). The orphan nuclear receptor REV-ERBalpha controls circadian transcription within the positive limb of the mammalian circadian oscillator. *Cell* **110,** 251–260.
Puigserver, P., and Spiegelman, B. M. (2003). Peroxisome proliferator-activated receptor-gamma coactivator 1 alpha (PGC-1 alpha): transcriptional coactivator and metabolic regulator. *Endocr. Rev.* **24,** 78–90.
Qi, L., and Boateng, S. Y. (2006). The circadian protein Clock localizes to the sarcomeric Z-disk and is a sensor of myofilament cross-bridge activity in cardiac myocytes. *Biochem. Biophys. Res. Commun.* **351,** 1054–1059.
Ralph, M. R., Foster, R. G., Davis, F. C., and Menaker, M. (1990). Transplanted suprachiasmatic nucleus determines circadian period. *Science* **247,** 975–978.
Ralph, M. R., and Menaker, M. (1988). A mutation of the circadian system in golden hamsters. *Science* **241,** 1225–1227.
Ramsey, K. M., Yoshino, J., Brace, C. S., Abrassart, D., Kobayashi, Y., Marcheva, B., Hong, H. K., Chong, J. L., Buhr, E. D., Lee, C., Takahashi, J. S., Imai, S., *et al.* (2009). Circadian clock feedback cycle through NAMPT-mediated NAD+ biosynthesis. *Science* **324,** 651–654.
Rayasam, G. V., Tulasi, V. K., Sodhi, R., Davis, J. A., and Ray, A. (2009). Glycogen synthase kinase 3: more than a namesake. *Br. J. Pharmacol.* **156,** 885–898.
Rosbash, M. (2009). The implications of multiple circadian clock origins. *PLoS Biol.* **7,** e62.
Rosenberg, R. S., Zee, P. C., and Turek, F. W. (1991). Phase response curves to light in young and old hamsters. *Am. J. Physiol.* **261,** R491–R495.
Rudnicki, M. A., Schnegelsberg, P. N., Stead, R. H., Braun, T., Arnold, H. H., and Jaenisch, R. (1993). MyoD or Myf-5 is required for the formation of skeletal muscle. *Cell* **75,** 1351–1359.
Sahar, S., Zocchi, L., Kinoshita, C., Borrelli, E., and Sassone-Corsi, P. (2010). Regulation of BMAL1 protein stability and circadian function by GSK3beta-mediated phosphorylation. *PLoS One* **5,** e8561.
Sanada, K., Harada, Y., Sakai, M., Todo, T., and Fukada, Y. (2004). Serine phosphorylation of mCRY1 and mCRY2 by mitogen-activated protein kinase. *Genes Cells* **9,** 697–708.
Sanada, K., Okano, T., and Fukada, Y. (2002). Mitogen-activated protein kinase phosphorylates and negatively regulates basic helix-loop-helix-PAS transcription factor BMAL1. *J. Biol. Chem.* **277,** 267–271.

Sangoram, A. M., Saez, L., Antoch, M. P., Gekakis, N., Staknis, D., Whiteley, A., Fruechte, E. M., Vitaterna, M. H., Shimomura, K., King, D. P., Young, M. W., Weitz, C. J., *et al.* (1998). Mammalian circadian autoregulatory loop: a timeless ortholog and mPer1 interact and negatively regulate CLOCK-BMAL1-induced transcription. *Neuron* **21,** 1101–1113.

Sato, T. K., Panda, S., Miraglia, L. J., Reyes, T. M., Rudic, R. D., McNamara, P., Naik, K. A., FitzGerald, G. A., Kay, S. A., and Hogenesch, J. B. (2004). A functional genomics strategy reveals Rora as a component of the mammalian circadian clock. *Neuron* **43,** 527–537.

Satoh, Y., Kawai, H., Kudo, N., Kawashima, Y., and Mitsumoto, A. (2006). Time-restricted feeding entrains daily rhythms of energy metabolism in mice. *Am. J. Physiol. Regul. Integr. Comp. Physiol.* **290,** R1276–R1283.

Schibler, U. (2009). The 2008 Pittendrigh/Aschoff lecture: peripheral phase coordination in the mammalian circadian timing system. *J. Biol. Rhythms* **24,** 3–15.

Schwartz, W. J., and Zimmerman, P. (1990). Circadian timekeeping in BALB/c and C57BL/6 inbred mouse strains. *J. Neurosci.* **10,** 3685–3694.

Shearman, L. P., Sriram, S., Weaver, D. R., Maywood, E. S., Chaves, I., Zheng, B., Kume, K., Lee, C. C., van der Horst, G. T., Hastings, M. H., and Reppert, S. M. (2000). Interacting molecular loops in the mammalian circadian clock. *Science* **288,** 1013–1019.

Shirogane, T., Jin, J., Ang, X. L., and Harper, J. W. (2005). SCFbeta-TRCP controls clock-dependent transcription via casein kinase 1-dependent degradation of the mammalian period-1 (Per1) protein. *J. Biol. Chem.* **280,** 26863–26872.

Siepka, S. M., Yoo, S. H., Park, J., Song, W., Kumar, V., Hu, Y., Lee, C., and Takahashi, J. S. (2007). Circadian mutant Overtime reveals F-box protein FBXL3 regulation of cryptochrome and period gene expression. *Cell* **129,** 1011–1023.

Spengler, M. L., Kuropatwinski, K. K., Schumer, M., and Antoch, M. P. (2009). A serine cluster mediates BMAL1-dependent CLOCK phosphorylation and degradation. *Cell Cycle* **8,** 4138–4146.

Stephan, F. K. (1983). Circadian rhythms in the rat: constant darkness, entrainment to T cycles and to skeleton photoperiods. *Physiol. Behav.* **30,** 451–462.

Stephan, F. K., Swann, J. M., and Sisk, C. L. (1979). Entrainment of circadian rhythms by feeding schedules in rats with suprachiasmatic lesions. *Behav. Neural. Biol.* **25,** 545–554.

Stephan, F. K., and Zucker, I. (1972). Circadian rhythms in drinking behavior and locomotor activity of rats are eliminated by hypothalamic lesions. *Proc. Natl. Acad. Sci. USA* **69,** 1583–1586.

Stokkan, K. A., Yamazaki, S., Tei, H., Sakaki, Y., and Menaker, M. (2001). Entrainment of the circadian clock in the liver by feeding. *Science* **291,** 490–493.

Storch, K. F., Lipan, O., Leykin, I., Viswanathan, N., Davis, F. C., Wong, W. H., and Weitz, C. J. (2002). Extensive and divergent circadian gene expression in liver and heart. *Nature* **417,** 78–83.

Tahara, Y., Hirao, A., Moriya, T., Kudo, T., Shibata, S., Effects of medial hypothalamic lesions on feeding-induced entrainment of locomotor activity and liver Per2 expression in Per2::luc mice. *J. Biol. Rhythms* **25,** 9–18.

Takahashi, J. S., DeCoursey, P. J., Bauman, L., and Menaker, M. (1984). Spectral sensitivity of a novel photoreceptive system mediating entrainment of mammalian circadian rhythms. *Nature* **308,** 186–188.

Takahashi, J. S., Hong, H. K., Ko, C. H., and McDearmon, E. L. (2008). The genetics of mammalian circadian order and disorder: implications for physiology and disease. *Nat. Rev. Genet.* **9,** 764–775.

Tapscott, S. J. (2005). The circuitry of a master switch: Myod and the regulation of skeletal muscle gene transcription. *Development* **132,** 2685–2695.

Turek, F. W., Penev, P., Zhang, Y., van Reeth, O., and Zee, P. (1995). Effects of age on the circadian system. *Neurosci. Biobehav. Rev.* **19,** 53–58.

Ueda, H. R., Chen, W., Adachi, A., Wakamatsu, H., Hayashi, S., Takasugi, T., Nagano, M., Nakahama, K., Suzuki, Y., Sugano, S., Iino, M., Shigeyoshi, Y., *et al.* (2002). A transcription factor response element for gene expression during circadian night. *Nature* **418,** 534–539.

Vanselow, K., Vanselow, J. T., Westermark, P. O., Reischl, S., Maier, B., Korte, T., Herrmann, A., Herzel, H., Schlosser, A., and Kramer, A. (2006). Differential effects of PER2 phosphorylation: molecular basis for the human familial advanced sleep phase syndrome (FASPS). *Genes Dev.* **20,** 2660–2672.

Vieira, E., Nilsson, E. C., Nerstedt, A., Ormestad, M., Long, Y. C., Garcia-Roves, P. M., Zierath, J. R., and Mahlapuu, M. (2008). Relationship between AMPK and the transcriptional balance of clock-related genes in skeletal muscle. *Am. J. Physiol. Endocrinol. Metab.* **295,** E1032–E1037.

Vielhaber, E., Eide, E., Rivers, A., Gao, Z. H., and Virshup, D. M. (2000). Nuclear entry of the circadian regulator mPER1 is controlled by mammalian casein kinase I epsilon. *Mol. Cell Biol.* **20,** 4888–4899.

Vinciguerra, M., Fulco, M., Ladurner, A., Sartorelli, V., and Rosenthal, N. (2010). SirT1 in muscle physiology and disease: lessons from mouse models. *Dis. Model Mech.* **3,** 298–303.

Vitaterna, M. H., King, D. P., Chang, A. M., Kornhauser, J. M., Lowrey, P. L., McDonald, J. D., Dove, W. F., Pinto, L. H., Turek, F. W., and Takahashi, J. S. (1994). Mutagenesis and mapping of a mouse gene, Clock, essential for circadian behavior. *Science* **264,** 719–725.

Welsh, D. K., Yoo, S. H., Liu, A. C., Takahashi, J. S., and Kay, S. A. (2004). Bioluminescence imaging of individual fibroblasts reveals persistent, independently phased circadian rhythms of clock gene expression. *Curr. Biol.* **14,** 2289–2295.

Wilson, M. M., Rice, R. W., and Critchlow, V. (1976). Evidence for a free-running circadian rhythm in pituitary-adrenal function in blinded adult female rats. *Neuroendocrinology* **20,** 289–295.

Wise, P. M., Cohen, I. R., Weiland, N. G., and London, E. D. (1988). Aging alters the circadian rhythm of glucose utilization in the suprachiasmatic nucleus. *Proc. Natl. Acad. Sci. USA* **85,** 5305–5309.

Witczak, C. A., Sharoff, C. G., and Goodyear, L. J. (2008). AMP-activated protein kinase in skeletal muscle: from structure and localization to its role as a master regulator of cellular metabolism. *Cell Mol. Life Sci.* **65,** 3737–3755.

Wyse, C. A., and Coogan, A. N. (2010). Impact of aging on diurnal expression patterns of CLOCK and BMAL1 in the mouse brain. *Brain Res.* **1337,** 21–31.

Yagita, K., Horie, K., Koinuma, S., Nakamura, W., Yamanaka, I., Urasaki, A., Shigeyoshi, Y., Kawakami, K., Shimada, S., Takeda, J., Uchiyama, Y., Development of the circadian oscillator during differentiation of mouse embryonic stem cells in vitro. *Proc. Natl. Acad. Sci. USA* **107,** 3846–3851.

Yamanaka, Y., Honma, S., and Honma, K. (2008). Scheduled exposures to a novel environment with a running-wheel differentially accelerate re-entrainment of mice peripheral clocks to new light-dark cycles. *Genes Cells* **13,** 497–507.

Yamazaki, S., Numano, R., Abe, M., Hida, A., Takahashi, R., Ueda, M., Block, G. D., Sakaki, Y., Menaker, M., and Tei, H. (2000). Resetting central and peripheral circadian oscillators in transgenic rats. *Science* **288,** 682–685.

Yamazaki, S., and Takahashi, J. S. (2005). Real-time luminescence reporting of circadian gene expression in mammals. *Methods Enzymol.* **393,** 288–301.

Yin, L., Joshi, S., Wu, N., Tong, X., and Lazar, M. A. (2010). E3 ligases Arf-bp1 and Pam mediate lithium-stimulated degradation of the circadian heme receptor Rev-erb alpha. *Proc. Natl. Acad. Sci. USA* **107,** 11614–11619.

Yin, L., Wang, J., Klein, P. S., and Lazar, M. A. (2006). Nuclear receptor Rev-erbalpha is a critical lithium-sensitive component of the circadian clock. *Science* **311,** 1002–1005.

Yoo, S. H., Yamazaki, S., Lowrey, P. L., Shimomura, K., Ko, C. H., Buhr, E. D., Siepka, S. M., Hong, H. K., Oh, W. J., Yoo, O. J., Menaker, M., and Takahashi, J. S. (2004). PERIOD2::LUCIFERASE real-time reporting of circadian dynamics reveals persistent circadian oscillations in mouse peripheral tissues. *Proc. Natl. Acad. Sci. USA* **101,** 5339–5346.

Zambon, A. C., McDearmon, E. L., Salomonis, N., Vranizan, K. M., Johansen, K. L., Adey, D., Takahashi, J. S., Schambelan, M., and Conklin, B. R. (2003). Time- and exercise-dependent gene regulation in human skeletal muscle. *Genome Biol.* **4,** R61.

Zhang, E. E., and Kay, S. A. (2010). Clocks not winding down: unravelling circadian networks. *Nat. Rev. Mol. Cell Biol.* **11,** 764–776.

Zhang, X., Dube, T. J., and Esser, K. A. (2009). Working around the clock: circadian rhythms and skeletal muscle. *J. Appl. Physiol.* **107,** 1647–1654.

Zheng, B., Larkin, D. W., Albrecht, U., Sun, Z. S., Sage, M., Eichele, G., Lee, C. C., and Bradley, A. (1999). The mPer2 gene encodes a functional component of the mammalian circadian clock. *Nature* **400,** 169–173.

Zylka, M. J., Shearman, L. P., Weaver, D. R., and Reppert, S. M. (1998). Three period homologs in mammals: differential light responses in the suprachiasmatic circadian clock and oscillating transcripts outside of brain. *Neuron* **20,** 1103–1110.

CHAPTER TEN

Regulation of Nucleocytoplasmic Transport in Skeletal Muscle

Monica N. Hall,* Anita H. Corbett,† *and* Grace K. Pavlath‡

Contents

Abstract

Proper skeletal muscle function is dependent on spatial and temporal control of gene expression in multinucleated myofibers. In addition, satellite cells, which are tissue-specific stem cells that contribute critically to repair and maintenance of skeletal muscle, are also required for normal muscle physiology. Gene expression in both myofibers and satellite cells is dependent upon nuclear proteins that require facilitated nuclear transport. A unique challenge for myofibers is controlling the transcriptional activity of hundreds of nuclei in a common cytoplasm yet achieving nuclear selectivity in transcription at specific locations such as neuromuscular synapses and myotendinous junctions. Nucleocytoplasmic transport of macromolecular cargoes is regulated by a complex interplay among various components of the nuclear transport machinery, namely nuclear pore complexes, nuclear envelope proteins, and various

* Graduate Program in Genetics and Molecular Biology, Emory University, Atlanta, Georgia, USA
† Department of Biochemistry, Emory University, Atlanta, Georgia, USA
‡ Department of Pharmacology, Emory University, Atlanta, Georgia, USA

Current Topics in Developmental Biology, Volume 96
ISSN 0070-2153, DOI: 10.1016/B978-0-12-385940-2.00010-3

soluble transport receptors. The focus of this review is to highlight what is known about the nuclear transport machinery and its regulation in skeletal muscle and to consider the unique challenges that multinucleated muscle cells as well as satellite cells encounter in regulating nucleocytoplasmic transport during cell differentiation and tissue adaptation. Understanding how regulated nucleocytoplasmic transport controls gene expression in skeletal muscle may lead to further insights into the mechanisms contributing to muscle growth and maintenance throughout the lifespan of an individual.

Abbreviations

cNLS	classical nuclear localization signal
NLS	nuclear localization signal
Kpna	karyopherin alpha
kpnb1	karyopherin beta1
NPC	nuclear pore complex
Nup	nucleoporin

1. Introduction

Skeletal muscle is a very plastic tissue that readily undergoes changes in mass and function in response to aging, injury, and disease. Such changes in muscle can impact breathing, locomotion, and metabolism and affect motility and lifespan. Proper skeletal muscle function is dependent on spatial and temporal control of gene expression mediated by proteins, such as transcription factors, that require facilitated transport to enter the nuclei. The subcellular localization of these regulatory proteins must be tightly controlled because altered import or export could result in aberrant muscle function.

Proper muscle function is dependent on myofibers, which are multinucleated cells containing many hundreds of nuclei distributed along the length of the cell in a common cytoplasm. Alongside each myofiber are adult muscle stem cells, called satellite cells, that lay beneath the basal lamina surrounding each myofiber. These satellite cells are normally quiescent but in response to muscle damage they are activated to begin proliferating and undergo differentiation and eventual fusion with each other or existing myofibers to repair muscles in a process called myogenesis. Myogenesis can be modeled *in vitro* by culturing myoblasts, the progeny of satellite cells, and inducing them to differentiate into multinucleated myotubes by changes in culture media. Myogenesis both *in vivo* and *in vitro* requires the coordinate activation and repression of many genes. Numerous nuclear proteins are required for proper gene expression and the nuclear repertoire of these proteins is very different along the myogenic continuum of quiescent satellite cells to mature postmitotic myofibers. How satellite cells differentially regulate nucleocytoplasmic transport of these nuclear proteins critical for regulating gene expression during quiescence, activation, and differentiation is unknown. In addition, how a myofiber with hundreds of nuclei coordinates and regulates nucleocytoplasmic transport is not clear.

2. Nuclear Envelope

The nuclear envelope of eukaryotic cells provides separation of the genetic material and transcriptional machinery within the nucleus from the translational machinery in the cytoplasm enhancing regulation of gene expression. The nuclear envelope is comprised of two lipid membrane bilayers, the outer nuclear membrane which is contiguous with the endoplasmic reticulum and the inner nuclear membrane which faces the nucleoplasm (Hetzer and Wente, 2009). The inner nuclear membrane contains integral transmembrane proteins that interact with lamins within a nuclear lamina meshwork of intermediate-type V filaments that lines the inner nuclear membrane. The nuclear lamina contributes to nuclear envelope stability and provides a platform for proteins involved in chromatin anchoring, DNA replication, and gene transcription (Kind and van Steensel, 2010).

Several studies have revealed a critical role for inner nuclear envelope proteins in regulating the expression of muscle-specific genes during muscle differentiation (Datta *et al.*, 2009; Huber *et al.*, 2009; Liu *et al.*, 2009; Ostlund *et al.*, 2009). For example, loss of function or mutation of lamins or lamin-associated inner nuclear membrane proteins can result in tissue-specific diseases which are referred to as nuclear envelopathies (Holaska, 2008; Mattout *et al.*, 2006). These diseases encompass a wide range of clinical phenotypes with different envelopathies affecting different tissues including muscle. Mutations in the nuclear envelope transmembrane protein emerin are associated with Emery-Dreifuss muscular dystrophy (Manilal *et al.*, 1996), while mutations in lamin A/C lead to two muscular dystrophies, Emery-Dreifuss muscular dystrophy and limb-girdle muscular dystrophy 1B (Muchir *et al.*, 2000). Emerin sequesters β-catenin and Lim domain protein, LMO7, at the inner nuclear periphery to regulate their participation in gene transcription; therefore, disease-causing mutations in emerin may disrupt access of these proteins to the transcriptional machinery (Holaska *et al.*, 2006; Markiewicz *et al.*, 2006). In addition, nuclear envelope transmembrane proteins, termed NETs, have been identified, a subset of which are hypothesized to have skeletal muscle-specific roles since they are highly expressed in skeletal muscle tissue compared with other mouse tissues (Chen *et al.*, 2006; Schirmer *et al.*, 2003). Specific roles for NETs appear to exist in signaling pathways during muscle differentiation (Datta *et al.*, 2009; Huber *et al.*, 2009; Liu *et al.*, 2009). For example, during differentiation of C2C12 cells, a mouse muscle cell line (Yaffe and Saxel, 1977), depletion of NET25 led to elevated mitogen-activated kinase (MAPK) signaling which delayed myogenesis (Huber *et al.*, 2009). In contrast, depletion of NET39 accelerated myogenesis through diminished mammalian target of rapamycin (mTOR) signaling and increased insulin-like growth factor 2 (IGF-2) production (Liu *et al.*, 2009). Together, these studies expose a crucial role for nuclear envelope proteins in regulating gene expression during muscle differentiation. Mutations in nuclear envelope transmembrane proteins may alter signaling at the nuclear envelope and may contribute to the altered gene expression observed in laminopathies affecting skeletal muscle. Further studies will likely uncover functional roles for other nuclear envelope proteins in regulating skeletal muscle gene expression.

3. Nuclear Pore Complexes

The nuclear envelope is perforated by nuclear pore complexes (NPCs) which fuse the outer nuclear membrane and inner nuclear membrane together to create channels for nucleocytoplasmic transport (Fig. 10.1; Lim and Fahrenkrog, 2006). NPCs are multiprotein suprastructures ($\sim$50 MDa) which provide channels for the nucleocytoplasmic exchange of ions and macromolecules (Alber *et al.*, 2007). While smaller ions and molecules can diffuse through the NPC, molecules larger than $\sim$40 kDa require a targeting signal and a soluble transport receptor to mediate transport through the NPC (Freitas and Cunha, 2009; Rabut *et al.*, 2004; Weis, 2003). The NPC is comprised of $\sim$30 types of nucleoporins or Nups, many of which are present in multiple copies, consistent with the eightfold symmetry of the NPC (Frenkiel-Krispin *et al.*, 2010). Based on the current model of the NPC, Nups are categorized as scaffold, transmembrane, or peripheral Nups (Fig. 10.1; Fernandez-Martinez and Rout, 2009). Scaffold Nups, also termed core Nups, provide structure to the NPC core by forming a cage-like scaffold, while transmembrane Nups located at the nuclear envelope–NPC interface, function in NPC biogenesis and nuclear envelope anchoring (Strambio-De-Castillia *et al.*, 2010). Peripheral Nups within the NPC function in cargo transport, chromatin anchoring, and gene transcription. Peripheral Nups lining the pore channel contain phenylalanine–glycine (FG) repeats that extend into the channel to function in NPC permeability and mediate facilitated transport of macromolecules

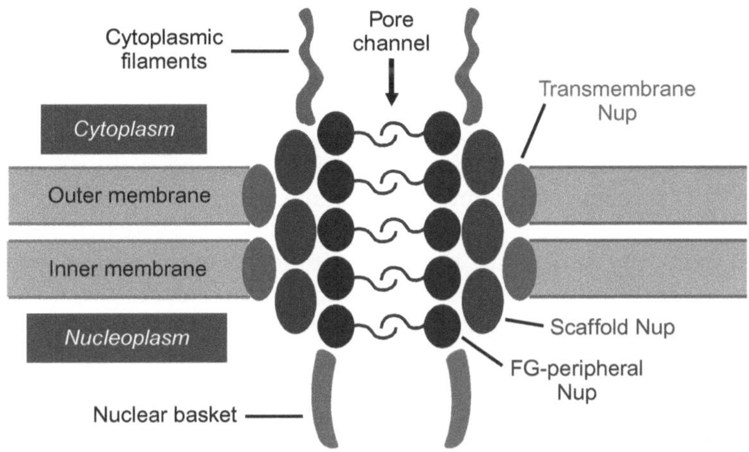

Figure 10.1 Schematic illustrating the relative location of various Nups within the NPC. The NPC resides within the nuclear envelope, a bilipid membrane comprised of an inner and outer nuclear membrane. The NPC has cytoplasmic filaments that extend into the cytoplasm and a nuclear basket that extends into the nucleoplasm. Peripheral Nups containing FG repeats line the pore channel to function in NPC permeability and the facilitated transport of macromolecules. Transmembrane Nups localize to the nuclear envelope–NPC interface, while scaffold Nups reside between transmembrane and peripheral Nups. (See Color Insert.)

(Strambio-De-Castillia *et al.*, 2010). The physical mechanics of NPC permeability and transport are still unclear with several models proposing varying arrangements of FG-Nups during transport; however, the functional role of individual FG-Nups in mediating transport and regulating different transport pathways has been well established (Walde and Kehlenbach, 2010).

NPCs within the nuclear envelope display variability in density and distribution between cell types and even within a single cell (Hetzer and Wente, 2009). For example, NPC density differs between *Xenopus* oocytes (>50 NPCs/mm^2) and C2C12 muscle cells (5 NPCs/mm^2) by 10-fold (D'Angelo *et al.*, 2009; Hetzer and Wente, 2009), meanwhile a 50% increase in NPC density was observed as mouse embryonic stem cells differentiated into cardiomyocytes (Perez-Terzic *et al.*, 2007). In addition, differences in NPC density across the nuclear envelope have been observed in *Saccharomyces cerevisiae* which suggests nuclear transport may be spatially regulated across the nuclear envelope (Winey *et al.*, 1997). NPC nucleoporin composition also varies between cell types where multiple Nups involved in nucleocytoplasmic transport display differential expression between tissues (Hetzer and Wente, 2009; Smitherman *et al.*, 2000; Tran and Wente, 2006). In skeletal muscle, the transcripts for numerous Nups are upregulated during satellite cell activation suggesting that an increase in NPC biogenesis occurs in proliferating satellite cells (Fukada *et al.*, 2007; Pallafacchina *et al.*, 2010). Meanwhile, *Tetrahymena thermophila*, a binucleated ciliated protozoa expresses a subset of Nups that differentially localize to either the macronucleus or the micronucleus to regulate transport of cargoes involved in nucleus-specific functions (Malone *et al.*, 2008). Variations in NPC density, distribution, and nucleoporin composition between cell types or nuclei sharing a common cytoplasm, suggest that NPCs and nucleoporins are regulated to accommodate for ever changing demands on nuclear transport in both mono- and multinucleated cells.

In proliferating eukaryotic cells, new NPCs are formed during mitosis and interphase which allows for regular replacement of Nups (Doucet and Hetzer, 2010). Therefore, *de novo* assembly of NPCs would not be predicted to occur in postmitotic cells. Experiments examining the synthesis of scaffold and peripheral Nups in postmitotic C2C12 myotubes, revealed that some peripheral Nups, such as NUP153 and NUP50, are continuously synthesized, whereas scaffold Nups, such as NUP107 and NUP160, are transcriptionally downregulated in myotubes and are consequently not replaced (D'Angelo *et al.*, 2009). Accumulation or loss of damaged scaffold Nups in nondividing cells, such as muscle, could result in dysfunctional NPCs and loss of integrity of the nucleocytoplasmic barrier. Indeed, oxidized scaffold Nups in neuronal nuclei from aged rats are associated with leakage of cytoplasmic proteins into the nucleus (D'Angelo *et al.*, 2009). These studies have significant implications for the functional integrity of the NPC in quiescent satellite cells and postmitotic multinucleated muscle cells. Further studies are required to determine if Nups in NPCs become damaged in skeletal muscle and whether the nucleocytoplasmic barrier is altered and contributes to muscle dysfunction during disease and aging.

4. Nuclear Import Pathways

Nuclear transport is a process whereby proteins or other macromolecules traverse the NPC either by directly interacting with peripheral FG-Nups within the NPC channel or by binding to import or export receptors that mediate transport through the NPC (Walde and Kehlenbach, 2010). The majority of nuclear transport receptors are karyopherin family members termed importins, transportins, or exportins which mediate transport across the NPC in an energy-dependent manner (Tran and Wente, 2006; Weis, 2003). Transport receptors tend to be divided into importins, which bind a nuclear localization signal (NLS) within a cargo protein to target it into the nucleus and exportins, which bind a nuclear export sequence (NES) within a cargo protein to target it for export to the cytoplasm (Cook *et al.*, 2007; Kalderon *et al.*, 1984a,b; Kutay and Guttinger, 2005; Lange *et al.*, 2007). Karyopherin transport receptors fall within two families, the karyopherin alpha (KPNA) family and the karyopherin beta family (Wagstaff and Jans, 2009). Increasing evidence suggests that both these karyopherin families have critical roles in controlling the nuclear import and export of key proteins involved in genetic reprogramming and cell adaptation in a large number of cell types and tissues (Kohler *et al.*, 1999; Quensel *et al.*, 2004; Talcott and Moore, 2000; Yasuhara *et al.*, 2007). Below we describe the mechanisms of karyopherin-dependent nuclear transport and detail what is currently known about these transport receptors and pathways in skeletal muscle.

4.1. Classical nuclear import: Karyopherin alpha family

Classical nuclear import, which is the best characterized of the nuclear transport pathways, is an active process that depends on KPNA and beta family members as well as a classical nuclear localization signal sequence (cNLS) defined by a string of basic residues contained within a protein (Hodel *et al.*, 2001; Kalderon *et al.*, 1984b; Robbins *et al.*, 1991). In *Mus musculus*, 35–55% of nuclear proteins may depend on classical nuclear import for nuclear targeting as determined using a bioinformatics approach (Marfori *et al.*, 2010). KPNAs recognize two types of cNLS, a monopartite signal comprised of a single string of basic amino acid residues or a bipartite signal containing two strings of basic variable residues flanking a 10–12 amino acid linker region (Kalderon *et al.*, 1984b; Robbins *et al.*, 1991). The prototypical sequence for the monopartite is the SV40 large-T antigen sequence, PKKKRKV, and for the bipartite, the nucleoplasmin sequence KRPAATKKAGQAKKKK (Hodel *et al.*, 2001; Kalderon *et al.*, 1984a; Robbins *et al.*, 1991). Classical nuclear protein import is mediated by a heterotrimeric complex of KPNA, which recognizes and binds the cNLS signal within a cargo protein, and karyopherin beta1 (KPNB1), which binds KPNA and interacts with FG-Nups within the NPC to mediate nuclear import (Fig 10.2; Matsuura *et al.*, 2003; Matsuura and Stewart, 2005). Once in the nucleus, a small GTPase, Ran-GTP, binds KPNB1, triggering disassembly of the trimeric complex and subsequent cargo release (Cook *et al.*, 2007). Disassembly of the import complex is also facilitated by CAS, the export receptor for KPNA and another member of the karyopherin family (Kutay *et al.*, 1997), and the nucleoporin, NPAP60L (Ogawa *et al.*, 2010). Upon cargo release, KPNA is recycled back to the cytoplasm by CAS, while KPNB1 is returned to the cytoplasm in complex with Ran-GTP (Hood and Silver, 1998; Kutay *et al.*, 1997). Thus, classical nuclear import cycles and directionality depend upon the GTPase, Ran, which facilitates assembly and disassembly of transport complexes (Cook *et al.*,

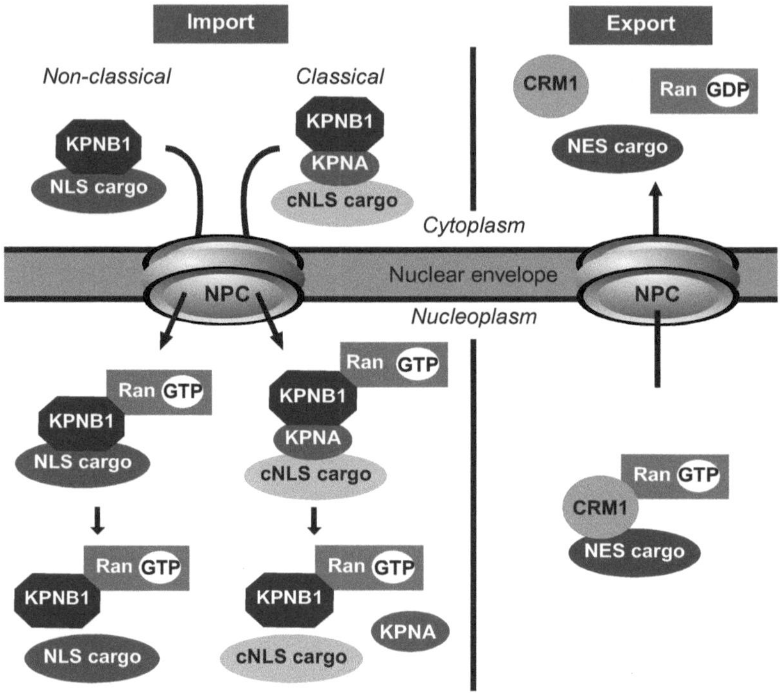

Figure 10.2 Basic model of karyopherin-mediated nuclear import and export pathways. The nuclear import pathway involves the import receptors karyopherin alpha (KPNA) and/or KPNB1 which can recognize proteins containing a classical (cNLS) or nonclassical (NLS) nuclear localization signal, respectively. The nuclear import of a cNLS-containing cargo involves both KPNA and KPNB1 import receptors, since KPNA recognizes the cNLS motif in the cargo protein and then KPNB1 mediates translocation of the import complex through the NPC by interacting with FG-Nups within the NPC. Once in the nucleus, Ran-GTP binding to KPNB1 results in the dissociation of the import complex and cargo release into the nucleus. The nuclear export pathway consists of an obligate trimeric complex consisting of the exportin (CRM1 for classical NES-containing cargo), export cargo, and Ran-GTP. Translocation of the complex through the NPC is mediated by interaction between the exportin and FG-Nups. Once in the cytoplasm, the hydrolysis of Ran-GTP to Ran-GDP results in the dissociation of the export complex and subsequent cargo release. (See Color Insert.)

2007). Ran-GTP, but not Ran-GDP, triggers cargo release in the nucleus upon binding to KPNB1, therefore, the directionality of import is driven by the presence of Ran-GTP in the nucleus and Ran-GDP in the cytoplasm (Lonhienne *et al.*, 2009). Ran-GTP levels are maintained in the nucleus through the nuclear import of Ran-GDP (Ribbeck *et al.*, 1998) and conversion to Ran-GTP by the nuclear localized Ran guanine nucleotide exchange factor (Ran GEF; Cook *et al.*, 2007; Smith *et al.*, 1998).

Meanwhile, in the cytoplasm, Ran-GTP is hydrolyzed to Ran-GDP by the Ran GTPase-activating protein (Ran GAP).

KPNB1 is the sole member of the karyopherin beta family to participate in classical nuclear import with KPNA (Liu and Liu, 2007). KPNB1 and other karyopherin beta family members participate in nonclassical nuclear import which involves either direct NLS binding or the use of non-KPNA adaptor proteins. Modeling studies for cNLS import reveal that the addition of the KPNA adaptor to the transport cycle, as opposed to direct protein import by KPNB1 alone, reduces import efficiency (Riddick and Macara, 2007); however, the loss of import efficiency is offset by the addition of multiple KPNA adaptor paralogs that allows for additional points of control over the nuclear localization of cNLS proteins.

Saccharomyces cerevisiae contains a single, essential KPNA import receptor. Six KPNA paralogs are found in mouse: KPNA1, KPNA2, KPNA3, KPNA4, KPNA6, and KPNA7 (Hu *et al.*, 2010; Tsuji *et al.*, 1997). Seven KPNA paralogs exist in human with which the mouse homologues share 90% amino acid identity (Kelley *et al.*, 2010; Kohler *et al.*, 1997, 1999; Tsuji *et al.*, 1997). Confusion regarding KPNA protein nomenclature between species exists in the literature. While KPNA gene names between human and mouse are consistent, protein nomenclature using the terms importin alpha or importins does not match between these two species (Table 10.1). In this chapter, we refer to KPNA paralogs using the *KPNA/kpna* nomenclature, instead of importin alpha, to minimize confusion when discussing KPNA paralog function. KPNA paralogs in mouse and human are categorized into three subtypes based on percentage of amino acid identity. Mouse subtypes are Subtype S: KPNA1 and KPNA6; Subtype P: KPNA2; and Subtype Q: KPNA3 and KPNA4. Subtype members share 80–90% amino acid identity, whereas different subtypes share 40–50% amino acid identity (Tsuji *et al.*, 1997). All KPNA paralogs function as nuclear import receptors, but paralogs may differ in their cNLS binding affinities and/or specificities for cNLS proteins (Hodel *et al.*, 2001; Kohler *et al.*, 1999; Quensel *et al.*, 2004; Talcott and Moore, 2000; Timney *et al.*, 2006; Yasuhara *et al.*, 2007). For example, human regulator of chromosome condensation 1, RCC1, which is the Ran GEF, depends solely on KPNA4 to access to the nucleus, while other cNLS proteins, such as RNA Helicase A, may utilize multiple KPNA paralogs to access the nucleus, but may have preference for one paralog over another (Aratani *et al.*, 2006; Quensel *et al.*, 2004). *In silico* experiments suggest that the rate of nuclear import of a cNLS cargo is limited by the levels of KPNA and Ran (Riddick and Macara, 2005).

The steady-state levels of different KPNA paralogs can vary both among tissue types and within a single tissue during differentiation suggesting distinct roles for individual KPNA paralogs in importing key factors required for cell function and differentiation (Goldfarb *et al.*, 2004; Mason and Goldfarb, 2009; Okada *et al.*, 2008; Poon and Jans, 2005). For example,

Table 10.1 Karyopherin alpha paralogs in different organisms

	Homo sapiens		*Mus musculus*		*Drosophila melanogaster*		*Saccharomyces cerevisiae*	
Subtype	Gene name	Importin designation	Gene name	Importin designation	Gene name	Importin designation	Gene name	Importin designation
S	*KPNA1*	alpha5	*Kpna1*	alpha1	*Kap-alpha1*	alpha1	*Srp1*	Kap60
	KPNA5	alpha6	–	–	–	–	–	–
	KPNA6	alpha7	*Kpna6*	alpha6	–	–	–	–
P	*KPNA2*	alpha1	*Kpna2*	alpha2	*Pen*	alpha2	–	–
	KPNA7	alpha7	*Kpna7*[a]	alpha7[a]	–	–	–	–
Q	*KPNA3*	alpha4	*Kpna3*	alpha3	–	–	–	–
	KPNA4	alpha3	*Kpna4*	alpha4	*Kap-alpha3*	alpha3	–	–

KPNA family members are categorized into three subtypes, S, P, and Q based on amino acid sequence homology.
Saccharomyces cerevisiae has a single karyopherin (SRP1), while *Homo sapiens*, *Mus musculus*, and *Drosophila melanogaster* each contain multiple karyopherin paralogs. The gene name and importin alpha designation are given for each species to clarify confusion regarding karyopherin/importin designations between species. A dash (–) indicates the absence of a karyopherin homologue in that species.

[a] Placement of recently discovered murine KPNA7 into subtype P is tentative.

during mouse spermatogenesis, KPNA paralogs are expressed with unique cellular and temporal expression profiles at discrete stages of development (Hogarth *et al.*, 2006). In contrast, during neural differentiation of mouse embryonic stem cells *in vitro*, the steady-state levels of one KPNA paralog increase, while that of another paralog decrease, thereby allowing for differential nuclear import of transcription factors involved in maintaining either the undifferentiated or differentiated state (Yasuhara *et al.*, 2007). This KPNA paralog switching was proposed by the authors of this study as a general mechanism that enables cells to coordinate differentiation by controlling the subcellular localization of transcription factors. In support of the subtype switching model, KPNA steady-state profiles in multiple differentiating human and mouse cell types are characterized by increases in expression of one KPNA paralog with concomitant decreases in another paralog (Kamei *et al.*, 1999; Kohler *et al.*, 1997, 2002; Okada *et al.*, 2008); however, in skeletal muscle cells, an increase in the steady-state levels of all five *Kpna*s was observed during differentiation (Hall *et al.*, unpublished data). The increase in all *Kpna*s may suggest an overall increase in demand for nuclear import during skeletal muscle differentiation. These studies suggest that the role of *KPNA*s in cell differentiation may differ between cell types that express different cNLS-containing cargo proteins.

Additional evidence for the nonredundant roles of individual KPNA paralogs in cellular physiology stems from loss-of-function experiments in model organisms. *Kpna1* null *Drosophila melanogaster* developed normally but displayed defects in gametogenesis resulting in sterility in both males and females (Ratan *et al.*, 2008). Similarly, male and female sterility occurred in *Kpna2* null flies (Mason *et al.*, 2002). The sterility in females could be rescued only by *Kpna2* transgenes, whereas the sterility in males could be rescued by *Kpna1*, *Kpna2*, or *Kpna3* transgenes suggesting distinct requirements for KPNA2 in male and female gametogenesis (Mason *et al.*, 2002). In contrast, *Kpna3* null flies displayed defects throughout development, whereas late stages of development and photoreceptor development could only be rescued with *Kpna3* but not *Kpna1* or *Kpna2* transgenes (Mason *et al.*, 2003). RNAi experiments in *Caenorhabditis elegans* demonstrated that KPNA3 but not KPNA2 is required for oocyte development (Geles and Adam, 2001). Together, these results in genetic model organisms support the notion that KPNA paralogs have evolved distinct functions in different cell types during development.

Loss-of-function experiments *in vitro* also provide support for distinct roles of KPNA paralogs in controlling cell proliferation and differentiation. During neural differentiation of mouse embryonic stem cells, depletion of KPNA1 by RNAi-mediated knockdown resulted in accelerated neural differentiation, while loss of KPNA5 delayed differentiation (Yasuhara *et al.*, 2007). In primary mouse muscle cells, the nonredundant roles for individual paralogs were revealed by siRNA experiments in which KPNA1 knockdown increased myoblast proliferation but KPNA2 knockdown decreased proliferation (Hall *et al.*, unpublished data). In contrast, no proliferation defect was observed with KPNA4 knockdown. KPNAs import negative regulators of proliferation, such as Rb and p27Kip1 in other cell types (Hu *et al.*, 2005; Shin *et al.*, 2005), so results of these RNAi experiments could suggest that KPNA1 imports a negative regulator of proliferation during myogenesis. Contrary to these findings, in Hela cells, RNAi-mediated knockdown of each KPNA, including KPNA1, resulted in a decrease in cell proliferation, which suggests that the role of individual KPNAs in cell proliferation may differ between cell types (Kohler *et al.*, 2002). Contrary to the results obtained *in vitro*, a recent study examining a *Kpna1* null mice revealed normal development of brain and other tissues, however, KPNA1 may have a different role in proliferation in the brain or compensation by other KPNAs may have occurred during development (Shmidt *et al.*, 2007). Further evidence for the nonredundant roles of KPNA paralogs in skeletal muscle comes from knockdown experiments where depletion of KPNA2, but not KPNA1 or KPNA4, resulted in reduced myotube size and decreased ability of muscle cells to migrate (Hall *et al.*, unpublished data). The small myotube phenotype observed may be due to the reduced import of multiple cNLS-containing proteins involved in regulating cell migration. Indeed, a similar migration defect was also observed upon loss of KPNA2 in a lung cancer cell line (Wang *et al.*, 2010). These studies suggest that myoblast proliferation and myotube growth rely on specific KPNA paralogs to regulate the nuclear import of key factors involved in proliferation and myotube growth during myogenesis.

Classical nuclear import has been implicated in transporting cargoes across large distances or from specific sites in the cell to the nucleus. This phenomenon has been most extensively studied in neurons (Lai *et al.*, 2008; Mikenberg *et al.*, 2007; Thompson *et al.*, 2004), but emerging evidence supports the presence of such spatial signaling in skeletal muscle. In rodent hippocampal neurons, classical nuclear import mediates the transport of cNLS cargoes from the synapse to the nucleus upon receptor activation (Thompson *et al.*, 2004). In skeletal muscle cells, KPNA mediates the nuclear import of myopodin, an actin bundling protein (Faul *et al.*, 2007), which has been shown to shuttle between the sarcomeric Z-disc and the nucleus in a differentiation and stress-dependent manner (Weins *et al.*, 2001). However, the role of myopodin nucleocytoplasmic shuttling during cell differentiation and stress is unclear. In contrast, a Z-disc-associated protein with known nuclear function is muscle limb protein (MLP), which acts as a mechanosensor in rat cardiomyocytes (Boateng *et al.*, 2009). Loss of cNLS-dependent import of MLP results in disarranged sarcomeres. Another cargo that undergoes similar shuttling is serum response factor (SRF), a transcription factor required for skeletal muscle growth that shuttles between the sarcomere and the nucleus (Li *et al.*, 2005). The nuclear import of SRF occurs via KPNA1/KPNB1 of the classical nuclear import pathway (McConville *et al.*, 2010). These findings suggest that nucleocytoplasmic import has a significant role in transmitting signals from sarcomeres to the nucleus in response to stimuli that induce muscle cell remodeling to adapt to cellular stress. Further studies should shed light on the role of classical nuclear import in overcoming the unique spatial challenges of signaling in a multinucleated muscle cell.

4.2. Karyopherin beta family members mediate import and export

Karyopherin beta family members comprise the majority of nuclear transport receptors which includes karyopherin beta import receptors (importins or transportins) involved in protein import and exportins involved in protein export (Cook and Conti, 2010; Cook *et al.*, 2007). Similar to KPNA-dependent nuclear import, the karyopherin beta nuclear import pathway is dependent upon energy and the Ran gradient for directionality (Lonhienne *et al.*, 2009). Karyopherin beta import receptors bind directly to nonclassical NLSs in cargo proteins to mediate import (Fig. 10.2) or may use a non-KPNA adaptor protein for NLS cargo recognition (Cook *et al.*, 2007; Kutay and Guttinger, 2005; Lange *et al.*, 2007). Currently, only karyopherin beta2-dependent cargoes have defined NLSs termed PY-NLS, while cargoes depending on other karyopherin beta family members do not have recognizable amino acid sequences that comprise NLS motifs (Marfori *et al.*, 2010).

In total, 14 karyopherin beta import receptors exist in *S. cerevisiae*, while humans have over 19 karyopherin betas (Chook and Suel, 2010). Karyopherin beta family members display protein homology ranging from 15% to 20% and function in the import of distinct sets of proteins, RNAs, and Nups (Chook and Suel, 2010; Marfori *et al.*, 2010). KPNB1, the best characterized member of the karyopherin family, is one of four essential karyopherin beta family members in *S. cerevisiae* (Chook and Suel, 2010). In contrast to classical nuclear import, KPNB1-dependent nuclear import does not occur through a conserved receptor–cargo binding conformation which provides KPNB1 with the flexibility to bind a wide variety of cargoes containing unique classes of NLS signals (Fiserova *et al.*, 2009; Marfori *et al.*, 2010).

Several studies provide evidence that karyopherin beta family members have critical roles in regulating the cellular localization of different NLS proteins involved in myogenesis and neuromuscular junction physiology (Giagtzoglou *et al.*, 2009; Higashi-Kovtun *et al.*, 2010; Mosca and Schwarz, 2010; van der Giessen and Gallouzi, 2007). In C2C12 myoblasts, the nuclear import of the RNA-binding protein HuR depends upon the karyopherin beta family member transportin-2 (van der Giessen and

Gallouzi, 2007). HuR is a RNA-binding protein involved in regulating the stability of mRNA transcripts encoding MyoD and Myogenin which are myogenic transcription factors required for differentiation (Figueroa *et al.*, 2003; van der Giessen and Gallouzi, 2007). During differentiation, cleavage of HuR prevents its nuclear import by transportin-2 which results in the stabilization of MyoD and Myogenin mRNAs and enhancement of myogenesis (Mazroui *et al.*, 2008). Several studies using a *Drosophila* model system have uncovered roles for karyopherin beta family members in regulating the import of proteins involved in postsynaptic membrane development and neurotransmitter release at the neuromuscular junction (Giagtzoglou *et al.*, 2009; Higashi-Kovtun *et al.*, 2010; Mosca and Schwarz, 2010). Wingless signaling at the neuromuscular junction causes cleavage and release of the C terminus of Frizzled2 (Fz2-C), which is then imported into the nucleus by KPNA2/KPNB1, or by the karyopherin beta family member karyopherin-beta11 (Mosca and Schwarz, 2010). In *Drosophila* mutants lacking either KPNA2 or karyopherin-beta11, a reduction in the nuclear import of Fz2-C and defects in the postsynaptic membrane were observed suggesting that multiple transport pathways are required for membrane development at the neuromuscular junction. Another karyopherin beta involved in neuromuscular junction physiology is *Drosophila* karyopherin beta13 which controls neurotransmitter release and intracellular Ca^{2+} levels at the neuromuscular junction (Giagtzoglou *et al.*, 2009). These data suggest that karyopherin beta family members regulate myogenesis, neuromuscular development and neurotransmitter release by importing a variety of proteins, including RNA-binding proteins and cell-surface receptor components.

Karyopherin beta family members also play roles in facilitated nuclear export of proteins to the cytoplasm (Cook and Conti, 2010; Wente and Rout, 2010). The best understood export pathway is the recognition of a classical nuclear export signal (NES) by the karyopherin beta, CRM1 (Fig. 10.2). The classical NES signal consists of a short string of hydrophobic leucine-rich residues, which are difficult to identify because they share sequence similarity with the hydrophobic cores of most proteins (Cook *et al.*, 2007). Prototypical sequences for an NES are the cyclin D NESs, RFLSLEPL, and TPTDVRDVDI as well as the mitogen-activated protein kinase kinase (MAPKK) NES, LQKKLEELEL (Kutay and Guttinger, 2005; Poon and Jans, 2005). Export receptors or exportins recognize and bind export cargoes while bound to the GTPase, Ran-GTP, in an obligate trimeric complex (Cook *et al.*, 2007; Kutay and Guttinger, 2005). As with facilitated nuclear import, directionality of export is driven by the compartmentalization of Ran-GTP in the nucleus and Ran-GDP in the cytoplasm.

At least six exportin genes are found in mouse and human (Okada *et al.*, 2008). These exportins facilitate the nuclear export of a variety of cargoes with some exportins displaying different specificity for individual cargoes. Similar to NLS-mediated import, NES-containing proteins may be exported by a single exportin or may utilize multiple exportins (Okada *et al.*, 2008). For example, the essential yeast exportin, CSE1, or CAS in vertebrates, has one export cargo, KPNA of the classical nuclear import pathway (Cook *et al.*, 2007), while the best characterized exportin, CRM1 (XPO1 or exportin-1), mediates export of at least 10 different classical NES-containing proteins (Cook *et al.*, 2007; Shen *et al.*, 2010; Wada *et al.*, 1998). Exportin family members also facilitate the export of tRNAs and pre-miRNAs (Cook *et al.*, 2007; Lund *et al.*, 2004). Together, these studies suggest exportins may have differing roles in the export of a wide variety of cargo proteins and RNAs required for proper cell function.

Nucleocytoplasmic shuttling via CRM1-dependent export and KPNA-dependent import coordinate the nuclear steady-state levels of proteins critical to muscle cell biology. For example, the nuclear localization of the transcription factor NF-κB is controlled by the subunit p65 which mediates both import and export of the complex through binding to KPNA/KPNB1 or CRM1, respectively (Micheli *et al.*, 2010; Zerfaoui *et al.*, 2010). The nuclear accumulation of p65 NF-κB suppresses MyoD transcription (Guttridge *et al.*, 2000) therefore modulation of nuclear import or export of p65 could control p65 NF-kB activity, MyoD expression, and ultimately myogenesis. The nuclear import and export of the forkhead box transcription factor FOXO3a is critical for skeletal muscle atrophy (Sandri *et al.*, 2004). In C2C12 cells, the nuclear import of Foxo3a was observed upon inhibition of the phosphatidylinositol 3-kinase, PI3K/Akt pathway, while nuclear export of FOXO3a was observed upon activation of the stress-activated protein kinase (SAPK) pathway (Clavel *et al.*, 2010). Control over the cellular localization of FOXO3a by two different signaling pathways may provide global control over atrophy by regulating the transcription of genes, such as *Atrogin-1*, that are involved in skeletal muscle atrophy (Clavel *et al.*, 2010). Control over the nucleoctyoplasmic shuttling of cargo proteins by different transport pathways may provide global control over gene expression during myogenesis.

5. Identifying Classical Nuclear Import-Dependent Cargoes

Identifying the specific cargo proteins that are transported via various nucleocytoplasmic transport pathways is key for understanding the regulatory networks that govern cell function. Here we focus on how cNLS-dependent cargoes are identified since this pathway is the best characterized nuclear transport pathway; however, many of the challenges in cargo identification presented here apply also to other receptor-mediated transport pathways. In *Mus musculus*, 30–55% of nuclear proteins are predicted to depend on classical nuclear import (Marfori *et al.*, 2010). However, only a few proteins with key functional roles in muscle are known to contain a functional cNLS, such as NOTCH (Huenniger *et al.*, 2010) and NFATc2 (Okamura *et al.*, 2000). While bioinformatics approaches exist to identify classical nuclear import signal sequences within cargoes, these putative cargoes still require functional testing to ensure that such signals actually mediate transport via this pathway.

Putative cNLS motif sequences within proteins can be identified with prediction software (Cokol *et al.*, 2000; Horton *et al.*, 2007; Kosugi *et al.*, 2009; Nguyen Ba *et al.*, 2009). The consensus sequence for the monopartite cNLS has been characterized in both structural and thermodynamic studies where the first residue is a lysine followed by a second and fourth basic residue as follows: K(K/R)X(K/R) (Conti and Kuriyan, 2000; Fontes *et al.*, 2000; Hodel *et al.*, 2001). The consensus sequence for the bipartite cNLS has also been characterized as KRX10-12KRRK (Fontes *et al.*, 2003). Small deviations from the consensus sequence may increase or decrease KPNA-cargo binding affinity, while large deviations likely result in failed import because KPNA-cargo binding is either too weak or too strong for efficient cargo import and release (Lange *et al.*, 2007). A drawback to cNLS prediction algorithms is that linear sequence is analyzed and these algorithms do not identify nonlinear synthetic cNLS signals created through intra- or interprotein interactions. For example, signal transducer and activator of transcription (STAT1) forms a homodimeric complex in which each dimer contributes basic resides to form a functional synthetic cNLS that is not detected by current cNLS algorithms (Fagerlund *et al.*, 2002). While cNLS prediction models identify consensus sequences, functional studies meeting several criteria must be performed before a cNLS is deemed functional (Lange *et al.*, 2007). A cNLS is functional if it is both necessary and sufficient for import of the cargo protein and import of the cargo depends upon the classical nuclear import machinery (Lange *et al.*, 2007).

An alternative approach to identifying cNLS-dependent cargoes is a candidate-based approach which involves transport receptor loss-of-function experiments, whereby a phenotype observed upon receptor depletion may offer hints to potential cargo. A candidate-based approach may prove difficult since the phenotypes observed during depletion of transport receptors are likely combinatorial due to the altered nuclear transport of many cargo proteins that are involved in regulating a large number of genes. Overall, identifying cargoes dependent upon classical nuclear import receptors will be critical to understanding the role of nucleoctyoplasmic transport in regulating cell function and fate in skeletal muscle.

6. Remodeling of the Nuclear Transport Machinery

Alterations in global nucleocytoplasmic transport provide another layer of control over gene expression. Global changes in the efficiency or rate of nuclear transport can occur through alterations or remodeling of key components of the nuclear transport machinery. For example, altering the expression or localization of karyopherin transport receptors, Ran and/or Ran-associated proteins, or Nups results in changes in transport efficiency (Hodel *et al.*, 2001, 2006; Riddick and Macara, 2005; Timney *et al.*, 2006; Wagstaff and Jans, 2009). A muscle cell could alter nucleocytoplasmic transport to adjust for changes in demand for nuclear transport over a wide range of cellular conditions such as cell quiescence, proliferation, differentiation, stress, aging, and disease.

Experimental evidence indicates that steady-state levels for different components of the nuclear transport machinery can vary during myogenesis. Microarray analyses suggest that satellite cell entry into the cell cycle is marked by global remodeling of the nuclear transport machinery (Fukada *et al.*, 2007; Pallafacchina *et al.*, 2010). A wide variety of mRNAs that encode components of the nuclear transport machinery, such as nucleoporins and various karyopherin transport receptors, were increased in proliferating satellite cells *in vivo* compared to quiescent satellite cells. This widespread upregulation of the nuclear transport machinery may be functionally required to allow for rapid changes in gene expression associated with the myogenic lineage progression of satellite cells. Nuclear pore composition can influence stem cell differentiation as evidenced from studies of NUP-133-deficient epiblast and embryonic stem cells in mice which differentiated inefficiently along the neural lineage (Lupu *et al.*, 2008). Nuclear transport machinery remodeling has also been observed during muscle differentiation, however the extent and type of remodeling differs among muscle cell types. In mouse primary muscle cells, the steady-state levels of all five KPNA import receptors increased during skeletal muscle differentiation suggesting an increase in demand for nuclear import during differentiation (Hall *et al.*, unpublished data). The differentiation of mouse embryonic stem cells into a cardiac lineage resulted in the down-regulation of karyopherins, exportins, Nups- and Ran-related proteins, while an increase in NPC density was observed along with the expansion of individual NPC diameter suggesting an increase in demand for nuclear transport during cardiac differentiation (Perez-Terzic *et al.*, 2007). In contrast, during the differentiation of C2C12s, the steady-state levels of Nup proteins and overall density of NPCs remained constant (D'Angelo *et al.*, 2009). Differences in nuclear transport machinery remodeling appear to be cell type-dependent and suggest that remodeling of the nuclear transport machinery is a key process that controls global nucleocytoplasmic transport during cellular differentiation. Characterizing the functional role of members of the nuclear transport machinery during skeletal muscle proliferation and differentiation will be essential to understanding the role of nucleocytoplasmic transport as a driver of muscle cell differentiation and function.

Remodeling of the nuclear transport machinery has also been observed during cellular response to chemical or mechanical stress in multiple cell types. Remodeling of the nuclear transport machinery to reduce or block transport during cellular stress may allow a cell to globally "pause" gene signaling pathways in order to redirect gene expression to respond to a particular cellular stress. Cellular stress, such as oxidative stress, can inhibit both import and export transport receptors and reduce the levels, localization and posttranslation modifications of several Nups involved in nuclear export (Crampton *et al.*, 2009; Kodiha *et al.*, 2008; Miyamoto *et al.*, 2004). In vascular smooth muscle cells (VSMCs), remodeling of the nuclear import machinery was observed during exposure to ceramide, a antiproliferative sphingolipid implicated in the final stage of atherosclerotic plaque formation. Ceramide treatment of cultured VSMCs resulted in reduced cell proliferation and inhibition of classical nuclear import due to mislocalization of KPNA and CAS to the cytoplasm (Faustino *et al.*, 2008). Nuclear transport machinery remodeling during mechanical stress was also observed during stretching of VSMCs which results in smooth muscle cell hyperplasia and hypertrophy (Richard *et al.*, 2007). Mechanical stretching of VSMCs *in vitro* resulted in an increase in nuclear import and the steady-state protein levels of Nups, along with alterations in MAPK signaling. Further studies are required to determine whether remodeling of the nuclear transport machinery also occurs with oxidative or mechanical stress in skeletal muscle cells and what role it may play in controlling the nuclear localization of critical proteins necessary for cellular responses to these physiologic perturbations.

Aging and disease are also associated with changes in the nuclear transport machinery. In myocardial microvascular endothelial cells and human fibroblasts, a reduction in expression of KPNA and therefore, classical nuclear import occurred with cellular aging (Ahluwalia *et al.*, 2010; Pujol *et al.*, 2002). Another study using Hela cells expressing a mutant form of Lamin A responsible for the premature aging disease, Hutchinson Gilford progeria syndrome, revealed a reduction in nuclear import efficiency along with alterations in the localization of NUP62, NUP153, and the exportin CRM1 (Busch *et al.*, 2009), suggesting that remodeling of the nuclear import machinery may have a role in disease pathology. In contrast, human cardiomyocytes from patients with heart failure, displayed an increase in karyopherin import receptors, exportins, Ran regulators, and Nups as well as differences in NPC configuration and morphology as compared to healthy cardiomyocytes (Cortes *et al.*, 2010) suggesting that remodeling may occur as a disease response. These findings suggest that changes in the nuclear transport machinery may occur either as result of aging and disease or may occur in response to disease to restore tissue function. Given the extensive loss of skeletal muscle mass that can occur with aging or disease, further studies are warranted to determine if remodeling of the nuclear transport machinery also occurs with age or disease in skeletal muscle and whether nucleocytoplasmic communication is impaired as a consequence.

7. Challenges in Studying Nucleocytoplasmic Transport in Multinucleated Cells

The basic mechanics for nucleocytoplasmic transport of proteins and RNA identified to date and described in this review have almost exclusively been examined in cells with a single nucleus. Skeletal muscle is the only permanent multinucleated cell type in the body and constitutes ~50% of body mass; yet, how these cells spatially and temporally regulate and coordinate nucleocytoplasmic transport among hundreds of nuclei is unknown.

Spatial and temporal regulation of nucleocytoplasmic transport between nuclei likely occurs within a single myofiber since transcriptional activity of specific gene loci can differ among nuclei within the same myofiber. Such differences in transcription could arise from specific regional requirements for cell function. Myonuclei at the neuromuscular junction express transcripts for subunits of the acetylcholine receptor at much higher levels than nonsynaptic nuclei leading to the accumulation of the acetylcholine receptor protein at the neuromuscular junction (Burden, 1993; Fontaine and Changeux, 1989; Sanes *et al.*, 1991; Simon *et al.*, 1992) and thereby facilitating coordinated neuronal activation of muscle contraction. In addition, the myonuclei located at the myotendinous junction in stretched myofibers express the transcript for sarcomeric myosin heavy chain at higher levels than other myonuclei (Dix and Eisenberg, 1990), thereby enhancing sarcomere addition and cell growth at the ends of myofibers in response to muscle stretching. Less clear are the reasons why myonuclei distributed along the length of a myofiber and at times even right next to each other exhibit differences in the transcription of both endogenous genes and transgenes (Newlands *et al.*, 1998). Transcriptional differences among nuclei also occur in myotubes *in vitro* (Berman *et al.*, 1990; Su *et al.*, 1995). The molecular mechanisms responsible for such nuclear diversity in transcriptional activity in skeletal muscle are unknown. Differences in transcriptional activity among nuclei in a common cytoplasm also occur in other cell types and organisms. For example, only a subset of nuclei in multinucleated mouse osteoclasts or human placental syncytiotrophoblasts is transcriptionally active (Ellery *et al.*, 2009; Youn *et al.*, 2010). In binucleated *Tetrahymena thermophila*, the macronucleus is transcriptionally active, whereas the micronucleus is transcriptionally inert (Karrer, 2000). Further, nuclei in the syncytial blastoderm of *Drosophila* and the syncytial germ line of *C. elegans* are also transcriptionally distinct (Burden, 1993). The molecular mechanisms responsible for the transcriptional differences among nuclei in these other cell types and organisms are not fully elucidated.

Nuclear proteins are localized to some nuclei and not others within a myofiber which further suggests differential nuclear targeting occurs in skeletal muscle. One such protein is endonuclease G, which is a mitochondrial protein that can translocate to nuclei and induce DNA fragmentation and apoptosis independent of caspase. At the initiation of disuse muscle atrophy, endonuclease G translocates to a subset of myofiber nuclei (Dupont-Versteegden *et al.*, 2006) and may serve as a means to control the loss of myonuclei commonly observed in disuse atrophy without cell death. Other examples of such proteins are specific nuclear envelope proteins which are more highly concentrated in synaptic myonuclei compared to nonsynaptic nuclei. These include Syne-1, Syne-2, and nesprin-1α which appear to participate in nuclear localization and/or anchoring (Apel *et al.*, 2000; Grady *et al.*, 2005; Puckelwartz *et al.*, 2010; Zhang *et al.*, 2007). Unequal nuclear localization of proteins is also observed in cultured myotubes. These include transcription factors with important roles in muscle differentiation and growth such as NFAT5 (O'Connor *et al.*, 2007), NFATc1 (Abbott *et al.*, 1998), and MyoD (Ferri *et al.*, 2009), the growth inhibitory protein, myostatin (Artaza *et al.*, 2002) as well as MYO18B (Salamon *et al.*, 2003), an unconventional myosin heavy chain with an unidentified role in muscle physiology. The mechanisms that govern differential protein targeting among neighboring myonuclei *in vivo* and *in vitro* are unknown but may be related to nucleus-specific transport mechanisms and merit further study.

Studies in *Tetrahymena* provide potential clues for how nucleus-specific transport mechanisms may be regulated in multinucleated myofibers. Certain nuclear proteins in *Tetrahymena* are selectively accumulated in either macro- or micro-nuclei (White *et al.*, 1989). Analyses of GFP-labeled KPNA proteins revealed that 9 of the 13 KPNA proteins localized exclusively to the micronucleus suggesting that nucleus-specific transport systems must exist (Malone *et al.*, 2008). Further studies demonstrated that the NPCs of macronuclei and micronuclei contain unique subsets of FG-containing nucleoporins which are responsible for this nuclear selectivity (Iwamoto *et al.*, 2009; Malone *et al.*, 2008). Interestingly, homologs of NUP98 contributed to nuclear selectivity: two NUP98 homologs localized exclusively to macronuclei, whereas the other two exclusively localized to micronuclei. Specific structural components of the NUP98 homologs were functionally required for the nuclear selectivity as shown by chimeric protein experiments (Iwamoto *et al.*, 2009). The NUP98 homologs that localized to the macronucleus contained amino acid repeats of GLFG, whereas homologs that localized to the micronucleus lacked GLFG and instead contained novel NIFN repeats. These results suggest that structural alterations of the NPC can contribute to nucleus-specific protein transport in a multinucleated cell. Such structural alterations may modulate the interaction of karyopherin transport receptors with specific components of the NPC and consequently alter nuclear accumulation of proteins. Immunofluorescence analyses of primary mouse myotubes *in vitro* reveal that the steady-state levels of KPNA2 differ among nuclei supporting the hypothesis that specific karyopherin transport receptors may also undergo selective nuclear targeting in skeletal muscle as in *Tetrahymena* (Hall *et al.*, unpublished data). Further studies are required to define the contribution of NPC composition and karyopherin transport receptors to nuclear differences in transcription and protein content in skeletal muscle.

8. Summary

Nucleocytoplasmic transport plays a key regulatory role in cellular physiology. While much is known about facilitated nuclear transport in other cell types, the study of nucleocytoplasmic transport in skeletal muscle is still in its infancy. Multinucleated myofibers are faced with unique challenges compared to most other mammalian cell types in controlling the function of hundreds of nuclei in a common cytoplasm. Although a fair bit is known about nuclear envelope proteins in skeletal muscle because of their association with several muscular dystrophies, very little is known about the NPC or karyopherin transport receptors. Further knowledge about the nuclear transport machinery is needed in skeletal muscle to enhance our understanding of how gene expression is controlled in normal, aged, and diseased muscle as well as to provide insight into satellite cell biology.

Acknowledgments

G. K. P. is supported by National Institute of Health grants AR051372, AR052730, AR047314, and NS059340.

References

Abbott, K. L., Friday, B. B., Thaloor, D., Murphy, T. J., and Pavlath, G. K. (1998). Activation and cellular localization of the cyclosporine A-sensitive transcription factor NF-AT in skeletal muscle cells. *Mol. Biol. Cell* **9,** 2905–2916.

Ahluwalia, A., Narula, J., Jones, M. K., Deng, X., and Tarnawski, A. S. (2010). Impaired angiogenesis in aging myocardial microvascular endothelial cells is associated with reduced importin alpha and decreased nuclear transport of HIF1 alpha: Mechanistic implications. *J. Physiol. Pharmacol.* **61,** 133–139.

Alber, F., Dokudovskaya, S., Veenhoff, L. M., Zhang, W., Kipper, J., Devos, D., Suprapto, A., Karni-Schmidt, O., Williams, R., Chait, B. T., Sali, A., and Rout, M. P. (2007). The molecular architecture of the nuclear pore complex. *Nature* **450,** 695–701.

Apel, E. D., Lewis, R. M., Grady, R. M., and Sanes, J. R. (2000). Syne-1, a dystrophin- and Klarsicht-related protein associated with synaptic nuclei at the neuromuscular junction. *J. Biol. Chem.* **275,** 31986–31995.

Aratani, S., Oishi, T., Fujita, H., Nakazawa, M., Fujii, R., Imamoto, N., Yoneda, Y., Fukamizu, A., and Nakajima, T. (2006). The nuclear import of RNA helicase A is mediated by importin-alpha3. *Biochem. Biophys. Res. Commun.* **340,** 125–133.

Artaza, J. N., Bhasin, S., Mallidis, C., Taylor, W., Ma, K., and Gonzalez-Cadavid, N. F. (2002). Endogenous expression and localization of myostatin and its relation to myosin heavy chain distribution in C2C12 skeletal muscle cells. *J. Cell. Physiol.* **190,** 170–179.

Berman, S. A., Bursztajn, S., Bowen, B., and Gilbert, W. (1990). Localization of an acetylcholine receptor intron to the nuclear membrane. *Science* **247,** 212–214.

Boateng, S. Y., Senyo, S. E., Qi, L., Goldspink, P. H., and Russell, B. (2009). Myocyte remodeling in response to hypertrophic stimuli requires nucleocytoplasmic shuttling of muscle LIM protein. *J. Mol. Cell. Cardiol.* **47,** 426–435.

Burden, S. J. (1993). Synapse-specific gene expression. *Trends Genet.* **9,** 12–16.

Busch, A., Kiel, T., Heupel, W. M., Wehnert, M., and Hubner, S. (2009). Nuclear protein import is reduced in cells expressing nuclear envelopathy-causing lamin A mutants. *Exp. Cell Res.* **315,** 2373–2385.

Chen, I. H., Huber, M., Guan, T., Bubeck, A., and Gerace, L. (2006). Nuclear envelope transmembrane proteins (NETs) that are up-regulated during myogenesis. *BMC Cell Biol.* **7,** 38.

Chook, Y. M., and Suel, K. E. (2010). Nuclear import by karyopherin-betas: Recognition and inhibition. *Biochim. Biophysic. Acta* [E pub ahead of print].

Clavel, S., Siffroi-Fernandez, S., Coldefy, A. S., Boulukos, K., Pisani, D. F., and Derijard, B. (2010). Regulation of the intracellular localization of Foxo3a by stress-activated protein kinase signaling pathways in skeletal muscle cells. *Mol. Cell. Biol.* **30,** 470–480.

Cokol, M., Nair, R., and Rost, B. (2000). Finding nuclear localization signals. *EMBO Rep.* **1,** 411–415.

Conti, E., and Kuriyan, J. (2000). Crystallographic analysis of the specific yet versatile recognition of distinct nuclear localization signals by karyopherin alpha. *Structure* **8,** 329–338.

Cook, A. G., and Conti, E. (2010). Nuclear export complexes in the frame. *Curr. Opin. Struct. Biol.* **20,** 247–252.

Cook, A., Bono, F., Jinek, M., and Conti, E. (2007). Structural biology of nucleocytoplasmic transport. *Annu. Rev. Biochem.* **76,** 647–671.

Cortes, R., Rosello-Lleti, E., Rivera, M., Martinez-Dolz, L., Salvador, A., Azorin, I., and Portoles, M. (2010). Influence of heart failure on nucleocytoplasmic transport in human cardiomyocytes. *Cardiovasc. Res.* **85,** 464–472.

Crampton, N., Kodiha, M., Shrivastava, S., Umar, R., and Stochaj, U. (2009). Oxidative stress inhibits nuclear protein export by multiple mechanisms that target FG nucleoporins and Crm1. *Mol. Biol. Cell* **20,** 5106–5116.

D'Angelo, M. A., Raices, M., Panowski, S. H., and Hetzer, M. W. (2009). Age-dependent deterioration of nuclear pore complexes causes a loss of nuclear integrity in postmitotic cells. *Cell* **136,** 284–295.

Datta, K., Guan, T., and Gerace, L. (2009). NET37, a nuclear envelope transmembrane protein with glycosidase homology, is involved in myoblast differentiation. *J. Biol. Chem.* **284,** 29666–29676.

Dix, D. J., and Eisenberg, B. R. (1990). Myosin mRNA accumulation and myofibrillogenesis at the myotendinous junction of stretched muscle fibers. *J. Cell Biol.* **111,** 1885–1894.
Doucet, C. M., and Hetzer, M. W. (2010). Nuclear pore biogenesis into an intact nuclear envelope. *Chromosoma* **119,** 469–477.
Dupont-Versteegden, E. E., Strotman, B. A., Gurley, C. M., Gaddy, D., Knox, M., Fluckey, J. D., and Peterson, C. A. (2006). Nuclear translocation of EndoG at the initiation of disuse muscle atrophy and apoptosis is specific to myonuclei. *Am. J. Physiol. Regul. Integr. Comp. Physiol.* **291,** R1730–R1740.
Ellery, P. M., Cindrova-Davies, T., Jauniaux, E., Ferguson-Smith, A. C., and Burton, G. J. (2009). Evidence for transcriptional activity in the syncytiotrophoblast of the human placenta. *Placenta* **30,** 329–334.
Fagerlund, R., Melen, K., Kinnunen, L., and Julkunen, I. (2002). Arginine/lysine-rich nuclear localization signals mediate interactions between dimeric STATs and importin alpha 5. *J. Biol. Chem.* **277,** 30072–30078.
Faul, C., Dhume, A., Schecter, A. D., and Mundel, P. (2007). Protein kinase A, Ca2+/calmodulin-dependent kinase II, and calcineurin regulate the intracellular trafficking of myopodin between the Z-disc and the nucleus of cardiac myocytes. *Mol. Cell. Biol.* **27,** 8215–8227.
Faustino, R. S., Cheung, P., Richard, M. N., Dibrov, E., Kneesch, A. L., Deniset, J. F., Chahine, M. N., Lee, K., Blackwood, D., and Pierce, G. N. (2008). Ceramide regulation of nuclear protein import. *J. Lipid Res.* **49,** 654–662.
Fernandez-Martinez, J., and Rout, M. P. (2009). Nuclear pore complex biogenesis. *Curr. Opin. Cell Biol.* **21,** 603–612.
Ferri, P., Barbieri, E., Burattini, S., Guescini, M., D'Emilio, A., Biagiotti, L., Del Grande, P., De Luca, A., Stocchi, V., and Falcieri, E. (2009). Expression and subcellular localization of myogenic regulatory factors during the differentiation of skeletal muscle C2C12 myoblasts. *J. Cell. Biochem.* **108,** 1302–1317.
Figueroa, A., Cuadrado, A., Fan, J., Atasoy, U., Muscat, G. E., Munoz-Canoves, P., Gorospe, M., and Munoz, A. (2003). Role of HuR in skeletal myogenesis through coordinate regulation of muscle differentiation genes. *Mol. Cell. Biol.* **23,** 4991–5004.
Fiserova, J., Kiseleva, E., and Goldberg, M. W. (2009). Nuclear envelope and nuclear pore complex structure and organization in tobacco BY-2 cells. *Plant J.* **59,** 243–255.
Fontaine, B., and Changeux, J. P. (1989). Localization of nicotinic acetylcholine receptor alpha-subunit transcripts during myogenesis and motor endplate development in the chick. *J. Cell Biol.* **108,** 1025–1037.
Fontes, M. R., Teh, T., and Kobe, B. (2000). Structural basis of recognition of monopartite and bipartite nuclear localization sequences by mammalian importin-alpha. *J. Mol. Biol.* **297,** 1183–1194.
Fontes, M. R., Teh, T., Jans, D., Brinkworth, R. I., and Kobe, B. (2003). Structural basis for the specificity of bipartite nuclear localization sequence binding by importin-alpha. *J. Biol. Chem.* **278,** 27981–27987.
Freitas, N., and Cunha, C. (2009). Mechanisms and signals for the nuclear import of proteins. *Curr. Genomics* **10,** 550–557.
Frenkiel-Krispin, D., Maco, B., Aebi, U., and Medalia, O. (2010). Structural analysis of a metazoan nuclear pore complex reveals a fused concentric ring architecture. *J. Mol. Biol.* **395,** 578–586.
Fukada, S. I., Uezumi, A., Ikemoto, M., Masuda, S., Segawa, M., Tanimura, N., Yamamoto, H., Miyagoe-Suzuki, Y., and Takeda, S. (2007). Molecular signature of quiescent satellite cells in adult skeletal muscle. *Stem Cells* **25,** 2448–2459.
Geles, K. G., and Adam, S. A. (2001). Germline and developmental roles of the nuclear transport factor importin alpha3 in *C. elegans*. *Development* **128,** 1817–1830.

Giagtzoglou, N., Lin, Y. Q., Haueter, C., and Bellen, H. J. (2009). Importin 13 regulates neurotransmitter release at the Drosophila neuromuscular junction. *J. Neurosci.* **29,** 5628–5639.

Goldfarb, D. S., Corbett, A. H., Mason, D. A., Harreman, M. T., and Adam, S. A. (2004). Importin alpha: A multipurpose nuclear-transport receptor. *Trends Cell Biol.* **14,** 505–514.

Grady, R. M., Starr, D. A., Ackerman, G. L., Sanes, J. R., and Han, M. (2005). Syne proteins anchor muscle nuclei at the neuromuscular junction. *Proc. Natl. Acad. Sci. USA* **102,** 4359–4364.

Guttridge, D. C., Mayo, M. W., Madrid, L. V., Wang, C. Y., and Baldwin, A. S., Jr. (2000). NF-κB-induced loss of MyoD messenger RNA: possible role in muscle decay and cachexia. *Science* **289,** 2363–2366.

Hetzer, M. W., and Wente, S. R. (2009). Border control at the nucleus: Biogenesis and organization of the nuclear membrane and pore complexes. *Dev. Cell* **17,** 606–616.

Higashi-Kovtun, M. E., Mosca, T. J., Dickman, D. K., Meinertzhagen, I. A., and Schwarz, T. L. (2010). Importin-beta11 regulates synaptic phosphorylated mothers against decapentaplegic, and thereby influences synaptic development and function at the Drosophila neuromuscular junction. *J. Neurosci.* **30,** 5253–5268.

Hodel, M. R., Corbett, A. H., and Hodel, A. E. (2001). Dissection of a nuclear localization signal. *J. Biol. Chem.* **276,** 1317–1325.

Hodel, A. E., Harreman, M. T., Pulliam, K. F., Harben, M. E., Holmes, J. S., Hodel, M. R., Berland, K. M., and Corbett, A. H. (2006). Nuclear localization signal receptor affinity correlates with in vivo localization in *Saccharomyces cerevisiae. J. Biol. Chem.* **281,** 23545–23556.

Hogarth, C. A., Calanni, S., Jans, D. A., and Loveland, K. L. (2006). Importin alpha mRNAs have distinct expression profiles during spermatogenesis. *Dev. Dyn.* **235,** 253–262.

Holaska, J. M. (2008). Emerin and the nuclear lamina in muscle and cardiac disease. *Circ. Res.* **103,** 16–23.

Holaska, J. M., Rais-Bahrami, S., and Wilson, K. L. (2006). Lmo7 is an emerin-binding protein that regulates the transcription of emerin and many other muscle-relevant genes. *Hum. Mol. Genet.* **15,** 3459–3472.

Hood, J. K., and Silver, P. A. (1998). Cse1p is required for export of Srp1p/importin-alpha from the nucleus in Saccharomyces cerevisiae. *J. Biol. Chem.* **273,** 35142–35146.

Horton, P., Park, K. J., Obayashi, T., Fujita, N., Harada, H., Adams-Collier, C. J., and Nakai, K. (2007). WoLF PSORT: Protein localization predictor. *Nucleic Acids Res.* **35,** W585–W587.

Hu, W., Kemp, B. E., and Jans, D. A. (2005). Kinetic properties of nuclear transport conferred by the retinoblastoma (Rb) NLS. *J. Cell. Biochem.* **95,** 782–793.

Hu, J., Wang, F., Yuan, Y., Zhu, X., Wang, Y., Zhang, Y., Kou, Z., Wang, S., and Gao, S. (2010). Novel importin-alpha family member Kpna7 is required for normal fertility and fecundity in the mouse. *J. Biol. Chem.* **285,** 33113–33122.

Huber, M. D., Guan, T., and Gerace, L. (2009). Overlapping functions of nuclear envelope proteins NET25 (Lem2) and emerin in regulation of extracellular signal-regulated kinase signaling in myoblast differentiation. *Mol. Cell. Biol.* **29,** 5718–5728.

Huenniger, K., Kramer, A., Soom, M., Chang, I., Kohler, M., Depping, R., Kehlenbach, R. H., and Kaether, C. (2010). Notch1 signaling is mediated by importins alpha 3, 4, and 7. *Cell. Mol. Life Sci.* **67,** 3187–3196.

Iwamoto, M., Mori, C., Kojidani, T., Bunai, F., Hori, T., Fukagawa, T., Hiraoka, Y., and Haraguchi, T. (2009). Two distinct repeat sequences of Nup98 nucleoporins characterize dual nuclei in the binucleated ciliate tetrahymena. *Curr. Biol.* **19,** 843–847.

Kalderon, D., Richardson, W. D., Markham, A. F., and Smith, A. E. (1984a). Sequence requirements for nuclear location of simian virus 40 large-T antigen. *Nature* **311,** 33–38.

Kalderon, D., Roberts, B. L., Richardson, W. D., and Smith, A. E. (1984b). A short amino acid sequence able to specify nuclear location. *Cell* **39,** 499–509.

Kamei, Y., Yuba, S., Nakayama, T., and Yoneda, Y. (1999). Three distinct classes of the alpha-subunit of the nuclear pore-targeting complex (importin-alpha) are differentially expressed in adult mouse tissues. *J. Histochem. Cytochem.* **47,** 363–372.

Karrer, K. M. (2000). Tetrahymena genetics: Two nuclei are better than one. *Methods Cell Biol.* **62,** 127–186.

Kelley, J. B., Talley, A. M., Spencer, A., Gioeli, D., and Paschal, B. M. (2010). Karyopherin alpha7 (KPNA7), a divergent member of the importin alpha family of nuclear import receptors. *BMC Cell Biol.* **11,** 63.

Kind, J., and van Steensel, B. (2010). Genome-nuclear lamina interactions and gene regulation. *Curr. Opin. Cell Biol.* **22,** 320–325.

Kodiha, M., Tran, D., Qian, C., Morogan, A., Presley, J. F., Brown, C. M., and Stochaj, U. (2008). Oxidative stress mislocalizes and retains transport factor importin-alpha and nucleoporins Nup153 and Nup88 in nuclei where they generate high molecular mass complexes. *Biochim. Biophys. Acta* **1783,** 405–418.

Kohler, M., Ansieau, S., Prehn, S., Leutz, A., Haller, H., and Hartmann, E. (1997). Cloning of two novel human importin-alpha subunits and analysis of the expression pattern of the importin-alpha protein family. *FEBS Lett.* **417,** 104–108.

Kohler, M., Speck, C., Christiansen, M., Bischoff, F. R., Prehn, S., Haller, H., Gorlich, D., and Hartmann, E. (1999). Evidence for distinct substrate specificities of importin alpha family members in nuclear protein import. *Mol. Cell. Biol.* **19,** 7782–7791.

Kohler, M., Fiebeler, A., Hartwig, M., Thiel, S., Prehn, S., Kettritz, R., Luft, F. C., and Hartmann, E. (2002). Differential expression of classical nuclear transport factors during cellular proliferation and differentiation. *Cell. Physiol. Biochem.* **12,** 335–344.

Kosugi, S., Hasebe, M., Tomita, M., and Yanagawa, H. (2009). Systematic identification of cell cycle-dependent yeast nucleocytoplasmic shuttling proteins by prediction of composite motifs. *Proc. Natl. Acad. Sci. USA* **106,** 10171–10176.

Kutay, U., and Guttinger, S. (2005). Leucine-rich nuclear-export signals: Born to be weak. *Trends Cell Biol.* **15,** 121–124.

Kutay, U., Bischoff, F. R., Kostka, S., Kraft, R., and Gorlich, D. (1997). Export of importin alpha from the nucleus is mediated by a specific nuclear transport factor. *Cell* **90,** 1061–1071.

Lai, K. O., Zhao, Y., Ch'ng, T. H., and Martin, K. C. (2008). Importin-mediated retrograde transport of CREB2 from distal processes to the nucleus in neurons. *Proc. Natl. Acad. Sci. USA* **105,** 17175–17180.

Lange, A., Mills, R. E., Lange, C. J., Stewart, M., Devine, S. E., and Corbett, A. H. (2007). Classical nuclear localization signals: Definition, function, and interaction with importin alpha. *J. Biol. Chem.* **282,** 5101–5105.

Li, S., Czubryt, M. P., McAnally, J., Bassel-Duby, R., Richardson, J. A., Wiebel, F. F., Nordheim, A., and Olson, E. N. (2005). Requirement for serum response factor for skeletal muscle growth and maturation revealed by tissue-specific gene deletion in mice. *Proc. Natl. Acad. Sci. USA* **102,** 1082–1087.

Lim, R. Y., and Fahrenkrog, B. (2006). The nuclear pore complex up close. *Curr. Opin. Cell Biol.* **18,** 342–347.

Liu, S. M., and Liu, W. M. (2007). Recent developments in the understanding of nuclear protein import. *Protein Pept. Lett.* **14,** 723–733.

Liu, G. H., Guan, T., Datta, K., Coppinger, J., Yates, J., 3rd, and Gerace, L. (2009). Regulation of myoblast differentiation by the nuclear envelope protein NET39. *Mol. Cell. Biol.* **29,** 5800–5812.

Lonhienne, T. G., Forwood, J. K., Marfori, M., Robin, G., Kobe, B., and Carroll, B. J. (2009). Importin-beta is a GDP-to-GTP exchange factor of Ran: Implications for the mechanism of nuclear import. *J. Biol. Chem.* **284,** 22549–22558.

Lund, E., Guttinger, S., Calado, A., Dahlberg, J. E., and Kutay, U. (2004). Nuclear export of microRNA precursors. *Science* **303,** 95–98.

Lupu, F., Alves, A., Anderson, K., Doye, V., and Lacy, E. (2008). Nuclear pore composition regulates neural stem/progenitor cell differentiation in the mouse embryo. *Dev. Cell* **14,** 831–842.

Malone, C. D., Falkowska, K. A., Li, A. Y., Galanti, S. E., Kanuru, R. C., LaMont, E. G., Mazzarella, K. C., Micev, A. J., Osman, M. M., Piotrowski, N. K., Suszko, J. W., Timm, A. C., *et al.* (2008). Nucleus-specific importin alpha proteins and nucleoporins regulate protein import and nuclear division in the binucleate Tetrahymena thermophila. *Eukaryot. Cell* **7,** 1487–1499.

Manilal, S., Nguyen, T. M., Sewry, C. A., and Morris, G. E. (1996). The Emery-Dreifuss muscular dystrophy protein, emerin, is a nuclear membrane protein. *Hum. Mol. Genet.* **5,** 801–808.

Marfori, M., Mynott, A., Ellis, J. J., Mehdi, A. M., Saunders, N. F., Curmi, P. M., Forwood, J. K., Boden, M., and Kobe, B. (2010). Molecular basis for specificity of nuclear import and prediction of nuclear localization. *Biochim. Biophys. Acta* [Epub ahead of print].

Markiewicz, E., Tilgner, K., Barker, N., van de Wetering, M., Clevers, H., Dorobek, M., Hausmanowa-Petrusewicz, I., Ramaekers, F. C., Broers, J. L., Blankesteijn, W. M., Salpingidou, G., Wilson, R. G., *et al.* (2006). The inner nuclear membrane protein emerin regulates beta-catenin activity by restricting its accumulation in the nucleus. *EMBO J.* **25,** 3275–3285.

Mason, D. A., and Goldfarb, D. S. (2009). The nuclear transport machinery as a regulator of Drosophila development. *Semin. Cell Dev. Biol.* **20,** 582–589.

Mason, D. A., Fleming, R. J., and Goldfarb, D. S. (2002). Drosophila melanogaster importin alpha1 and alpha3 can replace importin alpha2 during spermatogenesis but not oogenesis. *Genetics* **161,** 157–170.

Mason, D. A., Mathe, E., Fleming, R. J., and Goldfarb, D. S. (2003). The *Drosophila melanogaster* importin alpha3 locus encodes an essential gene required for the development of both larval and adult tissues. *Genetics* **165,** 1943–1958.

Matsuura, Y., and Stewart, M. (2005). Nup50/Npap60 function in nuclear protein import complex disassembly and importin recycling. *EMBO J.* **24,** 3681–3689.

Matsuura, Y., Lange, A., Harreman, M. T., Corbett, A. H., and Stewart, M. (2003). Structural basis for Nup2p function in cargo release and karyopherin recycling in nuclear import. *EMBO J.* **22,** 5358–5369.

Mattout, A., Dechat, T., Adam, S. A., Goldman, R. D., and Gruenbaum, Y. (2006). Nuclear lamins, diseases and aging. *Curr. Opin. Cell Biol.* **18,** 335–341.

Mazroui, R., Di Marco, S., Clair, E., von Roretz, C., Tenenbaum, S. A., Keene, J. D., Saleh, M., and Gallouzi, I. E. (2008). Caspase-mediated cleavage of HuR in the cytoplasm contributes to pp 32/PHAP-I regulation of apoptosis. *J. Cell Biol.* **180,** 113–127.

McConville, J. F., Fernandes, D. J., Churchill, J., Dewundara, S., Kogut, P., Shah, S., Fuchs, G., Kedainis, D., Bellam, S. K., Patel, N. M., McCauley, J., Dulin, N. O., *et al.* (2010). Nuclear import of serum response factor in airway smooth muscle. *Am. J. Respir. Cell Mol. Biol.* [Epub ahead of print].

Micheli, L., Leonardi, L., Conti, F., Maresca, G., Colazingari, S., Mattei, E., Lira, S. A., Farioli-Vecchioli, S., Caruso, M., and Tirone, F. (2010). PC4/Tis7/IFRD1 stimulates skeletal muscle regeneration and is involved in myoblast differentiation as a regulator of MyoD and NF-kB. *J. Biol. Chem.* **286,** 5691–5707.

Mikenberg, I., Widera, D., Kaus, A., Kaltschmidt, B., and Kaltschmidt, C. (2007). Transcription factor NF-kappaB is transported to the nucleus via cytoplasmic dynein/dynactin motor complex in hippocampal neurons. *PLoS ONE* **2,** e589.

Miyamoto, Y., Saiwaki, T., Yamashita, J., Yasuda, Y., Kotera, I., Shibata, S., Shigeta, M., Hiraoka, Y., Haraguchi, T., and Yoneda, Y. (2004). Cellular stresses induce the nuclear accumulation of importin alpha and cause a conventional nuclear import block. *J. Cell Biol.* **165,** 617–623.

Mosca, T. J., and Schwarz, T. L. (2010). The nuclear import of Frizzled2-C by Importins-beta11 and alpha2 promotes postsynaptic development. *Nat. Neurosci.* **13,** 935–943.

Muchir, A., Bonne, G., van der Kooi, A. J., van Meegen, M., Baas, F., Bolhuis, P. A., de Visser, M., and Schwartz, K. (2000). Identification of mutations in the gene encoding lamins A/C in autosomal dominant limb girdle muscular dystrophy with atrioventricular conduction disturbances (LGMD1B). *Hum. Mol. Genet.* **9,** 1453–1459.

Newlands, S., Levitt, L. K., Robinson, C. S., Karpf, A. B., Hodgson, V. R., Wade, R. P., and Hardeman, E. C. (1998). Transcription occurs in pulses in muscle fibers. *Genes Dev.* **12,** 2748–2758.

Nguyen Ba, A. N., Pogoutse, A., Provart, N., and Moses, A. M. (2009). NLStradamus: A simple hidden Markov model for nuclear localization signal prediction. *BMC Bioinform.* **10,** 202.

O'Connor, R. S., Mills, S. T., Jones, K. A., Ho, S. N., and Pavlath, G. K. (2007). A combinatorial role for NFAT5 in both myoblast migration and differentiation during skeletal muscle myogenesis. *J. Cell Sci.* **120,** 149–159.

Ogawa, Y., Miyamoto, Y., Asally, M., Oka, M., Yasuda, Y., and Yoneda, Y. (2010). Two isoforms of Npap60 (Nup50) differentially regulate nuclear protein import. *Mol. Biol. Cell* **21,** 630–638.

Okada, N., Ishigami, Y., Suzuki, T., Kaneko, A., Yasui, K., Fukutomi, R., and Isemura, M. (2008). Importins and exportins in cellular differentiation. *J. Cell. Mol. Med.* **12,** 1863–1871.

Okamura, H., Aramburu, J., Garcia-Rodriguez, C., Viola, J. P., Raghavan, A., Tahiliani, M., Zhang, X., Qin, J., Hogan, P. G., and Rao, A. (2000). Concerted dephosphorylation of the transcription factor NFAT1 induces a conformational switch that regulates transcriptonal activity. *Mol. Cell.* **6,** 539–550.

Ostlund, C., Guan, T., Figlewicz, D. A., Hays, A. P., Worman, H. J., Gerace, L., and Schirmer, E. C. (2009). Reduction of a 4q35-encoded nuclear envelope protein in muscle differentiation. *Biochem. Biophys. Res. Commun.* **389,** 279–283.

Pallafacchina, G., Francois, S., Regnault, B., Czarny, B., Dive, V., Cumano, A., Montarras, D., and Buckingham, M. (2010). An adult tissue-specific stem cell in its niche: A gene profiling analysis of in vivo quiescent and activated muscle satellite cells. *Stem Cell Res.* **4,** 77–91.

Perez-Terzic, C., Faustino, R. S., Boorsma, B. J., Arrell, D. K., Niederlander, N. J., Behfar, A., and Terzic, A. (2007). Stem cells transform into a cardiac phenotype with remodeling of the nuclear transport machinery. *Nat. Clin. Pract.* **4**(Suppl. 1), S68–S76.

Poon, I. K., and Jans, D. A. (2005). Regulation of nuclear transport: Central role in development and transformation? *Traffic* **6,** 173–186.

Puckelwartz, M. J., Kessler, E. J., Kim, G., Dewitt, M. M., Zhang, Y., Earley, J. U., Depreux, F. F., Holaska, J., Mewborn, S. K., Pytel, P., and McNally, E. M. (2010). Nesprin-1 mutations in human and murine cardiomyopathy. *J. Mol. Cell. Cardiol.* **48,** 600–608.

Pujol, G., Soderqvist, H., and Radu, A. (2002). Age-associated reduction of nuclear protein import in human fibroblasts. *Biochem. Biophys. Res. Commun.* **294,** 354–358.

Quensel, C., Friedrich, B., Sommer, T., Hartmann, E., and Kohler, M. (2004). In vivo analysis of importin alpha proteins reveals cellular proliferation inhibition and substrate specificity. *Mol. Cell. Biol.* **24,** 10246–10255.

Rabut, G., Lenart, P., and Ellenberg, J. (2004). Dynamics of nuclear pore complex organization through the cell cycle. *Curr. Opin. Cell Biol.* **16,** 314–321.

Ratan, R., Mason, D. A., Sinnot, B., Goldfarb, D. S., and Fleming, R. J. (2008). Drosophila importin alpha1 performs paralog-specific functions essential for gametogenesis. *Genetics* **178,** 839–850.

Ribbeck, K., Lipowsky, G., Kent, H. M., Stewart, M., and Gorlich, D. (1998). NTF2 mediates nuclear import of Ran. *EMBO J.* **17,** 6587–6598.

Richard, M. N., Deniset, J. F., Kneesh, A. L., Blackwood, D., and Pierce, G. N. (2007). Mechanical stretching stimulates smooth muscle cell growth, nuclear protein import, and nuclear pore expression through mitogen-activated protein kinase activation. *J. Biol. Chem.* **282,** 23081–23088.

Riddick, G., and Macara, I. G. (2005). A systems analysis of importin-{alpha}-{beta} mediated nuclear protein import. *J. Cell Biol.* **168,** 1027–1038.

Riddick, G., and Macara, I. G. (2007). The adapter importin-alpha provides flexible control of nuclear import at the expense of efficiency. *Mol. Syst. Biol.* **3,** 118.

Robbins, J., Dilworth, S. M., Laskey, R. A., and Dingwall, C. (1991). Two interdependent basic domains in nucleoplasmin nuclear targeting sequence: Identification of a class of bipartite nuclear targeting sequence. *Cell* **64,** 615–623.

Salamon, M., Millino, C., Raffaello, A., Mongillo, M., Sandri, C., Bean, C., Negrisolo, E., Pallavicini, A., Valle, G., Zaccolo, M., Schiaffino, S., and Lanfranchi, G. (2003). Human MYO18B, a novel unconventional myosin heavy chain expressed in striated muscles moves into the myonuclei upon differentiation. *J. Mol. Biol.* **326,** 137–149.

Sandri, M., Sandri, C., Gilbert, A., Skurk, C., Calabria, E., Picard, A., Walsh, K., Schiaffino, S., Lecker, S. H., and Goldberg, A. L. (2004). Foxo transcription factors induce the atrophy-related ubiquitin ligase atrogin-1 and cause skeletal muscle atrophy. *Cell* **117,** 399–412.

Sanes, J. R., Johnson, Y. R., Kotzbauer, P. T., Mudd, J., Hanley, T., Martinou, J. C., and Merlie, J. P. (1991). Selective expression of an acetylcholine receptor-lacZ transgene in synaptic nuclei of adult muscle fibers. *Development* **113,** 1181–1191.

Schirmer, E. C., Florens, L., Guan, T., Yates, J. R., 3rd, and Gerace, L. (2003). Nuclear membrane proteins with potential disease links found by subtractive proteomics. *Science (New York, NY)* **301,** 1380–1382.

Shen, T., Liu, Y., Contreras, M., Hernandez-Ochoa, E. O., Randall, W. R., and Schneider, M. F. (2010). DNA binding sites target nuclear NFATc1 to heterochromatin regions in adult skeletal muscle fibers. *Histochem. Cell Biol.* **134,** 387–402.

Shin, I., Rotty, J., Wu, F. Y., and Arteaga, C. L. (2005). Phosphorylation of p27Kip1 at Thr-157 interferes with its association with importin alpha during G1 and prevents nuclear re-entry. *J. Biol. Chem.* **280,** 6055–6063.

Shmidt, T., Hampich, F., Ridders, M., Schultrich, S., Hans, V. H., Tenner, K., Vilianovich, L., Qadri, F., Alenina, N., Hartmann, E., Kohler, M., and Bader, M. (2007). Normal brain development in importin-alpha5 deficient-mice. *Nat. Cell Biol.* **9,** 1337–1338 (author reply 1339).

Simon, A. M., Hoppe, P., and Burden, S. J. (1992). Spatial restriction of AChR gene expression to subsynaptic nuclei. *Development* **114,** 545–553.

Smith, A., Brownawell, A., and Macara, I. G. (1998). Nuclear import of Ran is mediated by the transport factor NTF2. *Curr. Biol.* **8,** 1403–1406.

Smitherman, M., Lee, K., Swanger, J., Kapur, R., and Clurman, B. E. (2000). Characterization and targeted disruption of murine Nup50, a p27(Kip1)-interacting component of the nuclear pore complex. *Mol. Cell. Biol.* **20,** 5631–5642.

Strambio-De-Castillia, C., Niepel, M., and Rout, M. P. (2010). The nuclear pore complex: Bridging nuclear transport and gene regulation. *Nat. Rev. Mol. Cell Biol.* **11,** 490–501.

Su, X., Berman, S. A., Sullivan, T., and Bursztajn, S. (1995). Myoblast and myotube nuclei display similar patterns of heterogeneous acetylcholine receptor subunit mRNA expression. *J. Cell. Biochem.* **58,** 22–38.

Talcott, B., and Moore, M. S. (2000). The nuclear import of RCC1 requires a specific nuclear localization sequence receptor, karyopherin alpha3/Qip. *J. Biol. Chem.* **275,** 10099–10104.

Thompson, K. R., Otis, K. O., Chen, D. Y., Zhao, Y., O'Dell, T. J., and Martin, K. C. (2004). Synapse to nucleus signaling during long-term synaptic plasticity: A role for the classical active nuclear import pathway. *Neuron* **44,** 997–1009.

Timney, B. L., Tetenbaum-Novatt, J., Agate, D. S., Williams, R., Zhang, W., Chait, B. T., and Rout, M. P. (2006). Simple kinetic relationships and nonspecific competition govern nuclear import rates in vivo. *J. Cell Biol.* **175,** 579–593.

Tran, E. J., and Wente, S. R. (2006). Dynamic nuclear pore complexes: Life on the edge. *Cell* **125,** 1041–1053.

Tsuji, L., Takumi, T., Imamoto, N., and Yoneda, Y. (1997). Identification of novel homologues of mouse importin alpha, the alpha subunit of the nuclear pore-targeting complex, and their tissue-specific expression. *FEBS Lett.* **416,** 30–34.

van der Giessen, K., and Gallouzi, I. E. (2007). Involvement of transportin 2-mediated HuR import in muscle cell differentiation. *Mol. Biol. Cell* **18,** 2619–2629.

Wada, A., Fukuda, M., Mishima, M., and Nishida, E. (1998). Nuclear export of actin: A novel mechanism regulating the subcellular localization of a major cytoskeletal protein. *EMBO J.* **17,** 1635–1641.

Wagstaff, K. M., and Jans, D. A. (2009). Importins and beyond: Non-conventional nuclear transport mechanisms. *Traffic* **10,** 1188–1198.

Walde, S., and Kehlenbach, R. H. (2010). The part and the whole: Functions of nucleoporins in nucleocytoplasmic transport. *Trends Cell Biol.* **20,** 461–469.

Wang, C. I., Wang, C. L., Wang, C. W., Chen, C. D., Wu, C. C., Liang, Y., Tsai, Y. H., Chang, Y. S., Yu, J. S., and Yu, C. J. (2010). Importin subunit alpha-2 is identified as a potential biomarker for non-small cell lung cancer by integration of the cancer cell secretome and tissue transcriptome. *Int. J. Canc.* [Epub ahead of print].

Weins, A., Schwarz, K., Faul, C., Barisoni, L., Linke, W. A., and Mundel, P. (2001). Differentiation- and stress-dependent nuclear cytoplasmic redistribution of myopodin, a novel actin-bundling protein. *J. Cell Biol.* **155,** 393–404.

Weis, K. (2003). Regulating access to the genome: Nucleocytoplasmic transport throughout the cell cycle. *Cell* **112,** 441–451.

Wente, S. R., and Rout, M. P. (2010). The nuclear pore complex and nuclear transport. *Cold Spring Harb. Perspect. Biol.* **2,** a000562.

White, E. M., Allis, C. D., Goldfarb, D. S., Srivastva, A., Weir, J. W., and Gorovsky, M. A. (1989). Nucleus-specific and temporally restricted localization of proteins in Tetrahymena macronuclei and micronuclei. *J. Cell Biol.* **109,** 1983–1992.

Winey, M., Yarar, D., Giddings, T. H., Jr., and Mastronarde, D. N. (1997). Nuclear pore complex number and distribution throughout the Saccharomyces cerevisiae cell cycle by three-dimensional reconstruction from electron micrographs of nuclear envelopes. *Mol. Biol. Cell* **8,** 2119–2132.

Yaffe, D., and Saxel, O. (1977). A myogenic cell line with altered serum requirements for differentiation. *Differentiation* **7,** 159–166.

Yasuhara, N., Shibazaki, N., Tanaka, S., Nagai, M., Kamikawa, Y., Oe, S., Asally, M., Kamachi, Y., Kondoh, H., and Yoneda, Y. (2007). Triggering neural differentiation of ES cells by subtype switching of importin-alpha. *Nat. Cell Biol.* **9,** 72–79.

Youn, M. Y., Takada, I., Imai, Y., Yasuda, H., and Kato, S. (2010). Transcriptionally active nuclei are selective in mature multinucleated osteoclasts. *Genes Cells* **15,** 1025–1035.

Zerfaoui, M., Errami, Y., Naura, A. S., Suzuki, Y., Kim, H., Ju, J., Liu, T., Hans, C. P., Kim, J. G., Abd Elmageed, Z. Y., Koochekpour, S., Catling, A., *et al.* (2010). Poly(ADP-ribose) polymerase-1 is a determining factor in Crm1-mediated nuclear export and retention of p65 NF-kappa B upon TLR4 stimulation. *J. Immunol.* **185,** 1894–1902.

Zhang, X., Xu, R., Zhu, B., Yang, X., Ding, X., Duan, S., Xu, T., Zhuang, Y., and Han, M. (2007). Syne-1 and Syne-2 play crucial roles in myonuclear anchorage and motor neuron innervation. *Development* **134,** 901–908.

Index

Contents of Previous Volumes

Volume 47

Volume 48

Volume 49

Volume 52

Volume 53

Volume 54

Volume 57

Volume 63

Volume 64

Volume 66

Volume 67

Volume 68

Volume 69

Volume 70

Volume 71

Volume 72

Volume 73

Volume 76

Volume 77

Volume 80

Volume 81

Volume 82

Volume 83

Volume 84

Volume 85

Volume 86

Volume 89

Volume 90

Volume 91

Volume 92

Volume 93

Volume 94

Volume 95

FSC
www.fsc.org
MIX
Papier | Fördert
gute Waldnutzung
FSC® C083411

Kolibri 360 GmbH
Ferdinand-Jühlke-Straße 7
99095 Erfurt, Deutschland
produktsicherheit@kolibri360.de